Molecular Genetics of Mammalian Cells

MOLECULAR GENETICS OF MAMMALIAN CELLS
A Primer in Developmental Biology

George M. Malacinski, Editor
Indiana University

Christian C. Simonsen, Consulting Editor
Invitron Corporation

Michael Shepard, Consulting Editor
Genentech, Inc.

Macmillan Publishing Company
NEW YORK

Collier Macmillan Publishers
LONDON

Macmillan Publishing Company
866 Third Avenue, New York, NY 10022

Collier Macmillan Canada, Inc.

Printed in the United States of America

printing number
1 2 3 4 5 6 7 8 9 10   year 6 7 8 9 0 1 2 3 4 5

Library of Congress Cataloging-in-Publication Data
Main entry under title:

Molecular genetics of mammalian cells

    (Primers in developmental biology)
    Includes bibliographies and index.
    1. Mammals —Genetics.   2. Molecular genetics.
3. Cytogenetics. I. Malacinski, George M.
II. Simonsen, Christian C. III. Shepard, Michael.
IV. Series   [DNLM: 1. Gene Expression Regulation.
QH 450 M7182]
QL738.5.M65 1985   599′.087328   85-29927
ISBN 0-02-947950-9

This book is respectfully dedicated to the scientists who worked so hard in the 1950s and 1960s to establish experimental conditions that permitted mammalian cells to be cultured like microorganisms.

Primers in Developmental Biology

George M. Malacinski, Series Editor
Indiana University

Available

Pattern Formation: A Primer in Developmental Biology, George M. Malacinski, Editor, and Susan V. Bryant, Consulting Editor

Molecular Genetics of Mammalian Cells: A Primer in Developmental Biology, George M. Malacinski, Editor, and Christian C. Simonsen and Michael Shepard, Consulting Editors

Forthcoming

Developmental Genetics of Animals and Plants: A Primer in Developmental Biology, George M. Malacinski, Editor

Contents

Preface

Primers in Developmental Biology was conceived as a series of textbooks in contemporary biological research areas with the following aims:

1. To *identify* rapidly emerging and potentially important disciplines and subdisciplines of developmental biology
2. To *organize* those disciplines for students and scholars who might be seeking both an introduction to current research problems and a broad overview of a discipline
3. To provide *insight* into the thought processes of several of the important contributors to the discipline

Our editorial approach to *Molecular Genetics of Mammalian Cells* was similar to that of the first volume in the series, *Pattern Formation.* Rather than solicit a small number of highly referenced, pedantic chapters that might represent only a limited cross section of a research area, we conducted a comprehensive search. First, the field of molecular genetics of mammalian cells was surveyed by examining the few textbooks and collections of review articles currently in print. Then we conducted computer searches employing key words and author names to gather our data base. With the aid of colleagues we compiled subject lists and matched these with the appropriate potential authors. Finally, before we extended the invitation to the volume contributors, we subjected our roster to further review.

We formulated the following guidelines for our contributors:

1. Chapters would be written at a level that would be the most meaningful to the graduate student, postdoctoral fellow, or new investigator in the field of molecular genetics of mammalian cells.
2. Authors would be urged to speculate and to provide their personal views or interpretations.

3. Literature references would be limited in scope, to encourage authors to generalize.
4. Chapters would be cross-referenced, wherever possible, so that a careful reading of the entire book would give the reader both a comprehensive and coherent view of contemporary studies in molecular genetics of mammalian cells.

To generate an "insider's view" of the discipline, the editors (especially myself) talked informally and candidly with several of the contributors. A summary of several of these discussions is included in the "Editor's Discussion with the Contributors." All contributors also responded to two general questions we asked about their particular subject. We have included these at the end of each chapter. Using these approaches, we developed an advanced-level textbook for molecular genetics of mammalian cells. It is designed, like a textbook, to be general in scope. Yet *Molecular Genetics of Mammalian Cells* also contains a great amount of detailed information, and so can serve also as a reference book.

We are employing a similar strategy to develop the third volume of the series, *Developmental Genetics of Animals and Plants*. It will focus mainly on contemporary or molecular approaches to a variety of problems in developmental biology. In addition, the volume will contain introductory descriptions of many of the more widely employed genetic systems.

Sarah Greene, our editor at Macmillan, provided expert publishing advice, and Diane Malacinski provided the fine art print.

George M. Malacinski

Contributing Authors

Susana Balcells — Department of Genetics, M. D. Anderson Hospital and Tumor Institute, Houston, Texas 77030

Renato Baserga — Department of Pathology and Fels Research Institute, Temple University School of Medicine, Philadelphia, Pennsylvania 19140

Roberto Cattaneo — Institute for Molecular Biology I, University of Zurich, Hoenggerberg, 8093 Zurich, Switzerland

Barbara Christy — Department of Biology, The Johns Hopkins University, Baltimore, Maryland 21218

Martin J. Cline — Division of Hematology-Oncology, University of California, Los Angeles, California 90024

Laurence D. Etkin — Department of Genetics, M. D. Anderson Hospital and Tumor Institute, Houston, Texas 77030

Nancy Federspiel — Becton-Dickenson Monoclonal Center, 2375 Garcia Ave., Mountain View, California 94043

Bernd Groner — Ludwig Institute for Cancer Research, Inselspital, 3010 Bern, Switzerland

Howard V. Hershey — Department of Biology, Indiana University, Bloomington, Indiana 47405

Ricky R. Hirschhorn — T. H. Morgan School of Biological Sciences, Thomas Hunt Morgan Building, University of Kentucky, Lexington, Kentucky 40506

Nancy Hynes — Ludwig Institute for Cancer Research, Inselspital, 3010 Bern, Switzerland

Randal J. Kaufman — Genetics Institute, Inc., 87 Cambridge Park Drive, Cambridge, Massachusetts 02140

George Khoury — Laboratory of Molecular Virology, National Cancer Institute, Bethesda, Maryland 20205

Ching-Juh Lai — Laboratory of Infectious Diseases, National Institute of Allergy and Infectious Diseases, Bethesda, Maryland 20205

Laimonis Laimins — Department of Molecular Genetics and Cell Biology, University of Chicago, 920 E. 58th Street, Chicago, Illinois 60637

Lance A. Liotta — Laboratory of Pathology, National Institutes of Health, Bethesda, Maryland 20205

Lewis J. Markoff — Laboratory of Infectious Diseases, National Institute of Allergy and Infectious Diseases, Bethesda, Maryland 20205

R. Scott McIvor — Genentech, Inc., South San Francisco, California 94080

G. Stanley McKnight — Department of Pharmacology, University of Washington, Seattle, Washington, 98195

Joo-Hung Park — Department of Biology, Indiana University, Bloomington, Indiana 47405

Helmut Ponta — Kernforschungszentrum Karlsruhe, Institute for Genetics and Toxicology, D-7500 Karlsruhe, West Germany

Bradley Pearman — Department of Genetics, M. D. Anderson Hospital and Tumor Institute, Houston, Texas 77030

Nadia Rosenthal — Department of Cardiology, Children's Hospital Medical Center, Harvard Medical School, Boston, Massachusetts 02115

Michel M. Sanders — Department of Pharmacology, University of Washington, Seattle, Washington 98195

George Scangos — Department of Biology, The Johns Hopkins University, Baltimore, Maryland 21218

Heinz Schaller — Microbiology and ZMBH, University of Heidelberg, Im Neuenheimer Feld, 230, 6900 Heidelberg, West Germany

Robert T. Schimke — Department of Biological Sciences, Stanford University, Herrin Laboratories, Stanford, California 94305

Christian C. Simonsen — Department of Molecular and Cellular Biology, Invitron Corp., San Carlos, California 94070

Rolf Sprengel — Microbiology and ZMBH, University of Heidelberg, Im Neuenheimer Feld, 230, 6900 Heidelberg, West Germany

James E. Talmadge — Preclinical Screening Laboratory, Frederick Cancer Research Facility, Frederick, Maryland 21701

Milton W. Taylor Department of Biology, Indiana University, Bloomington, Indiana 47405

Hans Will Microbiology and ZMBH, University of Heidelberg Im Neuenheimer Feld, 230, 6900 Heidelberg, West Germany

Berton Zbar Laboratory of Immunobiology, Nantional Cancer Institute–Frederick Cancer Research Facility, Frederick, Maryland 21701

Introduction

A MAJOR GOAL of many developmental biologists is to understand the regulatory mechanisms that guide mammalian (especially human)
embryogenesis and generate cell specialization (differentiation). Indeed, that
goal is also included among the health services priorities of many national
governments and private foundations. The substantial amount of research
funding provided by public agencies as well as private organizations has
provided a major impetus to the field of mammalian development. Especially
favored of late have been those projects that employ biochemical and molecular approaches.

Two conceptual approaches dominate experimental thinking in this area.
The *first* approach emphasizes the use of simple, nonmammalian models as
tools for establishing the general principles that govern pattern formation and
cell differentiation in developing animals. Various examples of that approach
are included in volume 1 of *A Primer in Developmental Biology: Pattern Formation.* In that volume, model systems as diverse as cellular slime molds, insects, amphibia, and avian species are used to elucidate the types of mechanisms that regulate embryonic pattern specification. There is a high expecta-
tion among researchers (especially embryologists) who study those systems
that their analyses of patterning and gene expression will be applicable to the
mammalian embryo either by elucidating mechanisms which regulate gene
expression that are shared by phylogenetically lower organisms (e.g., insects)
and mammals or by preparing and testing experimental protocols that can
eventually be modified for use with more complex mammalian embryos.

The *second* approach emphasizes the direct use of mammalian systems.
Some molecular biologists believe that gene expression circuitry is either
more complex or highly idiosyncratic in mammals, negating the use of lower
vertebrates as substitutes for mammalian systems. Other researchers argue

that sufficient background information is already available from simpler systems and that there is no need to delay making the transition to mammalian systems. Finally, a compromise approach is offered by some: use virus-infected mammalian cells to provide a simple experimental system (viral gene-expression patterns) in the context of the mammalian cell's complex housekeeping metabolism.

Once a commitment is made to employ the latter approach—the direct use of the mammalian system for studies in developmental biology—it is important to realize at the outset that the mammalian embryo itself is not amenable to many of the experimental approaches used with lower vertebrates. Conventional manipulations such as surgical intervention, nuclear transplantation, and cell lineage analyses are usually cumbersome and difficult to accomplish. Use of either tissue or cell cultures is required. Consequently, the early years of the postdescriptive era of mammalian embryology focused on the development of adequate cell culture methods. Initially, growth initiation as well as the maintenance of a stable phenotype proved difficult. Once methods for establishing suitable cell lines were developed, real progress in understanding mammalian cell differentiation began. As insights into gene-expression circuitry were obtained, it became apparent that the lack of appropriate mammalian cell genetic systems was limiting the rate of progress. Although the development of somatic cell hybrids proved valuable, researchers were hampered by an inability to proceed as rapidly as desired with mammalian cell genetics.

The advent of recombinant DNA technology and transfection systems promises, however, to change all that. An exciting future lies ahead for this research area. Horizons that seemed so circumscribed a decade ago are now limited only by our scientific imagination.

This volume of *Primers in Developmental Biology* reviews many of the fascinating ways in which gene expression in mammalian cells are presently being analyzed. It is clear, however, that in each instance parts (cells), rather than the whole (tissues, organ, or embryo), are being analyzed. Embryologists often assert that the whole (e.g., embryo) will display properties that are greater than the sum of its parts. No doubt that will be the case. Yet, in the absence of rigorous whole-organism genetics and considering the difficulty of manipulating mammalian embryos, it is only at the subcellular and molecular levels that precise experimentation can be carried out.

A theme that emerged from *Pattern Formation* is that truly meaningful knowledge concerning what is really important at the molecular level cannot be identified without first obtaining insight at the cellular and tissue levels. The viewpoint taken by most of the authors in this volume runs counter to that theme. One of the main purposes of *Molecular Genetics of Mammalian Cells* is to review several contemporary ways of studying gene expression in mammalian cells. Experimental approaches are described that can serve as a springboard to an eventual comprehensive and coherent view of the mechanisms that control mammalian embryogenesis and cell differentiation.

Setting the debate about levels of analysis aside, another purpose of *Mo-*

lecular Genetics of Mammalian Cells is to highlight potential practical applications of this area of research. There are at least two. First, disease control efforts can benefit directly from studies of the type described in this volume. Viral infection will be better understood after some of the detailed analyses described in this volume are completed. Then there is the potential for gene therapy, which is also addressed in several chapters in this volume. Second, mammalian cell genetics may be exploited for commercial purposes. That is, mammalian cells can be employed in much the same fashion as have been industrial microorganisms. The opportunity is now available to use mammalian cells that have been transfected with heterologous genes for the production of valuable protein products.

The climate is favorable for a multidisciplinary approach to research problems that promises to contribute to basic knowledge as well as solve practical problems. The combination of the disciplines of developmental biology, genetics, molecular biology, and medicine represented by this volume can serve as a model for a holistic view of mammalian embryogenesis and cell differentiation.

George M. Malacinski

Editor's Discussion with Contributors

George M. Malacinski

While developing this volume, we had an opportunity to engage some of the contributors in informal, "off-the-record" conversations about several aspects of the topics included in this book. Several of their replies are summarized here.

1. *What constitutes the discipline of mammalian cell molecular genetics?*

   In addition to the topics included in this volume, mutations, karyotype analyses and somatic cell hybridization could be included as part of the subject of mammalian cell molecular genetics. But since this volume emphasizes the most recent experimental approaches, as well as a slant toward applying the discipline to developmental biology (especially its practical aspects), it is easy to understand why not all possible topics are included.

2. *Can we be sure that the various so-called mutant cell lines routinely employed by workers in this field represent true gene mutations? That is, can cells be trained to express an altered, stably inherited phenotype without the involvement of nucleotide sequence modification?*

   Evidence at the DNA level has substantiated the notion that cell lines that have been labeled "mutants" are genuine genetic mutations. In contrast, the epigenetic origin of so-called mutant phenotypes has not been substantiated.

3. *To what extent do cell culture methods occupy the time of the researcher? Is mammalian cell culture so routine these days that it is only incidental to the core of the project?*

   Many of the same problems that have been around for 20 to 30 years linger. Contamination by mycoplasma remains a problem. Many of the original media have not been improved (although recently renewed interest in this area is evident). Fibroblasts still are the major cell type used in most experiments. So it is safe to say that cell culture continues to be a preoccupation of researchers in this field.

4. *Why does dihydrofolate reductase (DHFR) get so much press?*

A variety of reasons, including several historical ones. Originally, it was perhaps the first mammalian gene to display bona fide gene amplification. Methotrexate (MTX) employed as an anticancer drug, exhibited decreased effectiveness against certain tumors. It was subsequently discovered that DHFR gene amplification represented the mechanism of resistance. Eventually, its protein sequence, gene structure, and regulation were subjected to intense investigation by several laboratories.

5. *Why complicate matters by using viruses to study gene expression in mammalian cells?*

Actually, the use of the appropriate virus probably simplifies matters, especially if the virus is integrated into the host cell's chromosome. Against the broad background of host functions, viral gene expression is often easily monitored.

6. *What major discoveries that apply to nonviral systems (e.g., uninfected mammalian cells) has the use of viruses generated?*

The list reads like a "what's new in molecular biology": oncogenes (sarc); enhancers (SV40 virus); introns (adenovirus); somatic cell hybrids (sendai virus), and so on.

7. *Conspicuously absent from this volume is a chapter on somatic cell heterokaryons and hybrids. Has their general usefulness been superseded by transfection/transformation?*

Somatic cell hybrids continue to be employed for gene mapping studies. And of course hybridomas generate monoclonal antibodies.

8. *Now that heterologous genes can be introduced into mammalian cells and achieve full expression, what next can we expect from this field?*

Exploitation, for sure. Transfected cells can be used by commercial laboratories for the production of biologically interesting molecules (e.g., interferon). Genetic disease therapy can also be expected, although it is probably some years off. Finally, in the more near term we can expect to be able to construct "minichromosomes" (as is being done in yeast systems) for *in vivo* studies on regulation.

9. *What obstacles must be overcome before gene therapy becomes a reality?*

There are many, and they vary according to the gene and cell type involved. Since most approaches now on the drawing board would employ *in vitro* transfection and back transplantation into the host tissue there is the choice of the vector system appropriate to the cell type at issue. The vector must of course deliver a gene construct that will generate high levels of the relevant protein gene products in the implanted cells. Those cells must in turn maintain themselves in the tissue environment and function in a relatively normal fashion.

Should direct transformation of a tissue or organ be attempted, different problems lie ahead. First, an appropriate vector (e.g., retrovirus) must be developed that does not generate undesirable side effects. Second, a way to use that vector to transform only the target cells in the tissue or organ of interest must be figured out. Recall, most tissues represent a mosaic of different cell types.

Clearly, the molecular biology (e.g., gene isolation, preparation of suitable constructs) is under control. It is the basic cellular and developmental biology aspects that need help.

10. *Does it ever get simpler? Consider the discovery of promoters, enhancers, etc. Is gene-expression circuitry—when the last analysis is complete—going to be simple or complex?*

The latter. The closer we look, the more we discover it is complex. Who would have guessed that genes reorganize? Evolutionary hangovers are probably more prevalent than many molecular biologists are willing to admit. That is, the cell's molecular behavior is being constantly remodeled by evolutionary considerations. In many ways it represents a Rube Goldberg apparatus, rather than the streamlined machine a molecular biologist, starting from scratch, might build.

One might even go so far as to say that the reductionists are losing their long-standing battle with their opposite number—the gadflies who favor intricate, complicated, and manifold regulation patterns. The tally of known cis-acting DNA sequence elements continues to rise, as does the number of trans-acting factors (nuclear proteins?).

11. *Because of the potential commercial opportunities associated with exploiting mammalian cell genetics, industries (especially the venture capital variety) are active in the field. Can we expect industry to generate basic knowledge? Or will industry concentrate on product development?*

There is no doubt a danger that the commercial pressures to do proprietary research will limit the more applied aspects, but the admittedly young biotechnology companies that have enjoyed the greatest success have been the ones to publish most quickly and most often. It might be more accurate to suggest that the biotechnology companies will continue to generate basic knowledge in focused areas: cloning, sequencing, and expressing therapeutically important proteins. The understanding of how these proteins work in the animal (e.g., interferons, growth factors, atrial peptides, hormones) will most likely not be emphasized in industry.

The field of molecular genetics of animal cells is just now beginning to be defined. The major issues of the last 10 years (oncogenes, receptor proteins, RNA processing, gene regulation) are by no means finished. The next 10 years will be no less exciting.

Glossary

carcinoma malignant tumor

clone genetically identical copy of a cell; nucleotide sequence amplified in a suitable vector (e.g., plasmid).

construct genetically engineered gene, usually prepared in a plasmid vector. Often contains unusual nucleotide sequence elements (e.g., heterologous promoters).

enhancer nucleotide sequence that increases the transcription of genes into mRNA. Curiously, enhancers can function when located either upstream or downstream of the 5′ end of a gene.

exon DNA coding sequence. Several are usually present in a gene, and through RNA processing, the nucleotide sequences they represent are spliced into the mRNA transcript, which is subsequently translated into a polypeptide.

intron nucleotide sequence in a gene that is represented in an initial RNA transcript but is spliced out during the processing that generates the mRNA that is subsequently translated into a polypeptide. Also called "intervening sequence."

metastasis invasion of foreign tissues by growth and spreading of transformed cells. The normal control mechanisms that usually limit a specific cell type to a well-defined territory (e.g., tissue or organ) do not operate.

Northern blot method for identifying individual RNAs (e.g., specific mRNA) in a complex mixture. Typically, RNA preparation is separated by size in an acrylamide gel. That gel is then "blotted" onto nitrocellulose paper. A radiolabeled DNA probe (e.g., cDNA) is hybridized to the nitrocellulose blot and the signal is revealed by autoradiography.

oncogene endogenous gene which, when activated, is responsible for the change of a cell's phenotype from its normal state of differentiation to a neoplastic character. Oncogenes are often found in the genome of RNA tumor viruses.

plasmid extrachromosomal genetic element found in various bacteria. It consists of a double-stranded, closed, circular DNA molecule and is commonly employed as a vector for preparing clones of heterologous genes.

promoter nucleotide sequence usually associated with a structural gene. It binds RNA polymerase and facilitates the initiation of transcription.

retrovirus RNA tumor virus. The RNA genome is transcribed into DNA by the enzyme reverse transcriptase. Often used as a vector for transferring heterologous genes into mammalian cells.

Southern blot method for identifying specific nucleotide sequences (e.g., specific gene) in a DNA preparation. Typically, the DNA is "restricted" (digested) with enzymes and separated by size in an agarose gel. The gel is then blotted onto nitrocellulose paper. A radiolabeled probe (e.g., mRNA) is hybridized to the nitrocellulose blot and the signal revealed by autoradiography.

TATA box nucleotide sequence associated with many structural genes. It facilitates site-specific initiation of transcription.

transcript RNA synthesized enzymatically (e.g., by RNA polymerase) from a DNA template.

transfection DNA-mediated gene transfer. Often carried out with selectable vectors (e.g., plasmids containing antibiotic resistance markers). Usually the DNA is introduced into mammalian cells by incubating the cells with DNA in the presence of DEAE-dextran, or by adding a DNA calcium phosphate coprecipitate to a suspension of the cells.

transformation change in a cell's phenotype that usually includes increased growth rate, shape alterations, and cell surface modifications. Often caused by infection with tumor virus.

Far must thy researches go
Wouldst thou learn the world to know;
Thou must tempt the dark abyss
Wouldst thou prove what *Being* is;
Naught but firmness gains the prize,
Naught but fullness makes us wise,
Buried deep truth e'er lies.
            Friedrich Schiller, *Proverbs of Confucius*

Regulation of Gene Expression

CHAPTER 1

Control of Gene Expression by DNA Methylation

Barbara Christy and George Scangos

THE PRESENCE of modified bases in the DNA of both prokaryotes and eukaryotes has been known for some time. In prokaryotes,
modified adenine residues clearly are involved in the recognition of foreign
DNA in restriction/modification systems (1). In eukaryotes, the function of
methylated cytosine residues has not been entirely elucidated. Methylation
has been postulated to be involved in many cellular processes, including restriction and modification similar to that of prokaryotes, DNA replication,
recombination, mutation and repair systems, maintenance of chromosome
structure, cell cycle regulation, cellular differentiation, and control of gene
expression (20, 22, 67).

The only function for which there is strong experimental support is
that of control of gene expression. There is a well-established inverse correlation between the level of DNA methylation and the expression of many
genes, although there are a few genes for which a positive correlation exists
(2, 37, 85), and several genes for which a correlation between the state of
methylation and the state of expression has not been demonstrated. The
mechanisms by which methylation may affect gene expression are not known,
but in theory methylation can influence binding of regulatory proteins,
chromatin conformation, or accessibility of important sites on the DNA to
various enzymatic processes.

Overall levels of methylation in different tissues and cell lines have been
examined by several methods, including high-pressure liquid chromatography, nearest-neighbor analysis, antibodies directed against 5-methylcy-
tosine, genomic sequencing, and cleavage of DNA with methylation-sensitive
restriction enzymes. One of the most useful techniques for studying methylation patterns involves the use of restriction enzymes that are sensitive to
methylation within their recognition sites. The pattern of cleavage with these

enzymes can be analyzed on Southern blots (78), using the cloned gene as a probe, and conclusions can be drawn about the state of methylation by the relative extent of cleavage at these sites. Restriction enzymes that have proved to be particularly useful in the study of methylation are the isoschizomers HpaII and MspI. Both enzymes recognize the same site (CCGG), but HpaII can cleave only if the internal cytosine in the recognition sequence is unmethylated, whereas cleavage by MspI is not affected by methylation at the internal cytosine (52, 90). Since most methylated cytosines occur at cytosine residues that are part of CpG sequences (60), comparison between the two digestion patterns can be used to determine the pattern of methylation within the gene.

In this chapter, we will attempt to summarize and discuss some of the data, derived by a variety of methods, that suggest that DNA methylation is involved in the control of gene expression.

In Vivo Distribution of 5-Methylcytosines in DNA

In DNA of various mammalian species, approximately 2 to 8 percent of the cytosines are present as 5-methylcytosine (20, 22, 67). Higher plants contain a greater proportion of 5-methylcytosine, while no 5-methylcytosine has yet been detected in the genomic DNA of some lower eukaryotes, such as *Drosophila* and nematodes (20). In animal cells, more than 95 percent of modified cytosine residues are found within CpG sequences, although some also can be found in CpC and other sequences (60). In mammalian genomes, CpG sequences occur less often than expected on the basis of statistical probabilities (20), which may reflect an evolutionary selection against random occurrence of these modifiable sites. In plants, modified cytosines are found at sequences other than CpG, commonly C-X-G sequences, where X = C, A, or T (35). In most cases, methylation at C-G or C-X-G is symmetrical. The cytosines on both strands are methylated, which allows maintenance of methylation patterns following semiconservative replication of the DNA.

In some organisms, it appears that methyl moieties are highly nonrandomly distributed, in long clusters of fully methylated or fully unmethylated regions. In sea urchins, some genes are usually found in one fraction, while other genes are found in the other fraction (5, 6). Methylated cytosine residues are found in both single-copy and repetitive sequences, although repetitive sequences appear to be more heavily methylated in general (26).

Within one species, the levels and patterns of methylation can vary significantly during development and differentiation. Although mouse or human sperm DNA has a lower overall 5-methylcytosine content than does DNA from other tissues (21), specific regions of the genome can have higher or lower levels of methylation than is found in corresponding adult tissue. For example, some of the satellite sequences found in the mouse genome are undermethylated in DNA from both male and female germ cells (66, 73), while many unique sequences are more heavily methylated in sperm cells than in somatic cells in which they are being expressed (22). Additionally, some repetitive sequences in extraembryonic lineages are substantially undermethylated relative to the DNA of the embryo proper (10). In the adult genome, repetitive sequences are more heavily methylated than moderately repetitive or single copy sequences (26), and in general, the expressed genes in a particular cell type seem to be about 30 percent as methylated overall as the "average" cellular DNA (59). During the development of differentiated cell types in the organism, different patterns of methylation seem to be established in differentiated cell types, which may be an important mechanism for the establishment of patterns of gene expression during differentiation.

DNA methylation has been proposed to play a role in the inactivation of one of the two X chromosomes in female cells (68). The inactive X chromosome becomes heterochromatic, late replicating, and transcriptionally inactive at most (but not all) loci within the chromosome (68). DNA isolated from inactive X chromosomes is unable to transfer the HPRT$^+$ phenotype into HPRT$^-$ cells (50), suggesting that inactivation occurs at the level of the DNA. Recent studies demonstrated that the HPRT locus on inactive X chromosomes was more heavily methylated than that on active X chromosomes (99, 101). The methylation inhibitors 5-azacytidine and 5-azadeoxycytidine reactivated HPRT and other genes on inactive X chromosomes in somatic cell hybrids and in transformation experiments (41, 56, 89). Taken together, these results suggest that DNA methylation is involved in the process of X inactivation.

It has been suggested that patterns and extent of methylation may change over time as cells age. Romanov and Vanyushin reported decreases *in vivo* in 5-methylcytosine in the DNA of aging cattle and salmon (71). In cell culture studies, Wilson and Jones (98) showed that the 5-methylcytosine content of normal diploid fibroblasts decreased with time in culture. The species whose cells most rapidly senesced in culture (mouse and hamster) showed the greatest and most rapid loss of 5-methylcytosine.

In contrast, Ehrlich et al. (21) analyzed tissues from human individuals of various ages, and found no significant decrease in the overall levels of 5-methylcytosine. Additionally, they examined the level of methylation in diploid human fibroblasts over time in culture, but found no decrease (21). However, the cells were analyzed for only 10 to 15 doublings, compared to the more than 100 doublings analyzed by Wilson and Jones (98). The results of Wilson and Jones suggested that changes in methylation occurred slowly in normal human fibroblasts, so that 10 to 15 doublings may not have been sufficient time for changes to be noted.

Other studies using tumor cell lines and freshly explanted tumor tissue also demonstrated that the extent of methylation could change over time in culture. Kuhlmann and Doerfler (44), showed that when Ad 12 induced tumors from hamsters were explanted into culture, the Ad 12 sequences, which were poorly methylated at the time of explantation, became more methylated as the cells were passaged. In contrast, subclones that were maintained in conditioned medium retained a methylation pattern similar to that of the original tumor cells. Although the virus was poorly methylated in these lines, it was not extensively expressed, indicating that undermethylation was not sufficient for expression of the viral genomes. These data suggest that changes in DNA methylation seen in culture are related, at least in part, to the culture conditions. The levels of methylation in established cell lines in culture were more stable than those of freshly explanted tissue, and not vary significantly with further culture (98).

Recent studies indicate that there may be correlations between methylation and tumorigenesis. This possibility is attractive, since DNA modification constitutes a heritable alteration of the DNA and so could alter the normal pattern of gene expression. Gama-Sosa et al. measured the 5-methylcytosine content of 103 human tumors of various types (27). DNA methylation of fresh tumor tissues (benign, primary, and metastatic tumors) was compared to the methylation level of normal tissues of the same type. Most of the metastatic neoplasms had significantly lower genomic 5-methylcytosine contents than the benign neoplasms or normal tissues. The percentage of primary neoplasms having hypomethylated DNA was intermediate between those of metastatic tumors and benign neoplasmas (27). These data suggest that an inverse correlation exists between the extent of DNA methylation and tumor progression. Studies by Wilson and Jones (97) demonstrated that several chemical carcinogens were capable of inhibiting DNA methylation in cultured mouse cell lines, which also suggested that changes in DNA methylation may be involved in the tumorigenic process.

In addition to changes in overall genome methylation seen in tumors, changes in the methylation pattern of specific genes have been correlated with the state of transformation of cultured cells. In several cases, [e.g., Groffen et al. (31)], hypomethylation of oncogenes was correlated with the transformed phenotype. Feinberg and Vogelstein (23) found that in four of five patients studied (with two different types of cancer), there was substantial hypomethylation of several genes in tumor cells compared to adjacent normal tissues. The sequences examined included human growth hormone, α-globin, and γ-globin. One patient also showed hypomethylation in cells from a metastasis relative to cells from the primary tumor, which were in turn less methylated ated than the cells from the adjacent normal tissue. Taken together, these studies demonstrate an inverse correlation between DNA methylation and tumor progression. It is possible that decreased methylation leads to increased transcription of genes involved in tumor formation, although there are not yet any data to suggest such a causal relationship.

Inverse Correlations

The relationship between DNA methylation and gene expression has been extensively studied for a number of genetic systems. In general, an inverse correlation exists between the state of expression of a gene and the degree of methylation in or around that gene, with a few exceptions. Genes that are expressed in a given tissue or cell line generally are hypomethylated in that tissue relative to other tissues in which they are not expressed. Globin genes, whose methylation has been studied in humans, rabbits, and chickens, all are hypomethylated in tissues in which they are expressed, relative to those tissues in which they are not expressed (53, 55, 76, 86, 91, 93). Human β-globin genes are most heavily methylated in sperm DNA, methylated to a lesser degree in somatic tissues, and least methylated in erythroid cells which express β-globin (86). Methylation of the globin genes also varies with the stage of development. Following the switch from the embryonic and fetal patterns of expression to the adult pattern, methylation of the embryonic and fetal genes is increased, and they become less sensitive to DNase I digestion (93). These findings suggest that methylation plays a role in initiating or maintaining the inactive state of the embryonic and fetal globin genes in adult tissue.

Further evidence that methylation plays a role in regulation of globin gene transcription comes from studies utilizing agents that are known to inhibit DNA methylation, such as 5-azacytidine (40). Treatment of human sickle-cell anemia patients with 5-azacytidine led to an increased level of fetal hemoglobin along with hypomethylation of 2 of 15 assayable sites around the γ-ζ-β-globin complex. One of these two sites was 107-base pairs (bp) 5′ to the γ^G globin gene and the other was 107 bp 5′ to the γ^A gene (11). Similar results were obtained in baboons treated with 5-azacytidine (19).

More recently, Ginder et al. (29) have shown that adult chickens treated with 5-azacytidine had decreased methylation of globin genes, and an increased amount of embryonic ρ-globin mRNA in circulating erythrocytes. An additional increase in mRNA level was found in birds that were treated with both 5-azacytidine and sodium butyrate, but not with sodium butyrate alone. Butyrate treatment did not detectably alter the pattern of methylation, but in conjunction with 5-azacytidine treatment, did induce the appearance of a DNase-I hypersensitive site. The authors concluded that the 5-azacytidine induced demethylation was necessary but not sufficient for ρ-globin induction, and that a multistep process was involved.

Other *in vivo* systems in which hypomethylation of DNA has been correlated strongly with expression of specific genes are the chicken ovalbumin, conalbumin (ovotransferrin), and ovomucoid genes. Mandel and Chambon (51) demonstrated that the methylation patterns in and around all of these genes varied in a tissue-specific manner. These three genes code for egg white proteins, which are expressed in the oviduct of the laying hen. Among the tissues examined (oviduct, erythrocyte, liver, kidney, spleen, and brain from laying hens; liver and sperm from roosters), the level of methylation was lowest in the oviduct of the laying hen. Changes in methylation pattern in this system were specific, since it was found that some sites were methylated in all of the tissues examined, while some were partially methylated in most tissues, with a greater degree of hypomethylation observed in the tissues where they were expressed. When ovalbumin mRNA synthesis was terminated by stopping estradiol treatment, no change was seen in the methylation of the ovalbumin gene, indicating that hypomethylation was not sufficient for expression, and that there was another level of control imposed on the system.

Therefore, as in other inducible systems, methylation does not appear to play a role in the rapid modulation of gene expression, but appears to be involved more in long-term tissue-specific regulation.

In vitro tissue culture studies also have demonstrated that expression of many genes is inversely correlated with methylation. Yagi and Koshland (100) examined methylation and expression of the immunoglobulin J-chain genes in cell lines representing successive stages in B-cell development, and demonstrated that J-chain gene expression was correlated with hypomethylation of the genes. The J-chain gene was heavily methylated in cell types in which it was not expressed, and less methylated in cell lines representative of antigen-stimulated lymphocytes synthesizing J chain. Analysis of the C_μ and $C_{\gamma 2b}$ genes in the same set of cell lines demonstrated that the C_μ gene was undermethylated in all lines examined, while $C_{\gamma 2b}$ was undermethylated only in those cell types that were synthesizing IgG2b protein. One hypothesis to explain these findings is that the changes in J-chain gene transcription and methylation are a specific differentiation-induced response to contact of the B cells with antigen. These experiments seem to implicate a demethylation event in the differentiation process. In addition, demethylation accompanied class switching in the heavy chain genes. The state of methylation of the genes was inversely correlated with their expression during the different stages of B-cell differentiation (70).

Expression of the albumin gene in cultured rat hepatoma cells also was strongly correlated with undermethylation of certain sites or regions in the 5′ end of the gene (64). Four MspI/HpaII sites were analyzed within the 15-kilobase (kb) albumin gene. In male rat liver, a tissue in which albumin is expressed, all four sites were unmethylated. In the DNA of hepatoma cells, three sites within the gene were methylated, and the degree of methylation of the 5′-most site was strongly correlated with the degree of expression of albumin. The 5′-most site was heavily methylated in subclones of these lines that did not express albumin. Revertants of these subclones which synthesized albumin were undermethylated at the 5′-most site (in the 5′ portion of the first exon). However, one exception to this generalization was observed. In one subclone that did not express albumin, the 5′-most site was mostly unmethylated, suggesting again that undermethylation is not sufficient for expression. In summary, with albumin as well as other genes, methylation was inversely correlated with expression, and undermethylation of at least one site at the 5′ end of the gene was correlated with, but was not sufficient for, stable gene expression.

Selectable Genes

The extent of methylation of several genes which act as selectable markers during gene transfer into cultured cells has been correlated with expression. Several laboratories, including our own, have generated mouse cell lines that contain the Herpes simplex virus thymidine kinase (TK) gene in which expression of the gene is inversely correlated with the extent of its methylation (12–14, 63). The TK gene is particularly useful for studying mechanisms controlling gene expression because expression of TK can be a necessary or lethal event, depending on the culture conditions. When TK-positive lines are placed in medium in which TK expression is lethal, derivative cell lines that retain but do not express the TK genes can be selected. When the TK-deficient cell lines are returned to medium in which expression of TK again is required for survival, TK-positive "reexpressors" can be selected. In such cell lines, the TK genes in the TK-deficient cell lines often were more heavily methylated than those in their TK-positive parents or derivatives (12, 13, 63). Treatment of TK-deficient cells with 5-azacytidine (which inhibits methylation) resulted in an increased number of TK-positive derivatives (12, 14), which again contained TK genes with reduced levels of methylation relative to their TK-deficient parents. Analysis of a large number of cell lines identified three sites within the gene and flanking region at which the state of methylation was always correlated with expression—the sites were unmethylated in TK-positive cell lines and methylated in TK-deficient cells (13).

Other laboratories have obtained similar series of cell lines in which expression of the TK genes did not appear to be correlated with changes in the methylation pattern (18, 69). Davies et al. (18) characterized cell lines in which changes in expression were correlated with changes in the TK chromatin structures (as assayed by changes in the nuclease sensitivity). In these lines, no changes in methylation pattern of the TK genes could be detected when the state of expression changed. Additionally, Robins et al., characterized a series of TK-containing lines, many of which appeared to lose the TK sequences when subjected to backselection (69). These results indicate that the Herpes simplex virus thymidine kinase (HSV TK) gene is subject to various control mechanisms. Since the site of integration into the mouse cell genome is likely to be different for each cell line (75), the different responses may reflect constraints imposed by the integration site and indicate the types of responses that occur most frequently in that genomic region.

Viral Systems

In general, virion DNA in virus particles is not detectably methylated (20), although some exceptions, such as Epstein-Barr virus, Shope papilloma virus, and human papilloma virus type 1a have been reported (17, 42, 65, 72, 94). In general, integrated viral genomes or portions of genomes that are heavily methylated are transcriptionally inactive, while genomes or portions of genomes that are expressed are less methylated (20), suggesting that integrated viral genomes are subject to the same types of epigenetic control mechanisms as cellular genes.

Human adenoviruses have been extensively studied as a model system for the control of gene expression. DNA isolated from purified adenovirus virions has little or no methylation (84). Although the DNA of the human host cell lines used to propagate the adenovirus is methylated, the adenovirus DNA, which replicates extrachromosomally in the nucleus of the cells, is unmethylated. Adenovirus DNA may escape methylation because extrachromosomal molecules are unable to be methylated by the host methylation enzymes, as has been suggested for polyoma virus by Subramanian (83). It is possible that the enzymes involved in the methylation process are chromatin-bound, and are unable to diffuse onto the extrachromosomal molecules. Alternatively, the virus may have a mechanism for actively avoiding methylation of its DNA. Another hypothesis is that there is little *de novo* methylase activity present in the cells that are used in the infection, although they retain their "maintenance methylase" activity.

Adenovirus DNA also can be found integrated into the genome of nonpermissive transformed and tumor cell lines. In this integrated state it is often extensively methylated (44, 84, 88), suggesting that adenovirus DNA can be methylated *de novo* in these cells, probably during or following integration into genomic DNA. Inverse correlations between expression of adenoviral genes and their methylation patterns have been established in these lines. One clear example was described by Sutter and Doerfler (84), who investigated the expression of the E2a region of adenovirus 2 in three transformed hamster lines. In one line, the 72k DNA binding protein was expressed and was unmethylated at all HpaII sites. In the other two lines, the 72k DNA binding protein was not expressed, and the gene was methylated at the HpaII sites. *In vitro* methylation of this gene also demonstrated that its expression was attenuated by methylation (47).

The situation appears to be different for some of the other DNA tumor viruses that have been examined. Polyoma and SV40 virion DNAs also are not detectably methylated, but in contrast to adenovirus, polyoma and SV40 genomes integrated into transformed cells are not extensively methylated, although polyoma sequences in transformed lines may be some what methylated in the late region of the virus genome (48). *In vitro* methylation of SV40 with rat liver methylase followed by microinjection into permissive cells failed to demonstrate any effect of methylation on T-antigen expression (30). There also was no apparent effect when SV40 was methylated *in vitro* by the bacterial methylases HpaII methylase or EcoRI methylase and transfected or microinjected into culture cells (92). In each case, the viral DNA was able to transform cells with an efficiency equal to that found using unmethylated DNA. In contrast to these findings, Fradin et al. (25) found that SV40 DNA that had been methylated *in vitro* by HpaII methylase was transcribed less efficiently after injection into *Xenopus* oocytes. It may be that methylation had a more severe effect on expression in *Xenopus* oocytes than in cultured mammalian cells.

Herpes simplex virus type 1 (HSV-1) often becomes latent in humans and animals following a primary infection (102). Recently, an *in vitro* model was developed invoking methylation as a mechanism involved in latency (102). An HSV-1 infected lymphoblastoid T-cell line was described that underwent two nonproductive latent stages. During these latent stages no detectable virus was produced, and the viral DNA was heavily methylated. During periods of active viral expression, the virus was not detectably methylated, demonstrating an inverse correlation between expression of the viral genes and the state of methylation of the viral genome. However, an alternative interpretation is possible. During the productive stages, there were 40 to 80 copies of the virus per cell that were not detectably methylated. During the latent phases, there were only one to two copies of the virus per cell, which were heavily methylated. It may be that these two methylated copies were always present, but that their detection was obscured by the 20- to 80-fold greater amount of nonmethylated DNA present when active viral production was taking place.

Expression of several integrated endogenous retroviral genomes in mice and chickens also has been inversely correlated with the state of methylation of their genomes. In chickens, there are two endogenous retroviral loci, one of which (termed ev-3) is transcriptionally active, while the other (termed ev-1) is inactive. Compared to ev-1, ev-3 is hypomethylated, DNase I sensitive, and contains nuclease hypersensitive sites in its long terminal repeats (LTRs) (32). Treatment of chicken cells with 5-azacytidine resulted in transcriptional activation of the ev-1 locus, suggesting that undermethylation may be casually related to expression in this case. Following 5-azacytidine treatment, the ev-1 locus was less methylated, and had acquired at least one hypersensitive site. This study thus provides an example in which changes in the state of methylation were correlated with alterations in the state of the chromatin structure as assayed by the presence of DNase hypersensitive sites.

In mice, expression of several endogenous retroviruses can be correlated with methylation. For example, mouse mammary tumor virus (MMTV) exists as an endogenous stable genetic locus in all mouse strains studied (7, 15). Viral genomes can be passed to offspring by infection with virus in the mother's milk and by passage of endogenous viral sequences through the germ line. Mouse strains that carry milkborne virus as well as endogenous virus have a higher incidence of tumors than those strains that carry only endogenous virus (15). However, removal of milkborne virus reduced, but did not completely eliminate, the occurrence of tumors, suggesting that the endogenous virus also may be involved in tumor formation. The endogenous MMTV genomes were highly methylated, while the milkborne proviral copies were not detectably methylated. The unmethylated viruses were also transcriptionally active, while the methylated endogenous copies were transcriptionally inactive (15). Breznik and Cohen (7) showed that endogenous proviral sequences became unmethylated at specific sites during mammary tumor formation in uninfected Balb/c mice (which do not contain detectable milkborne virus) relative to DNA isolated from Balb/c liver or normal lactating mammary gland.

Inbred laboratory strains of mice also carry endogenous type C viruses that are inactive but can be induced by treatment with 5-azacytidine as well as other agents (62). The endogenous viral sequences in mouse cells are resistant to digestion by Hpa II, in contrast to exogenous infective virus, indicating that the repressed endogenous viral sequences are methylated (62). The genes of intracisternal A particles (IAP; retroviruslike elements found in about 1000 copies per haploid genome in *Mus musculus*) contain low levels of methylation in both murine plasmacytomas (where they are expressed) and NIH3T3 cells (where they are not) (58). Normal liver, which also does not contain detectable transcripts of IAP genes, contains only highly methylated copies. The fact that the IAP genes in NIH3T3 cells are hypomethylated, yet not expressed, indicates that hypomethylation of these genes is not sufficient for their expression. Since a kappa light chain gene also was found to be hypomethylated but not expressed in these cells (58), undermethylation may be a general phenomenon in NIH3T3 cells. This observation may partially explain the fact that 3T3 cells are more readily transformed in culture than are primary mouse cells and some other cell lines (45).

Jaenisch and coworkers studied the methylation patterns of an exogenous retrovirus, Moloney murine leukemia virus (M-MuLV), which had been introduced into the germ line of mice by infection of mouse embryos. In 13 mouse lines derived in this manner, the expression of the virus was correlated with the amount of methylation of the viral genomes (82). In one such line, Mov-3, the DNA was noninfectious, and was heavily methylated at HhaI sites. When the proviral sequences from Mov-3 were cloned, removing all 5-methylcytosine, they became infectious (36), indicating that the endogenous provirus was not inactive due to a defect in the viral sequences, and implicating DNA modification in the inactivation of the virus.

Little expression of M-MuLV was found after infection of embryonal carcinoma cells in culture (81), in contrast to the high-level expression obtained after infection of differentiated cells. In embryonal cells, the integrated viral DNA was methylated, while in differentiated cells it was not, suggesting that *de novo* methylation may be more frequent in embryonic cells in culture than in more differentiated cells. *In vivo*, integrated viral copies were more highly methylated when embryos were infected at early stages (preimplantation), than when later stage embryos were infected (39). Thus, it appears that both *in vitro* and *in vivo*, *de novo* methylation of integrated proviral DNA occurs more frequently in early embryonic cells, suggesting that more *de novo* methylase activity is present at these stages, and consistent with the suggestion that *de novo* methylation is important for differentiation (39, 81).

Positive Correlations

In general, most genes whose expression appears to be affected by DNA methylation demonstrate an inverse correlation between the extent of methylation and expression of the gene. However, there are a few examples in which DNA methylation appears to play a positive role in the regulation of transcription. One example is from a prokaryotic system, involving the mom gene of bacteriophage Mu. Bacteriophage Mu possesses a DNA modification enzyme which modifies about 15 percent of the adenine residues to a new unusual form (Ax or N_6-carboxymethyladenine), rendering the bacteriophage resistant to restriction (37). It has been shown that *Escherichia coli* host cell adenine methyltransferase activity is required for transcription of the phage mom gene, since mom RNA can be detected only when the bacteriophage is grown in dam+ strains of *E. coli*, and can only be detected after induction of the Mu prophage (37). Therefore, in this case, the DNA methylase enzyme exerts a positive regulatory influence on the transcription of a gene.

Two models were proposed by Hattman (37) to explain these findings. The first model proposes that the *E. coli* dam⁺ methylase acts by methylating a site or sites on Mu DNA which then act to stimulate transcription of the mom gene. The second model proposes that the methylase acts by binding to specific sites on the Mu DNA, and the bound enzyme exerts a regulatory effect. The first model is supported by the fact that a cloned T4 adenine methylase gene can be substituted for the *E. coli* methylase. Since the structural properties of the T4 methylase are different from the properties of the *E. coli* methylase even though they both methylate the same DNA sequence (GATC), it is unlikely that both enzymes exert their effect through binding to the same sites. Certain Mu deletions allow mom expression independent of dam⁺ activity, which suggests that a specific site in the Mu DNA is the target site for the dam⁺ effect, and that the dam⁺ effect works through relieving an inhibitory function.

A positive correlation between methylation and gene expression was demonstrated in a eukaryotic system by Tanka et al. (85) for the mouse H-2K gene (a class 1 major histocompatibility antigen gene) in F9 teratocarcinoma cells. F9 cells can be induced to differentiate by treatment with retinoic acid. Expression of the classical transplantation antigens becomes detectable concomitant with differentiation (85). Tanaka et al. (85) studied two differentiated cell lines that were derived from F9 cells following treatment with retinoic acid. One of the clones (called c19) was isolated after 4 days of treatment while the other (c11) was isolated following retinoic acid treatment for 6 months. The c19 cells were biochemically and phenotypically "more differentiated" (less embryonic) in character than the parental F9 cells, and the c11 cells in turn were "more differentiated" than the c19 cells.

The structure and expression pattern of the H-2K gene was analyzed in these three cell lines representing three stages of differentiation, using a probe specific for the H-2K gene, F9 stem cells did not express the H-2K antigen, while c19 cells expressed H-2K at a low level and c11 cells expressed H-2K at a high level. The level of H-2K protein was correlated with the amount of H-2K mRNA in the cell lines. The pattern of methylation of the H-2K genes was examined in the three lines, using methylation-sensitive restriction enzymes. In F9 cells, the H-2K gene, which was not expressed, contained a low level of methylation. The H-2K gene in c19 cells was more methylated than in F9 cells, while the gene in c11 cells, which contained the highest level of H-2K mRNA, was most heavily methylated.

These data demonstrate a positive correlation between the level of methylation and expression of the H-2K gene in the three cell lines studied. To rule out the possibility that the two cloned cell lines did not represent the overall population, the state of methylation of H-2K genes in mass cultures of F9 cells treated with retinoic acid for 7 to 14 days was studied. The longer the cells were carried in retinoic acid, the more methylated the H-2K genes became (85). Additionally, treatment of the differentiated c11 cells with 5-azacytidine led to an increased level of expression of several gene products, but a decrease in the expression of the H-2K gene product. These data suggest a positive correlation between the extent of methylation and the level of expression of the H-2K gene in teratocarcinoma cells, Interestingly, the inactive H-2 genes of teratocarcinoma stem cells also were shown to be more DNase I sensitive than the active genes of the differentiated cells (16).

Another positive correlation between expression and the level of methylation of a gene was reported recently by Battistuzzi et al. (2). Glucose-6-phosphate dehydrogenase is an X-linked housekeeping gene, expressed in almost all cells, although the levels of the enzyme vary in a tissue-specific manner. Methylation at specific cytosine residues in the 3′ portion of this gene was correlated positively with the level of expression observed in the various cell types. Interestingly, these authors concluded that different sites of methylation were involved in the inactivation of this gene on the inactive X chromosome than are involved in its regulation on the active X chromosome.

Is Methylation Causally Related to Expression?

Even though correlations between hypomethylation and expression of many genes have been demonstrated, it has been difficult to determine whether or not the methylation is casually related to the lack of expression. Indeed, certain studies seem to suggest that hypermethylation may be caused by a lack of transcription of a gene, or that hypomethylation is a consequence of the active transcription of certain genes. For example, Wilks et al. (96) showed that the one reproducible change in the methylation pattern of the chicken vitellogenin gene, whose expression was induced by estrogen in the liver of immature chickens, was a demethylation of one HpaII site 0.6 kb 5′ to the gene. However, when the kinetics of induction were examined, it was found that transcription started 6 to 10 hours following stimulation, while the change in methylation seemed to occur between 10 and 24 hours. These data suggest that the change in methylation actually was preceded by the onset of transcription.

Similar results were obtained for the chick vitellogenin gene by Burch and Weintraub (8), who also found that the demethylation of a single site occurred, and that the demethylation followed, rather than preceded, the transciption of the gene. The demethylation seen by Wilks et al. also was seen following estrogen stimulation in the oviduct, which does not express vitellogenin, but does contain estrogen receptors and expresses several other genes in an estrogen-inducible manner. This finding suggested that the demethylation was induced by estrogen whether or not the vitellogenin gene actually was activated, and that the specific demethylation was not sufficient to activate the vitellogenin gene.

Burch and Weintraub (8) found that the appearance of several DNase I hypersensitive sites was correlated with estrogen stimulation either in the liver (which expresses the gene following estrogen stimulation) or the oviduct (which does not express the gene), while certain other hypersensitive sites were found only in liver DNA during estrogen stimulation. It was shown that the appearance of the hypersensitive sites preceded the demethylation event in the liver, and that the generation of the hypersensitive sites did not require demethylation (8). From these studies, it is not clear whether the changes in methylation of these sequences is the cause or effect of the changes in the transcription status. Another group (28) found that the estrogen-inducible vitellogenin genes of *Xenopus laevis* were expressed when fully methylated at all HpaII and HhaI sites, in contrast to several other *Xenopus* genes. However, in this and many other studies, subtle demethylation events that occurred at specific sites not recognizable by restriction endonucleases would not have been detected.

In contrast to the above studies, there are several lines of evidence that suggest that hypomethylation, at least in some cases, is sufficient to induce expression of previously inactive genes. Studies using inhibitors of DNA-methylation suggest that, in some cases, inhibition of DNA methylation can cause the activation of genes. Perhaps the most commonly used inhibitor is 5-azacytidine, a cytidine analog which is modified in the 5 position, and therefore is unable to serve as a substrate for the maintenance methylase enzyme (40). Treatment of cells with 5-azacytidine can activate expression of some previously inactive genes, including the HSV TK gene, fetal globin genes, and some integrated (repressed) viral genomes (11, 12, 14, 32). Interpretation of these studies is complicated by possible effects of 5-azacytidine other than inhibition of methylation. 5-Azacytidine has been shown to inhibit DNA and RNA synthesis, protein synthesis, RNA processing, and *de novo* pyrimidine synthesis (40). 5-Azacytidine also is a potent carcinogen and is weakly mutagenic (46). Additionally, the decreased methylation seen following treatment with 5-azacytidine is greater than that expected for the amount of 5-azacytidine that is incorporated (40), suggesting that this compound exerts its effect by inhibition of the methylase enzyme, as well as by being unable to serve as a substrate for the enzyme. Recently, it was proposed that the inhibition may be explained partially by covalent bond formation between DNA containing 5-azacytidine and the methylase enzyme (74).

In Vitro Methylation

A more direct method for demonstrating the causal relationship between changes in methylation and changes in gene expression has been comparison of gene transfer frequencies of methylated and unmethylated DNA, since any changes in expression can be attributed to the methylation. Bacterial methylases generally have been used for *in vitro* methylation studies, because they are easily obtainable and modify specific sites in DNA. Bacterial methylase enzymes modify each of their recognition sites present in bacterial DNA or on cloned DNA carried on plasmids in the bacteria. For example, HpaII methylase, isolated from *Hemophilus parainfluenzae*, modifies the internal C in the sequence CCGG (52), which, as part of a CpG dinucleotide, has been shown to be correlated with expression in many genes. For some studies, prokaryotic adenine methylases such as EcoRI methylase have been used (92). Since there is no detectable methylated adenine in mammalian cells, interpretation of results obtained with this enzyme is difficult.

More recently, mammalian methylase enzymes have been purified and used for *in vitro* methylation studies. These enzymes are physiologically more relevant for the study of mammalian gene expression, although their activity *in vitro* does not fully mimic their activity *in vivo*. *In vitro*, some purified mammalian methylases methylate every CpG sequence, whereas this is not the case *in vivo* (34, 77). It may be that one or more factor(s) necessary for specificity are missing from the purified enzyme preparations, or that the specificity is conferred by a complex mechanism within the cell, influenced by the timing of replication, structure of the chromatin, and/or availability and activity of the methylase enzyme.

In vitro methylation of DNA with any of the methylases often is sufficient to inhibit gene expression. Typically, selectable genes are methylated *in vitro* using one of the methylase enzymes, transfected into cultured cells, and the ability of the DNA to confer the selectable phenotype is determined. The HSV and chicken TK genes, when methylated at HpaII sites, produced TK+ colonies at a frequency which was 2- to 20-fold lower than that obtained using unmethylated DNA (95). When the HSV TK gene was methylated using EcoRI methylase, which methylates a single adenine residue about 70 bp upstream from the cap site, and microinjected into cultured cells, the number of TK-positive cells (as assayed by the ability to incorporate [3H] thymidine) was reduced by a factor of about three (92). Stein et al. (79) cotransferred APRT genes, which had been methylated at all HpaII sites, with the unmodified HSV TK gene, which was used as the selectable gene. Nineteen of twenty-nine positive clones derived by cotransfection of unmethylated or mock-methylated APRT genes with TK were phenotypically APRT+, while only 1 of 18 clones derived by transfection with methylated APRT genes was phenotypically APRT+. The average number of APRT sequences integrated in these cells was similar for both groups. When the methylated APRT genes were used, the integrated copies retained high levels of methylation. Interestingly, when APRT+ revertants were isolated from these negative lines by growth in selective medium, some had undergone extensive demethylation, while others contained amplified copies of the genes, some of which were unmethylated.

These data suggest that some amplified copies of genes may not be methylated, a hypothesis which also has been suggested by the observation that the integrated (unamplified) rRNA genes in *Xenopus* are heavily methylated, while the amplified copies are not extensively methylated (4). One explanation for these results is that during the unscheduled replication, which probably occurs during the amplification process, methylase activity cannot keep up with the replication. Alternatively, the amplification process may take place at a time when methylase activity is not high.

Of several studies of *in vitro* methylation, little effect on expression and replication of papovaviral genomes has, in general, been found (25, 30, 83, 92). Waechter and Baserga (92) microinjected SV40 large T-antigen genes that had been treated with EcoRI methylase and found no effect on T-antigen expression as assayed by immunofluorescence. Graessmann et al. (30) injected either intact SV40 or polyoma viral DNA after methylation with rat liver methylase and found no effect on early and late gene expression, viral replication, transformation ability, or virus maturation. The virus extracted from transformed cells was partly or completely demethylated. Subramanian (83) found no effects of HpaII methylation on the infectivity of polyoma, and also found that there was a loss of methyl groups in the DNA of progeny virus. Fradin et al. (25) tested the effects of HpaII methylase on SV40 expression after injection into *Xenopus* oocytes. No differences were found in T-antigen expression after methylation, although a decrease in expression of the late protein VP1 was detected. Since replication normally is required for VP1 expression, and SV40 replicates poorly in *Xenopus* oocytes, it is difficult to interpret the results. In contrast, Liboi and Basilico (48) studied the expression of polyoma virus integrated into the cellular genome and found an inverse correlation between methylation and T-antigen expression. These data suggest that freely replicating and integrated viruses are under different types of control, and that once a viral genome is integrated, it is regulated by epigenetic phenomena in a fashion similar to cellular genes. These data further suggest that the mammalian cell maintenance methylase enzymes might not be able to methylate molecules that are maintained extrachromosomally in the nucleus following replication of the DNA. However, there are several indications that this is probably not the case. In a few cases, viral DNA which is probably episomal has been found to be partially methylated, including Epstein-Barr virus (65, 72), Shope papilloma virus (94), and human papilloma virus type 1a (17). In the case of Shope papilloma virus, the methylation patterns were postulated to correlate with the degree of neoplasia, since the overall extent of methylation was variable in papillomas but uniformly higher in carcinomas (94). When the DNA of Bovine papilloma virus (BPV) was methylated *in vitro* at the 12 HpaII sites, approximately a 3-fold reduction in the frequency of transformation was obtained in mouse fibroblasts. When the 10 HhaI sites in the virus were methylated instead, no decrease in the transformation frequency was seen (our unpublished results). These results suggest that specific sites or regions of methylation may be important for inhibiting viral gene expression, replication, or both. The HpaII sites within the viral genome are clustered around transcriptional and replication origins, so that methylation of these sites may affect DNA conformation and/or regulatory protein binding differently from methylation of the less clustered HhaI sites.

Retroviral genomes also have been used in a number of *in vitro* methylation studies. McGeady et al. (54) used HpaII or HhaI methylase to methylate Moloney sarcoma virus (MSV) DNA *in vitro*. The ability to transform NIH3T3 cells was inhibited by methylation with either enzyme, while methylation with both enzymes did not inhibit transformation further. The inhibition of transformation activity could be reversed by treatment of the transfected cells with 5-azacytidine. The infrequent transformed cells resulting from methylated DNA contained only unmethylated (integrated) DNA, suggesting that loss of methyl groups was necessary for transformation. Loss of sufficient methyl groups to allow viral production occurred within the first 24 hours following transfection. The v-mos region of the virus was deduced to be the region in which methylation caused the greatest inhibition of viral transforming ability. These data were obtained by cotransfecting cells with plasmids that contained only specific regions of the provirus. These findings have interesting implications for regulation of the endogenous c-mos gene, which in NIH3T3 cells, contains methylated HpaII sites in the 5′ region of the gene and is not transcribed at detectable levels (54). These data suggest that an endogenous methylated c-mos gene might not be able to be activated solely by an insertion of an LTR adjacent to it.

Methylation of the Moloney murine leukemia virus (M-MuLV) at the 35 HpaII sites had only a marginal effect on transformation efficiency (77). When the same viral DNA was methylated using rat liver methylase (which methylates every CpG sequence in the virus), the biological activity was reduced by about three orders of magnitude (77). These results suggest that specific sites of methylation may be important for regulation of viral gene expression, and that sequences other than those detectable by restriction enzymes may be important. These results also point out that data obtained with bacterial methylase or restriction enzymes should be interpreted with caution. The absence of detectable effects with bacterial enzymes may indicate only that sites other than those recognized by the enzyme are important for expression.

The most comprehensive studies of *in vitro* methylation of viral sequences have been done using cloned adenoviral genes. When the cloned Ad 2 E2a gene (which codes for a 72kd Ad 2 specific DNA binding protein) was methylated with HpaII methylase (14 sites), its transcription in *Xenopus* oocytes was inhibited. In contrast, the transcription was not inhibited when the DNA was methylated with BsuRI methylase, which modifies the internal cytosine in the sequence GGCC, rather than in the sequence CCGG (87). These data suggest that highly specific sites that can be methylated are involved in the regulation of the expression of the gene. More recently, Langner et al. have shown that methylation at three HpaII sites in the promoter and 5′ region of the Ad 2 E2a gene inactivated it (47). The transcriptional activities of partially methylated clones of the Ad 2 E2a gene were tested following microinjection into *Xenopus* oocytes. Molecules were constructed either with the 5′ and promoter region methylated and the body of the gene unmethylated, or with the 3′ portion of the gene methylated and the 5′ region and promoter unmethylated. Only the construct methylated at the three 5′ sites failed to be transcribed in the oocytes, although all constructs persisted for at least 24 hours. These results demonstrated that specific methylation sites in the promoter and 5′ region were sufficient for transcriptional inactivation of the E2a gene in *Xenopus* oocytes. In contrast, adenine methylation of these constructs did not affect expression.

Several additional *in vitro* methylation experiments were carried out using adenovirus/chloramphenicol acetyltransferase (CAT) fusion genes (43). Use of the CAT gene allows sensitive and quantitative determination of the promoter activity following methylation, since the amount of enzyme is easily quantitated. CAT constructs containing the SV40 promoter were not affected by either HpaII or HhaI methylation. There are four HpaII sites in the CAT coding region and none in the SV40 promoter region, so that the only conclusion to be drawn is that methylation of the four HpaII sites within the CAT coding region does not affect the expression of the gene. Since there are no HhaI sites in either the CAT coding region or the SV40 promoter, the effects of methylation on this construct cannot be determined with HhaI methylase. The promoter fragments of two different adenovirus 12 genes, the E1a gene and the protein IX gene, were substituted for the SV40 promoter on the CAT construct, so that the CAT gene was under the control of one or the other promoter. The E1a promoter has two HpaII sites and three HhaI sites, and the construct containing this promoter was rendered nonfunctional by either HpaII or HhaI methylation (43). The protein IX promoter has one HpaII and one HhaI site downstream from the TATA sequence and two HhaI sites more than 300 base pairs (bp) upstream from the TATA box (probably outside the promoter region). The protein IX promoter was not inactivated by methylation with either HpaII or HhaI methylase (43).

These data showed that specific sites or combinations of sites of methylation were probably responsible for the inactivation of the promoter region. These data also suggested that methylation within the 5′ portion of these genes was more important for regulating expression than methylation within the main body of the gene. The results obtained using the protein IX gene promoter suggest that sites for methylation either far upstream or downstream from the TATA box may not affect expression.

Results obtained with the γ-globin gene also have provided strong support for the hypothesis that methylation of the promoter and 5′ flanking region of some genes is the most important modification associated with changes in gene expression. Busslinger et al. (9) used a novel *in vitro* methylation technique to methylate specific segments of the gene, while leaving other regions unmethylated. This method involves synthesis of hemimethylated DNA *in vitro* using single-stranded M13 DNA as a template for the synthesis of the minus strand, with 4-methylcytosine as the sole source of cytosine. Therefore, the regions of DNA that are synthesized *in vitro* contain 5-methylcytosine in place of all C residues. When this hemimethylated molecule is transfected into cells, the unmethylated strand presumably will be methylated by cellular maintenance methylase enzymes, for which hemimethylated DNA is the preferred substrate. Using this method, it was found that cells transformed with DNA methylated throughout the M13 vector and the entire globin insert did not express the globin gene. When the γ-globin structural gene was methylated but the 5′ flanking sequences were left unmethylated, there was no inhibition of expression of the globin gene. In contrast, methylation of only the 5′ region of the gene (leaving the structural gene sequences unmodified) resulted in the inhibition of transcription (9). These experiments demonstrate that methylation of the 5′ flanking region of this gene (positions −760 to +100) alone can inhibit its transcription.

In addition to these *in vitro* methylation experiments, which suggest that methylation in the 5' regions of some genes is sufficient to inhibit their expression, other correlations between 5' region undermethylation and expression have been shown. The genes for APRT and DHFR, two housekeeping genes that are expressed in virtually all cell types, are relatively undermethylated in the 5' regions in mouse cells (80). The specific site in a rat hepatoma albumin gene whose undermethylation was correlated with albumin expression is located in the 5' region of the gene (64). In addition to these examples, studies using the HSV TK gene have implicated several specific sites whose methylation pattern is correlated with the state of expression of the gene (13). One of these sites is in the 5' flanking region of the gene, while two other sites are within the structural gene itself, one in the 5' portion and one in the 3' portion. Although two of these sites are within the structural gene sequences, these results also suggest that specific sites of methylation may be important in the control of expression.

Possible Mechanisms of Action

A model for the effects of methylation on gene expression can be envisioned in which methylation could affect gene expression in either a positive or negative manner. A change in the level or pattern of methylation of a certain region could affect the chromatin conformation of that region of DNA. There is some evidence that is consistent with this hypothesis. First, there are several systems in which correlations between chromatin structure (as assayed by nuclease sensitivity) and the relative amount of methylation have been demonstrated (16, 32, 33). Second, methylation of synthetic oligonucleotides can increase their ability to form Z-DNA, allowing the DNA to make the transition from B- to Z-DNA at a physiological salt concentration (3). Although histones bind to Z-DNA *in vitro*, the complex formed does not exhibit a normal nucleosome structure (61). In addition, it has been shown that 5-methylcytosine–guanosine base pairs are more stable than C-G base pairs.

Changes in chromatin structure brought about by changes in methylation theoretically could alter the binding of protein factors important in transcription or the regulation of transcription. These factors could be directly involved in the transcription process (such as RNA polymerase), or they could be positive or negative regulatory factors. There is some evidence that DNA modifications and alterations in conformation can affect protein-DNA interactions. Restriction enzymes that contain CpG sequences in their recognition or cutting sites can be inhibited by methylation of these CpG sites (1, 52). In prokaryotes, lac repressor binding can be altered by modifications that cause alterations in the major groove of the DNA (24, 49). Another example was reported recently for the binding of glucocorticoid receptor to the first intron of the human growth hormone gene, in which the binding was inhibited by methylation of two symmetrically arranged clusters of guanine residues within the binding site (57). The fact that inverse correlations between expression and methylation are seen more often than positive correlations may be due to the fact that modification of a protein binding site causes decreased binding more often than increased binding.

Conclusions

The large number of studies that indicate a role for methylation in the regulation of gene expression leave little doubt that methylation is a fundamental mechanism involved in the control of many genes. Methylation does not seem to be as involved in gene induction or in transient changes in expression as it is in the establishment of long-term patterns of gene expression set up during development and differentiation. A large number of studies demonstrate strong inverse correlations between the extent of methylation and the level of expression of many genes. Similarly, some studies have suggested that levels of methylation can be inversely correlated with the degree of tumor progression, suggesting that demethylation of certain genes, leading to alterations in the level of their expression, may contribute to tumor initiation and progression. Studies employing *in vitro* methylation and 5-azactidine-induced demethylation suggests that methylation is causally related to gene expression. Certainly it seems that methylation of some genes is sufficient to abolish or reduce the level of their transcription. Other studies in which the temporal pattern of transcription and demethylation were studied seem to suggest that methylation may be a marker of genes that are not transcribed rather than a cause of the absence of transcription.

All of these interpretations may be correct, of course. Methylation may act differently for different genes so that in some cases demethylation may be a cause of transcription while in other cases it may be a consequence. It is clear that most genes are not actively transcribed when heavily methylated, and are not methylated significantly when actively transcribed.

There are genes whose expression is positively correlated with the extent of methylation—a mouse histocompatibility antigen gene being the best-documented example. The existence of such genes demonstrates that although the preponderance of genes seem to be negatively modulated by methylation, some genes are positively affected. One could envision models of action that invoke changes in chromatin structure or alterations in protein binding to explain these data. It could be that binding of proteins necessary for gene expression is affected by methylation. Several examples of proteins whose binding is known to be affected by methylation are given above. Alternatively, methylation might affect the chromatin conformation, making genes less accessible to transcription. The demonstration that alternating stretches of purines and pyrimidines undergo transition into the Z state at physiologic salt concentrations when methylated indicates that such alterations may in fact occur.

Cytosine methylation seems to be a gene regulatory mechanism found primarily in higher eukaryotes. Even organisms as advanced as *Drosophila* do not seem to utilize methylation as a mechanism of gene regulation, suggesting that methylation arose relatively recently as a genetic regulatory mechanism. Perhaps the demands placed on genetic mechanisms in organisms as complicated as mammals simply could not be adequately met by normal genetic processes. Methylation may in fact act as a safety net, representing one way of keeping genes off in tissues where their expression would be detrimental. This interpretation is consistent with the findings that demethylation is necessary but not sufficient for activation of silent genes. It may be that in the absence of methylation, the other genetic processes do not operate at sufficient levels to keep the genes off. By superimposing another level of gene regulation, cells are additionally protected against inappropriate gene expression.

Given the wealth of data, it is difficult to contend that cytosine methylation is not involved in gene regulation in vertebrates. Expression of many genes clearly is affected by the level of methylation—most gene expression is affected negatively and in a few cases, positively. Other genes may not be subject to regulation by DNA methylation, suggesting that methylation is not a universal regulatory mechanism. The mechanisms by which methylation affects gene expression (or vice versa) and more precise relationships between methylation and expression of specific genes should be elucidated during the next few years.

General References

DOERFLER, W. DNA methylation and gene activity, *Annu. Rev. Biochem.* 52:93–124.
EHRLICH, M. and WANG, R. Y. -H. 1981, 5-Methylcytosine in eukaryotic DNA, *Science* 212:1350–1357.

References

1. Arber, W., and Linn, S. 1969. DNA modification and restriction, *Annu. Rev. Biochem.* 38:467–500.
2. Battistuzzi, G., D'Urso, M., Toniolo, D., Persico, G. M., and Luzzatto, L. 1985. Tissue-specific levels of human glucose-6-phosphate dehydrogenase correlate with methylation of specific sites at the 3' end of the gene, *Proc. Nat. Acad. Sci. U.S.A.* 82:1465–1469.
3. Behe, M., and Felsenfeld, G. 1981. Effects of methylation on a synthetic polynucleotide: The B-Z transition in poly (dG-m5dC)-poly (dG-m5dC), *Proc. Nat. Acad. Sci. U.S.A.* 78:1619–1623.
4. Bird, A. P., and Southern, E. M. 1978. Use of restriction enzymes to study eukaryotic DNA methylation: I. The methylation pattern in ribosomal DNA from *Xenopus laevis, J. Mol. Biol.* 118:27–47.
5. Bird, A. P., and Taggart, M. H. 1980. Variable patterns of total DNA and rDNA methylation in animals, *Nucleic Acids Res.* 8:1485–1497.
6. Bird, A. P., Taggart, M. H., and Smith, B. A. 1979. Methylated and unmethylated DNA compartments in the sea urchin genome, *Cell* 17:889–901.
7. Breznik, T., and Cohen, J. C. 1982. Altered methylation of endogenous viral promoter sequences during mammary carcinogenesis, *Nature* (*London*) 295:255–257.

8. Burch, J. B. E., and Weintraub, H. 1983. Temporal order of chromatin structural changes associated with activation of the major chicken vitellogenin gene, *Cell* 33:65–76.

9. Busslinger, M., Hurst, J., and Flavell, R.A. 1983. DNA methylation and the regulation of globin gene expression, *Cell* 34:197–206.

10. Chapman, V., Forrester, L., Sanford, J., Hastie, N., and Rossant, J. 1984. Cell lineage-specific undermethylation of mouse repetitive DNA, *Nature (London)* 304:284–286.

11. Charache, S., Dover, G., Smith, K., Talbot, C. C., Moyer, M., and Boyer, S. 1983. Treatment of sickle cell anemia with 5-azacytidine results in increased fetal hemoglobin production and is associated with nonrandom hypomethylation of DNA around the γ-δ-β-globin gene complex, *Proc. Nat. Acad. Sci. U.S.A.* 80:4842–4846.

12. Christy, B., and Scangos, G. 1982. Expression of transferred thymidine kinase genes is controlled by methylation, *Proc. Nat. Acad. Sci. U.S.A.* 79:6299–6303.

13. Christy, B., and Scangos, G. 1984. Changes in structure and methylation pattern in a cluster of thymidine kinase genes, *Mol. Cell. Biol.* 4:611–617.

14. Clough, D. W., Kunkel, L. M., and Davidson, R. L. 1982. 5-Azacytidine-induced reactivation of a herpes simplex thymidine kinase gene, *Science* 216:70–73.

15. Cohen, J. C. 1980. Methylation of milk-borne and genetically transmitted mouse mammary tumor virus proviral DNA, *Cell* 19:653–662.

16. Croce, C. M., Linnenbach, A., Huebner, K., Parnes, J. R., Margulies, D. H., Appella, E. and Seidman, J. G. 1981. Control of expression of histocompatibility antigens (H-2) and B2-microglobulin in F9 teratocarcinoma stem cells *Proc. Nat. Acad. Sci. U.S.A.* 78:5754–5758.

17. Danos, O., Katinka, M., and Yaniv, M. 1980. Molecular cloning, refined physical map and heterogeneity of methylation sites of papilloma virus type 1a DNA, *Eur. J. Biochem.* 109:457–461.

18. Davies, R. L., Fuhrer-Krusi, S., and Kucherlapati, R. S. 1982. Modulation of transfected gene expression mediated by changes in chromatin structure, *Cell* 31:521–529.

19. DeSimone, J., Heller, P., Hall, L., and Zwiers, D. 1982. 5-Azacytidine stimulates fetal hemoglobin synthesis in anemic baboons, *Proc. Nat. Acad. Sci. U.S.A.* 79:4428–4431.

20. Doerfler, W. 1983. DNA methylation and gene activity, *Annu. Rev. Biochem.* 52:93–124.

21. Ehrlich, M., Gama-Sosa, M. A., Huang, L. H., Midgett, R. M., Lo, K.C., McCune, R. A., and Gehrke, C. 1982. Amount and distribution of 5-methylcytosine in human DNA from different types of tissues or cells, *Nucleic Acids Res.* 10:2709–2721.

22. Ehrlich, M., and Wang, R. Y. -H. 1981. 5-Methylcytosine in eukaryotic DNA, *Science* 212:1350–1357.

23. Feinberg, A. P., and Vogelstein, B. 1983. Hypomethylation distinguishes genes of some human cancers from their normal counterparts, *Nature (London)* 301:89–92.

24. Fisher, E. F., and Caruthers, M. H. 1979. Studies on gene control regions XII. The functional significance of a lac operator constitutive mutation, *Nucleic Acids Res.* 7:401–408.

25. Fradin, A., Manley, J. L., and Prives, C. L. 1982. Methylation of simian virus 40

HpaII site affects late, but not early, viral gene expression, *Proc. Nat. Acad. Sci. U.S.A.* 79:5142–5146.

26. Gama-Sosa, M., Midgett, R. M., Slagel, V. A., Githens, S., Kuo, K. C., Gehrke, C. W., and Ehrlich, M. 1983. Tissue-specific differences in DNA methylation in various mammals, *Biochim. Biophys. Acta* 740:212–219.

27. Gama-Sosa, M., Slagel, V. A., Trewyn, R. W., Oxenhandler, R., Kuo, K. C., Gehrke, C. W., and Ehrlich, M. 1983. The 5-methylcytosine content of DNA from human tumors, *Nucleic Acids Res.* 11:6883–6894.

28. Gerber-Huber, S., May, F. E. B., Westley, B. R., Felber, B. K., Hosbach, H. A., Andres, A.-C., and Ryffel, G. U. 1983. In contrast to other *Xenopus* genes the estrogen-inducible vitellogenin genes are expressed when totally methylated, *Cell* 33:43–51.

29. Ginder, G. D., Whitters, M. J., and Pohlman, J. K. 1984. Activation of a chicken embryonic globin gene in adult erythroid cells by 5-azacytidine and sodium butyrate, *Proc. Nat. Acad. Sci. U.S.A.* 81:3954–3958.

30. Graessman, M., Graessmann, A., Wagner, H., Werner, E., and Simon, D. 1983. Complete DNA methylation does not prevent polyoma and simian virus 40 early gene expression, *Proc. Nat. Acad. Sci. U.S.A.* 80:6470–6474.

31. Groffen, J., Heisterkamp, N., Blennerhassett, G., and Stephenson, J. R. 1983. Regulation of viral and cellular oncogene expression by cytosine methylation, *Virology* 126:213–227.

32. Groudine, M., Eisenman, R., and Weintraub, H. 1981. Chromatin structure of endogenous retroviral genes and activation by an inhibitor of DNA methylation, *Nature (London)* 292:311–317.

33. Groudine, M., and Weintraub, H. 1981. Activation of globin genes during chicken development, *Cell* 24:393–401.

34. Gruenbaum, Y., Cedar, H., and Razin, A. 1982. Substrate and sequence specificity of a eukaryotic DNA methylase, *Nature (London)* 295:620–622.

35. Gruenbaum, Y., Naveh-Many, T., Cedar, H., and Razin, A. 1981. Sequence specificity of methylation in higher plant DNA, *Nature (London)* 292:860–862.

36. Harbers, K., Schnieke, A., Stuhlmann, H., Jahner, D., and Jaenisch, R. 1981. DNA methylation and gene expression: Endogenous retroviral genome becomes infectious after molecular cloning, *Proc. Nat. Acad. Sci. U.S.A.* 78:7609–7613.

37. Hattman, S. 1982. DNA, methyltransferase-dependent transcription of the phage Mu mom gene, *Proc. Nat. Acad. Sci. U.S.A.* 79:5518–5521.

38. Holiday, R., and Pugh, J. E. 1975. DNA modification mechanisms and gene activity during development, *Science* 187:226–232.

39. Jahner, D., Stuhlmann, H., Stewart, C. L., Harbers, K., Lohler, J., Simon, I., and Jaenisch, R. 1982. De novo methylation and expression of retroviral genomes during mouse embryogenesis, *Nature (London)* 298:623–628.

40. Jones, P. A., and Taylor, S. M. 1980. Cellular differentiation, cytidine analogs and DNA methylation, *Cell* 20:85–93.

41. Jones, P. A., Taylor, S. M., Mohandas, T., and Shapiro, L. J. 1982. Cell cycle-specific reactivation of an inactive X-chromosome locus by 5-azadeoxycytidine, *Proc. Nat. Acad. Sci. U.S.A.* 79:1215–1219.

42. Kintner, C., and Sugden, B. 1981. Conservation and progressive methylation of Epstein-Barr viral DNA sequences in transformed cells, *J. Virol.* 38:305–316.

43. Kruczek, I., and Doerfler, W. 1983. Expression of the chloramphenicol acetyl-

transferase gene in mammalian cells under the control of adenovirus type 12 promoters: Effect of promoter methylation on gene expression, *Proc. Nat. Acad. Sci. U.S.A.* 80:7586–7590.

44. Kuhlmann, I., and Doerfler, W. 1982. Shifts in the extent and patterns of DNA methylation upon explantation and subcultivation of adenovirus type 12-induced hamster tumor cells, *Virology* 118:169–180.

45. Land, H., Parada, L. F., and Weinberg, R. A. 1983. Tumorigenic conversion of primary embryo fibroblasts requires at least two cooperating oncogenes, *Nature (London)* 304:596–602.

46. Landolph, J. R., and Jones, P. A. 1982. Mutagenicity of 5-azacytidine and related nucleosides in C3H/10T 1/2 clone 8 and V79 cells, *Cancer Res.* 42:817–823.

47. Langner, K. -D., Vardimon, L., Renz, D., and Doerfler, W. 1984. DNA methylation of three 5' C-C-G-G 3' sites in the promoter and 5' region inactivate the E2a gene of adenovirus type 2, *Proc. Nat. Acad. Sci. U.S.A.* 81:2950–2954.

48. Liboi, E., and Basilico, C. 1984. Inhibition of polyoma gene expression in transformed mouse cells by hypermethylation, *Virology* 135:440–451.

49. Lin, S.-Y., and Riggs, A. D. 1972. Lac operator analogues: Bromodeoxyuridine substitution in the lac operator affects the rate of dissociation of the lac repressor, *Proc. Nat. Acad. Sci. U.S.A.* 69:2574–2576.

50. Liskay, R. M., and Evans, R. J. 1980. Inactive X chromosome DNA does not function in DNA-mediated cell transformation for the hypoxanthine phosphoribosyltransferase gene, *Proc. Nat. Acad. Sci. U.S.A.* 77:4895–4898.

51. Mandel, J. L., and Chambon, P. 1979. DNA methylation: Organ-specific variations in the methylation pattern within and around ovalbumin and other chicken genes, *Nucleic Acids Res.* 7:2081–2103.

52. Mann, M. B., and Smith, H. O. 1977. Specificity of Hpa II and Hae III DNA methylases, *Nucleic Acids Res.* 4:4211–4221.

53. Mavilio, F., Giampaolo, A., Care, A., Magliaccio, G., Calandrini, M., Russo, G., Pagliardi, G. L., Mastroberardino, G., Marcinucci, M., and Peschle, C. 1983. Molecular mechanisms of human hemoglobin switching: Selective undermethylation and expression of globin genes in embryonic, fetal and adult erythroblasts, *Proc. Nat. Acad. Sci. U.S.A.* 80:6907–6911.

54. McGeady, M. L., Jhappan, C., Ascione, R., and Vande Woude, G. F. 1983. In vitro methylation of specific regions of the cloned Moloney sarcoma virus genome inhibits its transforming activity, *Mol. Cell. Biol.* 3:305–314.

55. McGhee, J. D., and Ginder, G. D. 1979. Specific DNA methylation sites in the vicinity of the chicken β-globin genes, *Nature (London)* 280:419–420.

56. Mohandas, T., Sparkis, R. S., and Shapiro, L. J. 1981. Reactivation of an inactive human X chromosome: Evidence for X inactivation by DNA methylation, *Science* 211:393–396.

57. Moore, D. D., Marks, A. R., Buckley, D. I., Kapler, G., Payvar, F., and Goodman, H. M. 1985. The first intron of the human growth hormone gene contains a binding site for glucocorticoid receptor, *Proc. Nat, Acad. Sci.* U.S.A. 82:699–702.

58. Morgan, R. A., and Huang, R. C. H. 1984. Correlation of undermethylation of intracisternal A-particle genes with expression in murine plasmacytomas but not in NIH/3T3 embryo fibroblasts, *Cancer Res.* 44:5234–5241.

59. Naveh-Many, T., and Cedar, H. 1981. Active gene sequences are undermethylated, *Proc. Nat. Acad. Sci. U.S.A.* 78:4246–4250.

60. Naveh-Many, T., and Cedar, H. 1982. Topographical distribution of 5-methylcytosine in animal and plant DNA, *Mol. Cell. Biol.* 2:758–562.

61. Nickol, A., Behe, M., and Felsenfeld, G. 1982. Effect of the B-Z transition in poly(dG-m5dC)-poly(dG-m5dC) on nucleosome formation, *Proc. Nat. Acad. Sci. U.S.A.* 79:1771–1775.

62. Niwa, O., and Sugahara, T. 1981. 5-Azacytidine induction of mouse endogenous type C virus and suppression of DNA methylation, *Proc. Nat. Acad. Sci. U.S.A.* 78:6290–6294.

63. Ostrander, M., Vogel, S., and Silverstein, S. 1982. Phenotypic switching in cells transformed with the herpes simplex virus thymidine kinase gene, *Mol. Cell. Biol.* 2:708–714.

64. Ott, M. -O., Sperling, L., Cassio, D., Levilliers, J., Sala-Trepat, J., and Weiss, M. C. 1982. Undermethylation at the 5′ end of the albumin gene is necessary but not sufficient for albumin production by rat hepatoma cells in culture, *Cell* 30:825–833.

65. Perlmann, C., Saemundsen, A. K., and Klein, G. 1982. A fraction of Epstein-Barr virus virion DNA is methylated in and around the EcoRI-J fragment, *Virology* 123:217–221.

66. Ponzetto-Zimmerman, C., and Wolgemuth, D. J. 1984. Methylation of satellite sequences in mouse spermatogenic and somatic DNAs, *Nucleic Acids Res.* 12:2807–2822.

67. Razin, A., and Riggs, A. D. 1980. DNA methylation and gene function, *Science* 219:604–610.

68. Riggs, A. D. 1975. X-inactivation, differentiation, and DNA methylation, *Cytogenet. Cell. Genet.* 14:9–25.

69. Robins, D. M., Axel, R., and Henderson, A. 1981. Chromosome structure and DNA sequence alterations associated with mutation of transformed genes, *J. Mol. Appl. Genet.* 1:191–203.

70. Rogers, J., and Wall, R. 1981. Immunoglobin heavy chain genes: Demethylation accompanies class switching, *Proc. Nat. Acad. Sci. U.S.A.* 78:7497–7501.

71. Romanov, G. A., and Vanyushin, B. F. 1981. Methylation of reiterated sequences in mammalian DNAs. Effects of the tissue type, age, malignancy and hormonal induction, *Biochim. Biophys. Acta* 653:204–218.

72. Saemundsen, A. K., Perlman, C., and Klein, G. 1983. Intracellular Epstein-Barr virus DNA is methylated in and around the EcoRI-J fragment in both producer and nonproducer cell lines, *Virology* 126:701–706.

73. Sanford, J., Forrester, L., Chapman, V., Chandley, A., and Hastie, N. 1984. Methylation patterns of repetitive DNA sequences in germ cells of *Mus musculus Nucleic Acids. Res.* 12:2823–2835.

74. Santi, D. V., Norment, A., and Garrett, C. E. 1984. Covalent bond formation between a DNA-cytosine methyltransferase and DNA containing 5-azacytosine, *Proc. Nat. Acad. Sci. U.S.A.* 81:6993–6997.

75. Scangos, G., and Ruddle, F. H. 1981. Mechanisms and applications of DNA-mediated gene transfer in mammalian cells—a review, *Gene* 14:1–10.

76. Shen, S. T., and Maniatis, T. 1980. Tissue-specific DNA methylation in a cluster of rabbit β-like globin genes, *Proc. Nat. Acad. Sci. U.S.A.* 77:6634–6638.

77. Simon, D., Stuhlmann, H., Jahner, D., Wagner, H., Werner, E., and Jaenisch, R. 1983. Retrovirus genomes methylated by mammalian but not bacterial methylase are non-infectious, *Nature (London)* 304:275–277.

78. Southern, E. M. 1975. Detection of specific sequences among DNA fragments separated by gel electrophoresis, *J. Mol. Biol.* 98:503–517.
79. Stein, R., Razin, A., and Cedar, H. 1982. In vitro methylation of the hamster adenine phosphoribosyltransferase gene inhibits its expression in mouse L cells, *Proc. Nat. Acad. Sci. U.S.A.* 79:3418–3422.
80. Stein, R., Sciaky-Gallili, N., Razin, A., and Cedar, H. 1983. Pattern of methylation of two genes coding for housekeeping functions, *Proc. Nat. Acad. Sci. U.S.A.* 80:2422–2426.
81. Stewart, C. L., Stuhlmann, H., Jahner, D., and Jaenisch, R. 1982. De novo methylation, expression, and infectivity of retroviral genomes introduced into embryonal carcinoma cells, *Proc. Nat. Acad. Sci. U.S.A.* 79:4098–4102.
82. Stuhlmann, H., Jahner, D., and Jaenisch, R. 1981. Infectivity and methylation of retroviral genomes is correlated with expression in the animal, *Cell* 26:221–232.
83. Subramanian, K. N. 1982. Effect of in vitro methylation at CpG sites on gene expression in a genome functioning autonomously in a vertebrate host, *Nucleic Acids Res.* 10:3475–3486.
84. Sutter, D., and Doerfler, W. 1980. Methylation of integrated adenovirus type 12 DNA sequences in transformed cells is inversely correlated with viral gene expression, *Proc. Nat. Acad. Sci. U.S.A.* 77:253–256.
85. Tanaka, K., Appella, E., and Jay, G. 1983. Developmental activation of the H-2K gene is correlated with an increase in DNA methylation, *Cell* 35:457–465.
86. van der Ploeg, L. H. T., and Flavell, R. A. 1980. DNA methylation in the human γ, δ, β-globin locus in erythroid and nonerythroid tissues, *Cell* 19:947–958.
87. Vardimon, L., Kressman, A., Cedar, H., Maechler, M., and Doerfler, W. 1982. Expression of a cloned adenovirus gene is inhibited by in vitro methylation, *Proc. Nat. Acad. Sci. U.S.A.* 79:1073–1077.
88. Vardimon, L., Neumann, R., Kuhlmann, I., Sutter, D., and Doerfler, W. 1980. DNA methylation and viral gene expression in adenovirus-transformed and -infected cells, *Nucleic Acids Res.* 8:2461–2473.
89. Venolia, L., Gartler, S. M., Wassman, E. R., Yen, P., Mohandas, T., and Shapiro, L. J. 1982. Transformation with DNA from 5-azacytidine-reactivated X chromosomes, *Proc. Nat. Acad. Sci. U.S.A.* 79:2352–2354.
90. Waalwyck, C., and Flavell, R. A. 1978. Msp I, an isoschizomer of Hpa II which cleaves both unmethylated and methylated Hpa II sites, *Nucleic Acids Res.* 5:3231–3236.
91. Waalwyck, C., and Flavell, R. A. 1978. DNA methylation at a CCGG sequence in the large intron of the rabbit β-globin gene: Tissue specific variations, *Nucleic Acids Res.* 5:4631–4641.
92. Waechter, D. E., and Baserga, R. 1982. Effect of methylation on expression of microinjected genes, *Proc. Nat. Acad. Sci. U.S.A.* 79:1106–1110.
93. Weintraub, H., Larsen, A., and Groudine, M. 1981. α-Globin-gene switching during the development of chicken embryos: Expression and chromosome structure, *Cell* 24:333–344.
94. Wettstein, F. O., and Stevens, J. G. 1983. Shope papilloma virus DNA is extensively methylated in non-virus-producing neoplasms, *Virology* 126:493–504.
95. Wigler, M., Levy, D., and Perucho, M. 1981. The somatic replication of DNA methylation, *Cell* 24:33–40.
96. Wilks, A. F., Cozens, P. J., Mattaj, I. W., and Jost, J. -P. 1982. Estrogen induces a

demethylation at the 5′ end region of the chicken vitellogenin gene, *Proc. Nat. Acad. Sci. U.S.A.* 79:4252–4255.

97. Wilson, V. L., and Jones, P. A. 1983. Inhibition of DNA methylation by chemical carcinogens in vitro, *Cell* 32:239–246.

98. Wilson, V. L., and Jones, P. A. 1983. DNA methylation decreases in aging but not in immortal cells. *Science* 220:1055–1057.

99. Wolf, S. F., Jolly, D. J., Lunnen, K. D., Friedmann, T., and Migeon, B. R. 1984. Methylation of the hypoxanthine phosphoribosyltransferase locus on the human X chromosome: Implications for X-chromosome inactivation, *Proc. Nat. Acad. Sci. U.S.A.* 81:2806–2810.

100. Yagi, M. and Koshland, M. E. 1981. Expression of the J chain gene during B cell differentiation is inversely correlated with DNA methylation, *Proc. Nat. Acad. Sci. U.S.A.* 78:4907–4911.

101. Yen, P. H., Patel, P. Chinault, A. C. Mohandes, T., and Shapiro, L. J. 1984. Differential methylation of hypoxanthine phosphoribosyltransferase genes on active and ina tive human X chromosomes, *Proc. Nat. Acad. Sci. U.S.A.* 81:1759–1763.

102. Youssoufian, H., Hammer, S. M., Hirsch, M. S., and Mulder, C. 1982. Methylation of the viral genome in an in vitro model of herpes simplex virus latency, *Proc. Nat. Acad. Sci. U.S.A.* 79:2207–2210.

Questions for Discussion with the Editors

1. *Please speculate on the complexity of the methylation machinery in a typical eukaryotic organism. Is the expression of only a single gene (e.g., coding for the "methylase" enzyme) required for sequence specific methylation? Do cell or tissue specific isoforms of the methylation enzyme(s) exist in an organism (e.g., mouse)?*

It is important to distinguish between establishment and maintenance of methylation patterns. It is not necessary to postulate different isoforms of maintenance methylase enzymes, since a simple protein could maintain patterns in all tissues merely by methylating hemimethylated sites after DNA replication. Such an enzyme would maintain a given pattern of methylation regardless of the initial distribution of methylated residues.

More complex is the question of how different patterns of methylation are established in differentiated tissue during development and differentiation. Here one must postulate either tissue-specific isoforms of *de novo* methylase enzymes which methylate different sites in different tissues, or other tissue-specific factors which render specific sites within chromatin sensitive to methylation or other factors which control the specificity of the enzyme. It is clear that the level and pattern of methylation changes drastically during meiosis and differentiation, but little is known of the mechanisms by which these changes occur.

2. *Do all cells in a given differentiated tissue or organ display the same extent of methylation of a specific gene? That is, have quantitative studies generated data to indicate that the extent of methylation (either hypo or hyper) is uniform throughout all cells which constitute a highly differentiated tissue?*

The question as to whether all cells in a given differentiated tissue or organ display a homogeneous methylation pattern is a technically difficult one to address. However, there have been some studies that have looked at methylation patterns in tissues which suggest that the methylation patterns at some sites are heterogenous within a givin tissue. This heterogeneity is suggested by the finding that certain sites are only partially methylated in some tissues, that is, that the site is methylated in less than 100 percent of the molecules. Some examples of genes that appear to be partially methylated at some sites are the ovalbumin gene in chicken tissues[1] and the albumin gene in rat liver.[2] some of this variability probably reflects the heterogenous cell composition of tissues, but some may also reflect differences in methylation pattern among cells of the same type. There is also the problem of contamination of organs with other tissue types, especially blood. It will be important but technically difficult to distinguish between these possibilities.

[1]Mandel, J. L., and Chambon, P. 1979. DNA methylation: Organ-specific variations in the methylation pattern within and around ovalbumin and other chicken genes, *Nucleic Acids Res.*, 7:2081–2103.

[2]Ott, M. -O., Sperling, L., Cassio, D., Levilliers, J., Sala-Trepat, J., and Weiss, M. C. 1982. Undermethylation at the 5′ end of the albumin gene is necessary but not sufficient for albumin production by rat hepatoma cells in culture, *Cell* 30:825–833.

CHAPTER 2

Gene Expression in Cells with an Altered RNA Polymerase II

Ricky R. Hirschhorn and Renato Baserga

GENE EXPRESSION in higher eukaryotes has been studied in many systems, mostly *in vitro*. Unfortunately, many *in vitro* studies do not correlate well with *in vivo* studies. For instance, the absolute *in vitro* requirement for a TATA box for transcription does not strictly correlate with an *in vivo* requirement. In fact, several genes shown not to contain a TATA box are transcribed successfully *in vivo* (6). Clearly, *in vivo* experiments offer a more direct approach for studying physiologically significant processes.

Perhaps one of the more useful *in vivo* systems is that of tsAF8 cells, a baby hamster kidney (BHK) derived cell line (24) in which RNA polymerase II activity is temperature sensitive (27, 29, 31, 36). Extensive studies of the growth characteristics of these cells show that at the nonpermissive temperature of 40.6°C the cells arrest specifically in the G_1 phase of the cell cycle, while at the permissive temperature of 34°C, they grow normally (7). Additionally, if tsAF8 cells are made quiescent by serum deprivation and are subsequently stimulated with 10 percent serum, they enter S phase at the permissive temperature with a median lag time of 28 hours, but do not enter S phase at the nonpermissive temperature (2, 35). The execution point of the temperature sensitive mutation is about 8 hours before S phase. Viability is good for 3 to 5 days at the nonpermissive temperature and the cells can be easily rescued by returning to the permissive temperature (24). During the first 24 to 30 hours at the nonpermissive temperature, neither ribosomal RNA synthesis (3, 27) nor overall protein synthesis (29) are affected.

In this chapter we will first summarize the data that indicate that tsAF8 cells have a defective RNA polymerase II and then we will describe how this cell line can be utilized to study the expression of either endogeneous genes or genes introduced by various gene transfer methods.

tsAF8 Cells Contain a Defective RNA Polymerase II

The original observation by Rossini and Baserga (27) showed that RNA polymerase II activity, as measured in isolated nuclei, decreases with a half-life of about 10 hours in tsAF8 cells shifted to the nonpermissive temperature. At the permissive temperature, RNA polymerase II activity remains constant for at least 40 hours after serum stimulation. Under these same conditions, RNA polymerase I activity (measured in isolated nucleoli) remains high at both the permissive and nonpermissive temperatures (27) for at least 30 hours after serum stimulation. Binding studies using [$_3$H]-α-amanitin showed that the α-amanitin binding subunit of the RNA polymerase II molecule disappeared in tsAF8 cells at the nonpermissive temperature with a half-life of 10 hours. Neither RNA polymerase II activity nor α-amanitin binding ability of cell extracts are affected when the parent cell line, BHK cells, is incubated at 41°C (29). Indeed, BHK cells grow very well at 41°C, which makes temperature-sensitive mutants of this cell line particularly attractive. The same cannot be said of other cell lines from different species. Chinese hamster ovary (CHO) cells grow poorly at 41°C, human cells don't do well above 40°C, and 3T3 cells simply stop growing at 39°C. The defect in RNA polymerase II activity is therefore specific for tsAF8 shifted to the restrictive temperature. Under these same conditions overall protein synthesis is not affected for up to 24 hours at the nonpermissive temperature (29). The temperature-sensitive defect in tsAF8 cells can be corrected by transfection with DNA isolated from α-amanitin-resistant non–temperature-sensitive CHO cells (20) and the resulting temperature-resistant transformant shows an α-amanitin-resistant RNA polymerase II activity. Finally, when partially purified RNA polymerase II is manually microinjected into the nuclei of tsAF8 cells, the temperature-sensitive defect is corrected and the cells enter DNA synthesis at the nonpermissive temperature (26). All of these studies, taken together, indicate that the temperature-sensitive defect in tsAF8 cells is either in the synthesis, the assembly, or turnover of RNA polymerase II. That these cells arrest specifically in the G_1 phase of the cell cycle at the nonpermissive temperature indicates the requirement for unique copy gene transcription for the transition of cells from a resting to a growing state, and has been reviewed elsewhere (4).

Gene Transfer Studies in tsAF8 Cells

Gene transfer has been used both as a method of studying gene expression in higher eukaryotes as well as a method to introduce genes which confer a selective advantage to the recipient cell. There are several methods for introducing genes into viable eukaryotic cells including cell fusion, manual microinjection, transfection, liposome delivery, loaded red blood cell ghosts, and virus infection. All of these techniques can be used to study short-term (phenotypic) or long-term (genotypic) modification of cells. In this section we will concentrate on manual microinjection, transfection, and virus infection as methods of introducing DNA into viable tsAF8 cells. In most cases, we will look at short-term or transient expression of the transferred genes.

The manual microinjection technique was developed by Graessman and coworkers (12, 13) and Diacumakos and colleagues (1, 9) to permit individual nuclei (or cytoplasm) of cultured cells to be loaded with macromolecules including DNA, cRNA, mRNA, and proteins. An experienced operator can microinject approximately 300 to 400 cells in 1 hour. If DNA is the macromolecule of interest, and is at a concentration of 1 mg/ml, the cell can be microinjected with close to 2000 copies of a cloned gene depending, of course, on the size of the plasmid. The survival rate after this method of gene transfer is greater than 85 percent. Unfortunately, only a small number of cells can be microinjected in a reasonable time, which limits the types of experiments that can be done by this method.

The method of transfection whereby cells take up DNA as a coprecipitate with calcium phosphate was developed by Graham and Van der Eb (14) as a way of introducing purified viral DNA into cells. Since their original experiments, DNA transfection has been used to introduce a wide variety of genes into cells (8, 18, 19, 22, 25, 26, 32, 33, 37, 38). The original procedure was designed to study long-term genotypic effects and therefore, the relatively low efficiency of DNA transfection was unimportant. Because our laboratory is mostly interested in short-term changes caused by a variety of genes, we developed a modified transfection procedure which, under optimal conditions, permits up to 70 percent of the transfected cells to transiently express the introduced gene (32).

In the experiment shown in Table 1, cells were either transfected or microinjected with a recombinant plasmid DNA, pSV2G, which contains all of the early region of SV40 cloned into the plasmid vector pBR322 (11). This

TABLE 1 Efficiency of transfection or microinjection in different cell lines[a]

Cell line[b]	Percent T$^+$ cells	
	Microinjection	Transfection
TK$^-$ts13	63.8	25.2
tsAF8	60.1	21.5
ts13	85.0	9.7
TC-7	92.0	8.2
NIH3T3	73.8	3.0

[a]Transfection and microinjection with pSV2G DNA are described by Shen et al. (31). pSV2G contains the entire early region of SV40, including the region coding for the large tumor (T) antigen, cloned in pBR322 (11). Efficiency of T-antigen expression was determined by indirect immunofluorescence.

[b]The cell lines used are described in the text and conditions for their growth have been previously described (32).

Source: Adapted from Shen et al. (32).

region of SV40 codes for the large T viral antigen which can be dected by indirect immunofluorescence. Three of the cell lines used—TK-ts13, tsAF8, and ts13—were all originally derived from BHK cells. TC-7 cells are a subclone of the African green monkey kidney cell line CV-1. NIH3T3 is a mouse cell line. Notice that the efficiency of expression of pSV2G DNA introduced by microinjection is high in all cell lines tested. When this same gene is introduced into the same cells by transfection, the efficiency varies greatly. While the efficiency of transfection is above 20 percent for the hamster-derived cells, TK-ts13 and tsAF8, it is only 3 percent for the mouse-derived NIH3T3 cells. In no case was the efficiency of transfection as high as that of microinjection even though the gene was allowed to express for the same amount of time. For the purpose of this chapter, it is important to note that tsAF8 cells were highly transfectable with pSV2G DNA and this is also true for other cloned genes.

In the next experiment, the effect of shifting to the nonpermissive temperature on the expression of pSV2G DNA, introduced by microinjection or transfection, was studied (see Table 2). At various times after transfection or microinjection, tsAF8 cells were shifted up to the nonpermissive temperature, which inactivates the RNA polymerase II in the cell and essentially stops gene transcription. Forty-eight hours after transfection the cells were fixed and the percentage of cells expressing the SV40 T antigen was determined by indirect immunofluorescence. The different methods of gene transfer have a very different profile in expression of the transferred gene. Microinjection is clearly faster and more efficient. If the cells are immediately shifted up to the nonpermissive temperature following gene transfer, neither the microinjected nor the transfected cells express the SV40 genome. When the microinjected cells are shifted up 6 hours after gene transfer, sufficient

TABLE 2   Effect of nonpermissive temperature on the expression of SV40 T antigen in tsAF8 cells[a]

Time of shift up (hours)	Percent T+ cells	
	Transfection	Microinjection
0	0	ND[b]
3	0	6.6
6	0.8	24.1
8	ND	31.4
17	1.8	ND
24	3.5	30.7
48	5.8	74.0

[a]After microinjection or transfection of tsAF8 cells with pSV2G DNA, the cells were shifted to the nonpermissive temperature at the times indicated. After a total of 48 hours, the cells were fixed and the percent of cells expressing T antigen was determined by indirect immunofluorescence.

[b]ND, not done.

Source: Adapted from Baserga et al. (5).

SV40 mRNA has been transcribed for 24 percent of the cells to contain SV40 T antigen 42 hours later. Less than 1 percent of the transfected cells are T positive at this same time point. If the cells are allowed to remain at the permissive temperature for the entire time, 74 percent of the microinjected cells express the SV40 T antigen compared to only 6 percent of the transfected cells. Since transcription is required for the ultimate expression of the SV40 T antigen, it appears that direct microinjection permits transcription more efficiently and in less time than transfection.

Another method of gene transfer is that of virus infection. However, this is not a universal technique as many types of cells are resistant to infection by certain viruses. BHK-derived cell lines, such as tsAF8, are resistant to infection by SV40, but not by adenovirus 2. Therefore, adenovirus 2 information can be introduced into tsAF8 by infection but SV40 cannot. Since it is known that oncogenic DNA viruses, such as SV40 and adenovirus 2, can induce DNA synthesis in resting cells, gene expression can be monitored by simple autoradiographic assay for DNA synthesis in cultures labeled with [^{3}H]thymidine. Table 3 compares the effect of microinjection of SV40 to infection with adenovirus 2 on the ability of quiescent tsAF8 cells to enter S phase at the permissive and nonpermissive temperatures.

Before these data are discussed, it is useful to introduce information on another BHK-derived G_1-specific temperature-sensitive mutant, known as ts13 (34). Similar to tsAF8 cells, the nonpermissive temperature stops ts13 cells in G_1 (10, 34). Other macromolecular synthesis, including RNA and overall protein syntheses, are not severely affected for the first 24 hours at the nonpermissive temperature (34). In fact, microinjected genes are expressed

---

TABLE 3 Stimulation of DNA synthesis in tsAF8 cells at the permissive and nonpermissive temperatures after gene transfer

Temperature (°C)	SV40[a]	T[+b]	Adenovirus[c]
34(pt⁻)	65%	86%	60%
40.6 (npt)		0	52
34 (pt), 40.6 (npt)[d]	59	78	ND

Note: pt, Permissive temperature; npt, nonpermissive temperature; ND, not done.

[a]Quiescent tsAF8 cells were microinjected with pSV2G DNA and incubated at the temperatures indicated. The microinjected cells were labeled with [^{3}H]thymidine for 24 hours, fixed, and processed for autoradiography to determine the percentage of cells in DNA synthesis.

[b]The percentage of cells expressing T antigen was determined by indirect immunofluorescence.

[c]Quiescent tsAF8 cells were infected with adenovirus 2 and * incubated at the temperatures indicated. The percentage of cells in DNA synthesis 24 hours later was determined as described in footnote *a*.

[d]Following gene transfer, cells were incubated at 34°C for 6 hours and then shifted up to 40.6°C.

Source: Adapted from Floros et al. (11).

equivalently at both the permissive and nonpermissive temperatures (11). Unlike tsAF8 cells, the viability of ts13 cells declines rapidly if the cells are kept at the nonpermissive temperature for longer than 24 hours. The execution point of the temperature-sensitive mutation is distinct from that of tsAF8 cells and is about 3 hours prior to S (10).

At the permissive temperature, both SV40 and adenovirus 2 can stimulate quiescent tsAF8 cells into DNA synthesis with equal efficiency. At the nonpermissive temperature, SV40 cannot cross the temperature-sensitive block of tsAF8, yet it can cross the temperature sensitive block of ts13 cells and induce cells into DNA synthesis (11). Adenovirus 2 can stimulate both quiescent tsAF8 cells and quiescent ts13 cells at the nonpermissive temperature with equal efficiency (28). There are no T-positive cells present when tsAF8 cells microinjected with SV40 DNA are incubated at the nonpermissive temperature. When a shift up to the nonpermissive temperature is delayed for 6 hours after microinjection of tsAF8 cells, T antigen is expressed and the cells enter DNA synthesis. This is a curious result since it must be assumed that both SV40 and adenovirus 2 must be expressed as RNA to induce cellular DNA synthesis, yet adenovirus can do so while SV40 cannot. Since the RNA polymerase II activity declines rapidly in tsAF8 cells at the nonpermissive temperature, it appears that adenovirus 2 can sequester sufficient functional enzyme to permit the expression of the regions necessary for the induction of DNA synthesis while SV40 cannot. This ability could result from a difference in the strength of promoters or the quality and quantity of enhancers present in these two different viral genomes.

mRNA Stability in tsAF8 Cells

When tsAF8 cells are shifted to the nonpermissive temperature, transcription mediated by RNA polymerase II rapdily decreases. The implication is that one can then study the half-life of any mRNA species present in tsAF8 at the time of shift up for which there is a probe. Figure 1 shows the results of such an experiment. Total cytoplasmic RNA is isolated from exponentially growing tsAF8 cells that were shifted up to the nonpermissive temperature for different lengths of time. The RNA is then separated on a formaldehyde-agarose gel and blotted onto nitrocellulose. Figure 1A shows the results when this blot is hybridized with a cDNA clone, p2A9, whose expression has been characterized as being G_1-specific (16). The probe hybridizes to a single band of 750 bases at all time points studied, but with decreasing intensity, even though equivalent amounts of RNA were loaded onto the gel. When this probe is removed and the blot is reprobed with p11D6, a non–cell-cycle-related cDNA clone, the intensity of hybridization is the same for up to 6 hours at the nonpermissive temperature. When the hybridization is quantitated by scanning densitometry, it is clear that the RNA homologous to p2A9 decays with a half-life of less than 2 hours. The half-life of p11D6 is much longer and is approximated to be at least 12 hours as determined from later time points not shown in Fig. 1. This system has been used to study the

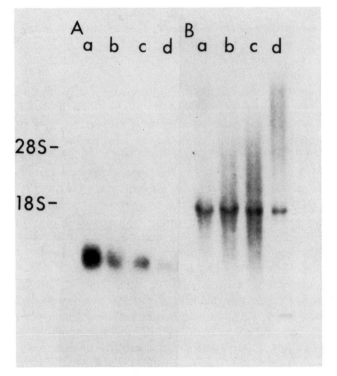

Figure 1 mRNA stability studies using total RNA isolated from tsAF8 cells. Total cytoplasmic RNA was isolated from tsAF8 cells shifted up to the nonpermissive temperature for different lengths of time. Following separation on a formaldehyde-agarose gel, the RNA was blotted onto nitrocellulose and hybridized as previously described (16). (A) p2A9 as probe. (B) p11D6 as probe. (a) 0 hours; (b) 3 hours; (c) 6 hours; and (d) 12 hours after shift up to the nonpermissive temperature. The same blot was used for both hybridizations.

mRNA stability of several other G_1-specific cDNA clones described by Hirschhorn et al. (16). In all cases, these mRNA species decay with a half-life of less than 2 hours. Other cloned genes, including c-myc and ornithine decarboxylase, decay with a relatively short half-life. Conversely, another non–cell-cycle-related cDNA clone, p7B5, hybridized with equal efficiency to all time points and was thus designated a long-lived species. These types of experiments point out the usefulness of tsAF8 cells in the study of mRNA stability for both long- and short-lived species. mRNA stability can be determined in these cells in the absence of drugs or radioactive precursors, both of which have been known to cause artifactual results (30). Drugs such as actinomycin D are known to have phenotypic effects. Incorporation of radio-active precursors into RNA can be very complicated to interpret since the altered precursor pool sizes can affect many cellular functions other than RNA synthesis and since complete removal of the labeled precursor is impossible for a true chase. The system described above using tsAF8 cells avoids these problems and allows the stability of specific mRNAs to be determined.

Gene Expression of Cell-Cycle-Related Sequences in tsAF8 Cells

In addition to the types of studies mentioned above, one can study, by various methods, the expression of endogenous sequences represented in the genome of tsAF8. Since our laboratory is predominantly interested in the cell cycle progression of eukaryotic cells, the types of studies we will describe here deal with the expression of sequences that are related to the eukaryotic cell cycle.

Recently, we identified several cDNA clones derived from a hamster cell line, whose expression was shown to specifically increase during the G_1-phase of the eukaryotic cell cycle (16). These clones were selected from a cDNA library prepared from poly(A^+)mRNA isolated from ts13 cells 6 hours after serum stimulation at the nonpermissive temperature. Replicas of the library were then hybridized with cDNA probes synthesized using poly(A^+)mRNA from serum-deprived (G_0) ts13 cells or from cells stimulated for 6 hours with serum (G_1). Less than 1 percent of the clones repeatedly hybridized to the G_1 probe and were considered to be derived from G_1-specific mRNA. The other colonies hybridized to both probes, or not at all. The level of expression of RNA homologous to these cDNA clones was studied in tsAF8 cells and ts13 cells at the permissive and nonpermissive temperatures. RNA isolated from either serum-deprived cells or cells stimulated with serum at 34°C or at 40.6° (tsAF8) or 39.6° (ts13) was dotted onto nitrocellulose and hybridized with cell-cycle-related clones, or a non–cell-cycle-dependent clone, p7B5. The resulting autoradiographs were then scanned with a soft laser densitometer.

An example of this type of experiment is shown in Fig. 2. In this experiment, the levels of expression of a cell-cycle-related cDNA clone, p2F1, and a non–cell-cycle-dependent cDNA clone, p7B5, were studied in tsAF8 cells at the permissive and nonpermissive temperatures. It is clear that the levels of RNA homologous to p7B5 do not change regardless of the physiological state of the cell and the temperature. Yet the levels of RNA homologous to p2F1 increase over that of quiescent cells when tsAF8 cells are stimulated at the permissive temperature, yet decrease dramatically when the cells are stimulated at the nonpermissive temperatures. In ts13 cells, these same RNA sequences are elevated at the nonpermissive temperature. A summary of these results for several cell-cycle-related cDNA clones is presented in Table 4. All of the cell-cycle-related sequences increase in G_1 in both tsAF8 and ts13 cells at the permissive temperature while p7B5 does not in either cell line. Some of these sequences increase in ts13 cells at the nonpermissive temperature to levels higher than those seen at the permissive temperature. Yet all of the cDNA clones that were shown to be G_1-specific decreased in tsAF8 cells at the nonpermissive temperature. This suggests that these cell-cycle-related sequences have a rapid turnover while some non–cell-cycle-related sequences such as p7B5 are long-lived.

Another interesting cell-cycle-related group of genes are the histones. Although many studies have shown that core histone RNA is present in detectable amounts only in S phase cells, studies in tsAF8 cells were required to

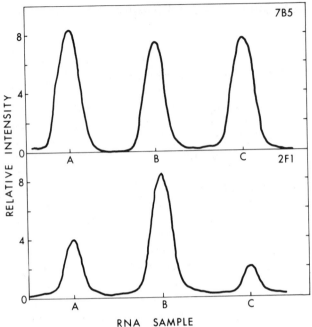

Figure 2 Densitometry tracing of dot blots made with cytoplasmic RNA from tsAF8 cells. Total cytoplasmic RNA was isolated from (A) quiescent tsAF8 cells; (B) tsAF8 cells serum stimulated for 6 hours at the permissive temperature; and (C) tsAF8 cells serum stimulated for 6 hours at the nonpermissive temperature. The RNA was dotted onto nitrocellulose and hybridized to p7B5 and p2F1 as described (16). The same blot was used for both hybridizations.

show conclusively that core histone expression depended absolutely on the entry of tsAF8 cells into the S phase of the cell cycle (17). In these experiments, summarized in Table 5, the levels of core histone RNA was determined in tsAF8 cells under different conditions including nonstimulated cells, cells stimulated for 24 hours at the permissive or nonpermissive temperatures, or when shift up to the nonpermissive temperature was delayed for 6, 16, and 20 hours. From this group of experiments, it is clear that the expression of the core histones was directly related to the number of cells capable of entering S phase. When entry into S is prevented by the nonpermissive temperature, no core histone expression is detectable. There is also a direct correlation between histone expression, the ability of cells to enter DNA synthesis, and the ability of the cells to divide (17).

Other cell-cycle-related genes are thymidine kinase (TK) and dihydrofolate reductase (DHFR) which are related to cell-cycle progression and are known to be associated with the onset of DNA synthesis. The expression of TK and DHFR was determined in tsAF8 and ts13 cells at the permissive and nonpermissive temperatures by determining both the enzymatic activities

TABLE 4 Expression of G_1-specific cDNA clones[a]

	Permissive temperature		Nonpermissive temperature	
Clone	tsAF8	ts13	tsAF8	ts13
p2A8	↑	↑	↓	=
p2A9	↑	↑	↓	↑
p2F1	↑	↑	↓	↑
p4F1	↑	↑	↓	↑
p7B5	=	=	=	=

Note: =, same as G_0; ↑, increased; ↓, decreased.

[a]Total cytoplasmic RNA was isolated from tsAF8 and ts13 cells stimulated with serum at both the permissive and nonpermissive temperatures, for varying lengths of time. RNA was dotted onto nitrocellulose and hybridized to nick-translated inserts isolated from the cDNA clones (listed in the first column) as previously described (16). The autoradiographic result was quantitated by densitometry.

TABLE 5 Levels of histone RNA in tsAF8 cells stimulated at the permissive and nonpermissive temperatures[a]

Temperature	H2A/H2B	H3	H4	Percent labeled cells
NS	—	—	—	3
pt, 24 hours	+	++	+	60
npt, 24 hours	—	—	—	2
↑ npt, 6 hours	—	—	—	5
↑ npt, 16 hours	+	+	+	16
↑ npt, 20 hours	++	++	++	27

Note: Not detectable; +, weak signal; ++, strong signal.

[a]Cells were stimulated at time zero and then incubated at either the permissive (pt) or nonpermissive (npt) temperature for 24 hours. Other cells were serum stimulated at time zero then incubated at the permissive temperature for either 6, 16, or 20 hours before shift up (arrow). NS represents cells that were not stimulated and were maintained at the permissive temperature. Total cellular RNA was isolated after 24 hours and separated on agarose-formaldehyde gels. Levels of histone RNA were determined by hybridization to ^{32}P-labeled histone-specific probes. DNA synthesis was determined by incubation in the presence of [^{3}H]thymidine for 24 hours.

TABLE 6 Expression of thymidine kinase (TK) and dihydrofolate reductase (DHFR) in tsAF8 and ts13 cells at the permissive and nonpermissive temperature[a]

	taAF8			ts13		
	Quiescence	34°C	40.6°C	Quiescence	34°C	39.6°C
Enzymatic activity:						
TK (cpm/μg)	80	1800	200	50	1000	50
DHFR (cpm/mg)	2200	8500	7800	2300	8200	1800
RNA levels:						
TK	—	+	—	—	+	—
DHFR	+	++	+	—	++	—

Note: —, Not detectable; +, weak signal; ++, strong signal.
[a]TK activity in tsAF8 and ts13 cells was determined by the method of Harris (15). DHFR activity in tsAF8 and ts13 cells was determined by the method of Johnson et al. (21). In both cases, activity was normalized to the amount of total protein present. RNA levels were determined by dot blot and Northern analysis as described by Hirschhorn et al. (16).

and the relative RNA levels (23). TK activity was determined by incorporation of [³H-] thymidine into cell extracts according to the method of Harris (15). DHFR was determined by incorporation of [³H]-methotrexate according to the method of Johnson et al. (21). RNA levels were determined by Northern blot analysis as previously described (16). These results are summarized in Table 6. When tsAF8 and ts13 cells are stimulated at the permissive temperature, the levels of TK mRNA and enzyme activity increased markedly. At the nonpermissive temperatures, both TK activity and mRNA levels remained low as in quiescent cells. DHFR activity and mRNA levels also increased when both cell lines were stimulated at the permissive temperatures. The levels of DHFR mRNA and enzyme activity in ts13 cells stimulated at the nonpermissive temperature are both low. However, DHFR enzyme activity and mRNA levels both increase in tsAF8 cells stimulated at the nonpermissive temperature. It may be that the levels of DHFR mRNA in tsAF8 cells at the nonpermissive temperature reflect the stability of the DHFR message. In any case, it appears that the expression of different genes is affected differently by the different temperature-sensitive blocks, even when the genes are growth related.

Conclusions

We have attempted to describe some of the experiments one can carry out using tsAF8 cells, in which one can take advantage of the fact that these cells contain a nonfunctional RNA polymerase II at the restrictive temperature and that they arrest specifically in G_1. Both endogenous and transferred genes can be studied since tsAF8 cells are easily microinjected and are highly transfectable. Use of these cells at both the permissive and nonpermissive temperatures can afford keen insight into the manner in which specific genes are regulated.

General References

BASERGA, R. 1985. *The Biology of Cell Reproduction*, Harvard University Press, Cambridge, Mass.

BASERGA, R., CROCE, C., and ROVERA, G. (eds.). 1980. *Introduction of Macromolecules into Viable Mammalian Cells*, Alan R. Liss, New York.

References

1. Anderson, W. F., Killos, L., Sanders-Haigh, L., Kretschmer, P. J., and Diacumakos, E. G. 1980. Replication and expression of thymidine kinase and human globin genes microinjected into mouse fibroblasts. *Proc. Nat. Acad. Sci. U.S.A.* 77:5399–5403.

2. Ashihara, T., Chang, S. D., and Baserga, R. 1978. Constancy of the shiftup point in two temperature-sensitive mammalian cell lines that arrest in G_1 *J. Cell. Physiol.* 96:15–22.

3. Ashihara, T., Traganos, F., Baserga, R., and Darzynkiewicz, Z. 1978. A comparison of cell cycle-related changes in post mitotic and quiescent AF8 cells as measured by cytofluorometry after acridine orange staining, *Cancer Res.* 38:2514–2518.

4. Baserga, R., Waechter, D. E., Soprano, K. J., and Galanti, N. 1982. Molecular biology of cell division, in: *Annals of the New York Academy of Sciences*, vol. 397 (R. Baserga, ed.), pp. 110–120, New York Academy of Sciences, New York.

5. Baserga, R., Shen, Y.-M., and Yuan, Z.-A. 1984. Cellular transformation systems, in: *Progress in Cancer Research and Therapy*, vol. 29 (S. R. Wolman and A. J. Mastromarino, eds.), pp. 377–385, Raven Press, New York.

6. Breathnach, R., and Chambon, P. 1981. Organization and expression of eucaryotic split genes coding for proteins, *Annu. Rev. Biochem.* 50:349–383.

7. Burstin, S. J., Meiss, H. K., and Basilico, C. 1974. A temperature-sensitive mutant of the BHK cell line, *J. Cell. Physiol.* 84:397–408.

8. Corsaro, C. M., and Pearson, M. L. 1981. Enhancing the efficiency of DNA-mediated gene transfer in mammalian cells, *Somatic Cell Genet.* 7:603–616.

9. Diacumakos, E. G. 1980. Introduction of macromolecules into viable mammalian cells by precise physical microinjection, in: *Introduction of Macromolecules into Viable Mammalian Cells* (R. Baserga, C. Croce, and G. Rovera, eds.), pp. 85–98, Alan R. Liss, New York.

10. Floros, J., Ashihara, T., and Baserga, R. 1978. Characterization of ts13 cells: A temperature-sensitive mutant of the G_1 phase of the cell cycle, *Cell Biol. Int. Rep.* 2:259–269.

11. Floros, J., Jonak, G., Galanti, N., and Baserga, R. 1981. Induction of cell DNA replication in G_1-specific ts mutants by microinjection of SV40 DNA, *Exp. Cell Res.* 132:215–223.

12. Graessmann, M., and Graessmann, A. 1976. "Early" simian-virus-40-specific RNA contains information for tumor antigen formation and chromatin replication, *Proc. Nat. Acad. Sci. U.S.A.* 73:366–370.

13. Graessmann, A., Graessmann, M., and Mueller, C. 1977. Regulatory function of simian virus 40 DNA replication for late viral gene expression, *Proc. Nat. Acad. Sci. U.S.A.* 74:4831–4834.

14. Graham, F. L., and Van der Eb, A. J. 1973. A new technique for the assay of infectivity of human adenovirus 5 DNA, *Virology* 52:456–467.

15. Harris, M. 1975. Non-mendelian segregation in hybrids between Chinese hamster cells, *J. Cell. Physiol.* 86:413–430.

16. Hirschhorn, R. R., Aller, P., Yuan, Z.-A., Gibson, C. W., and Baserga, R. 1984. Cell-cycle-specific cDNAs from mammalian cells temperature sensitive for growth, *Proc. Nat. Acad. Sci. U.S.A.* 81:6004–6008.

17. Hirschhorn, R. R., Marashi, F., Baserga, R., Stein, J., and Stein, G. 1984. Expression of histone genes in a G_1-specific temperature-sensitive mutant of the cell cycle, *Biochemistry* 23:3731–3735.

18. Huttner, K. M., Scangos, G. A., and Ruddle, F. H. 1979. DNA-mediated gene transfer of a circular plasmid into murine cells, *Proc. Nat. Acad. Sci. U.S.A.* 76:5820–5824.

19. Huttner, K. M., Barbosa, J. A., Scangos, G. A., Pratcheva, D. D., and Ruddle, F. H. 1981. DNA-mediated gene transfer without carrier DNA, *J. Cell. Biol.* 91:153–156.

20. Ingles, C. J., and Shales, M. 1982. DNA-mediated transfer of an RNA polymerase II gene: Reversion of the temperature-sensitive hamster cell cycle mutant tsAF8 by mammalian DNA, *Mol. Cell. Biol.* 2:666–673.

21. Johnson, L. F., Fuhrman, C. L., and Wiedemann, L. M. 1978. Regulation of dihydrofolate reductase gene expression in mouse fibroblasts during the transition from the resting to growing state, *J. Cell. Physiol.* 97:397–406.

22. Kit, S., Otsuka, H., Qavi, H., Trkula, D., Dubbs, D. R., and Hazen, M. 1980. Biochemical transformation or thymidine kinase (TK)-deficient mouse cells by herpes simplex virus type 1 DNA fragments purified from hybrid plasmids, *Nucleic Acids Res.* 8:5233–5253.

23. Liu, H. -T., Gibson, C. W., Hirschhorn, R. R., Rittling, S., Baserga, R., and Mercer, W. E. 1985. Expression of thymidine kinase and dihydrofolate reductase genes in mammalian ts mutants of the cell cycle. *J. Biol. Chem.* 260:3269–3277.

24. Meiss, H. K., and Basilico, C. 1972. Temperature sensitive mutants of BHK 21 cells, *Nature New Biol.* 239:66–68.

25. Miller, G., Wertheim, P., Wilson, G., Robinson, J., Geelen, J. L. M. C., van der Noordaa, J., and Van der Eb, A. J. 1979. Transfection of the human lymphoblastoid cells with herpes simplex viral DNA, *Proc. Nat. Acad. Sci. U.S.A.* 76:949–953.

26. Pellicer, A., Wigler, M., Axel, R., and Silverstein, S. 1978. The transfer and stable integration of the HSV thymidine kinase gene into mouse cells, *Cell* 14:133–141.

27. Rossini, M., and Baserga, R. 1978. RNA synthesis in a cell cycle-specific temperature sensitive mutant from a hamster cell line, *Biochemistry* 17:858–863.

28. Rossini, M., Weinmann, R., and Baserga, R. 1979. DNA synthesis in temperature-sensitive mutants of the cell cycle infected by polyoma virus and adenovirus, *Proc. Nat. Acad. Sci. U.S.A.* 76:4441–4445.

29. Rossini, M., Baserga, S., Huang, C. H., Ingles, C. J., and Baserga, R. 1980. Changes in RNA polymerase II in a cell cycle-specific temperature-sensitive mutant of hamster cells. *J. Cell. Physiol.* 103:97–103.

30. Rovera, G., Berman, S. and Baserga, R. 1970. Pulse labeling at RNA of mammalian cells, *Proc. Nat. Acad. Sci. U.S.A.* 65:876–883.

31. Shales, M., Bergsagel, J., and Ingles, C. J. 1980. Defective RNA polymerase II in the G$_1$-specific temperature sensitive hamster cell mutant tsAF8, *J. Cell. Physiol.* 105:527–532.

32. Shen, Y.-M., Hirschhorn, R. R., Mercer, W. E., Surmacz, E., Tsutsui, Y., Soprano, K. J., and Baserga, R. 1982. Gene transfer: DNA microinjection compared with DNA transfection with a very high efficiency, *Mol. Cell. Biol.* 2:1145–1154.

33. Shih, C., Shilo, B.-Z., Goldfarb, M. P., Donnenberg, A., and Weinberg, R. A. 1979. Passage of phenotypes of chemically transformed cells via transfection of DNA and chromatin, *Proc. Nat. Acad. Sci. U.S.A.* 76:5714–5718.

34. Talavera, A., and Basilico, C. 1977. Temperature sensitive mutants of BHK cells affected in cell cycle progression, *J. Cell. Physiol.* 92:425–436.

35. Talavera, A., and Basilico, C. 1978. Requirements of BHK cells for the exit from different quiescent states, *J. Cell. Physiol.* 97:429–440.

36. Waechter, D. E., Avignolo, C., Freund, E., Riggenbach, C. M., Mercer, W. E., McGuire, M., and Baserga, R. 1984. Microinjection of RNA polymerase II corrects the temperature-sensitive defect of tsAF8 cells, *Mol. Cell. Biol.* 60:77–82.

37. Wigler, M., Pellicer, A., Silverstein, S., Axel, R., Urlaub, G., and Chasin, L. 1979. DNA mediated transfer of the adenine phosphoribosyl transferase locus into mammalian cells, *Proc. Nat. Acad. Sci. U.S.A.* 76:1373–1376.

38. Wold, B., Wigler, M., Lacy, E., Maniatis, T., Silverstein, S., and Axel, R. 1979. Introduction and expression of a rabbit B-globin gene in mouse fibroblasts, *Proc. Nat. Acad. Sci. U.S.A.* 76:5684–5688.

Questions for Discussion with the Editors

1. *Concerning the altered RNA polymerase: What is your best guess concerning the molecular basis of the reduced enzyme level? Do tsAF8 cells display other (e.g., pleotrophic) effects?*

It can not be determined from the existing data if the reduced enzyme level present in tsAF8 cells at the nonpermissive temperature is a consequence of an alteration in the synthesis, assembly, or degradation of RNA polymerase II.

Further studies with this enzyme are required to pinpoint the defect. A point mutation somewhere in the RNA polymerase II locus could lead to any of these changes. However, the location of the mutation cannot be determined until the gene is isolated from both tsAF8 cells and BHK cells and then compared. The most striking consequence of a decreased RNA polymerase II activity is that of a specific G$_1$ arrest resulting from the cessation of unique copy gene transcription. This makes the tsAF8 cell line attractive not only as a recipient for gene transfer studies but also for studies of the eukaryotic cell cycle.

2. *Is the tsAF8 line unique, or would you expect a systematic search of various other lines to yield similar transcription alterable phenotypes?*

The tsAF8 cell line is unique. No other cell line has yet been characterized as having a defective RNA polymerase II activity. However, with a slowly increasing number of conditionally lethal mutants, other Syrian hamster cell lines may be shown to fall in the same complementation group as tsAF8. It is also possible that a cell line derived from a different species will be developed that has a defective RNA polymerase II. At present, no such lines have been reported.

CHAPTER **3**

Enhancer Elements and Tissue-Specific Gene Expression

Nadia Rosenthal, Laimonis Laimins, and George Khoury

DURING THE DEVELOPMENT of higher organisms, the specialized characteristics exhibited by differentiating cells depend ultimately on the ability to express genes selectively, in response either to intercellular or environmental signals. A primary goal of current research in developmental biology, therefore, is to describe the mechanism by which the expression of a single gene is controlled in a specific cell. This is a necessary prerequisite for understanding the complex regulation underlying the developmental programs of eukaryotic gene expression.

At present, the mechanisms governing gene expression are poorly understood. Extensive studies in prokaryotic systems have narrowed the major sites of transcriptional regulation to DNA sequences preceding active genes, such as the "Pribnow box" and the −35 sequence, which function as promoter elements for transcription initiation. More recently, researchers have identified sequences associated with known eukaryotic genes that function in an analogous way to promote transcription. Like their prokaryotic counterparts, eukaryotic promoters appear to contain a number of regulatory regions (elements) that share common consensus sequences and are present upstream of the majority of eukaryotic genes thus far studied (see Fig. 1). These include the Goldberg-Hogness (TATA) box, an AT-rich sequence, located 20 to 30 nucleotides before the site of transcription initiation, which appears to set the 5′ start site (initiating nucleotide) for transcription (22, 43, 45). In addition, a set of GC-rich hexanucleotides has been found associated with certain eukaryotic genes (35, 85). Recent studies have shown that this sequence in SV40 is responsible for binding a protein (SPI) (29) suspected of playing an

PROKARYOTIC UPSTREAM ELEMENTS

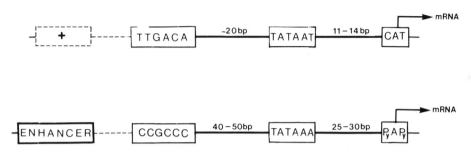

EUKARYOTIC UPSTREAM ELEMENTS

Figure 1 Consensus sequences of prokaryotic and eukaryotic promoter regions. Numbers are in base pairs relative to the cap site of the transcription unit.

important role in the transcriptional process (see below). Finally, a number of genes share an upstream element referred to as the CAAT box which has a consensus sequence CCAAT (30). Although it is presumed to bind to a regulatory factor, its role in gene regulation remains obscure.

A breakthrough in the search for additional eukaryotic control sequences was made by groups studying small DNA viruses. The work of these researchers led to the elucidation of enhancer elements. In retrospect, these viruses were found to possess a genetic system ideally suited to the identification of regulatory sequences. The genome sizes were small and in several cases their complete DNA sequence was known, making the generation of mutants relatively simple. Testing these mutants was also facilitated by the availability of well-developed tissue culture systems.

Early experiments in which the effect of these novel viral regulatory enhancer elements was noted were actually designed to study the replication origin of the well-characterized simian papovavirus, SV40. Several groups linked a viral DNA fragment containing the replication origin, including an adjacent group of 72-bp tandem repeated sequences, to the Herpes virus TK gene, and introduced the construction into mouse TK⁻ cells [(15) and Reddy et al., unpublished]. The purpose of these experiments was to show that the presence of the replication origin would increase the transformation frequency of the TK⁻ phenotype to TK⁺, either by increasing the TK template number through replication, by facilitating integration of the gene into the chromosomal DNA of the recipient cell, or both. The results of the initial experiments indicated that the linked SV40 replication origin fragment dramatically increased the transforming capacity of the TK gene. In retrospect, however, it was probably the presence of the SV40 repeated sequences, as yet undefined as a transcriptional regulatory element, that was responsible for increased TK gene expression.

While these studies were in progress, other investigators focused their attention on the regions of SV40 responsible for its transcriptional control. At least two lines of evidence had suggested that a significant fraction of the SV40 minichromosomes in infected monkey kidney cells (the permissive host cell for SV40) contained regions in which nucleosomes were frequently absent. This was based both on the sensitivity of minichromosomes to nucleases and on electron microscopic visualization of these structures (38, 62, 100). Restriction enzyme mapping in both sets of studies confirmed that the nucleosome-free regions were situated just to the late side of the SV40 DNA replication origin. A closer examination of this region reveals an AT-rich Goldberg-Hogness box for the early SV40 transcription unit, three 21-bp repeats, each containing two copies of a GC-rich hexanucleotide, and two 72-bp tandem repeats (116) (see Fig. 2). Deletion of the 72-bp repeats resulted in the loss of viral viability because the transcription of T antigen, a gene product required early in the viral life cycle, was drastically reduced (4, 52). This established

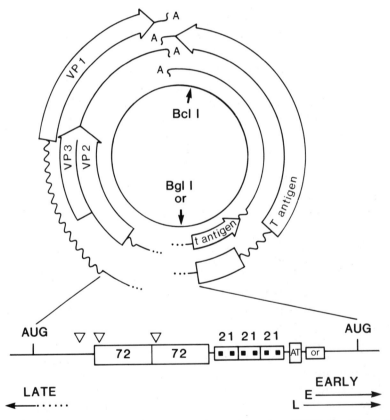

Figure 2 Regulatory sequences in the papovavirus SV40. Positions of core nucleotides in the 72-bp repeats and an upstream homologous region are indicated by open triangles. GC-rich hexamers in the 21-bp repeats are indicated by black boxes. The early gene TATA box lies in an AT-rich region between the 21-bp repeats and the origin of replication (or).

TABLE 1 Viral enhancers

Gene	Repeat no.	Size (bp)	Distance to mRNA initiator (bp)
SV40 (early)	2	72	110
BKV (early)	3	68	100
JCV (early)	2	98	100
LPV (early)	2	60	250
MSV (LTR)	2	72	110
Wild-type polyoma (early)	2	7 and 8	200
Embryonal carcinoma (EC)–polyoma (early)	2	30 to 100	200

the 72-bp repeats as an essential element for early gene expression in the virus.

In other studies, the repertoire of genes whose transcription could be augmented by the SV40 repeats was expanded. By linking different gene coding regions to the isolated SV40 repeats and introducing them into animal cells, several groups demonstrated that the level of transcription of heterologous sequences as diverse as β-globin (3), conalbumin (87), or the adenovirus major late transcription unit (87) were dramatically increased.

Other animal virus genomes were soon scanned for repeated sequences that could act as transcriptional enhancers (see Table 1). When Levinson et al. (78) replaced the 72-bp repeats in the SV40 genome with a nonrelated 72-bp tandem element present in the long terminal repeat (LTR) of the Moloney sarcoma virus (MSV), the resulting chimeric virus was viable. This suggested that a retroviral sequence, namely the MSV repeats, had a function similar to that of the SV40 element. Furthermore, like the SV40 enhancer, this element could activate the transcription of a heterologous gene (SV40 T antigen).

Experiments in several laboratories (39, 87) established the ability of the SV40 72-bp repeats to enhance correctly initiated gene transcription from a considerable distance (several kilobases), and in either direction. Thus, the enhancer could function at the 5' or 3' ends of a gene or even when it was located within an intervening sequence (see Fig. 3). In addition, it was active in either orientation relative to the test gene. These observations have critical

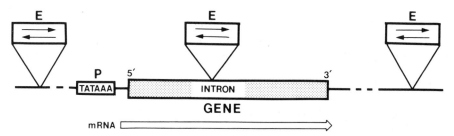

Figure 3 Functional positions of enhancer elements relative to regulated genes.

TABLE 2 Enhancer core consensus sequence

Enhancer	Sequence
SV40	G G T G T G G A A A G T C C
BKV	G T C A T G G T T T G G C T
JCV	C T C A T G C T T G G C T G
Py	C G T G T G G T T T T G C A
Py(2)	G G C C T G G A A T G T T T
BPV	G G A G T G G T G T G T A C
MSV	T C T G T G G T A A G C A G
ALA	A A C A T G G T A A C G A T
RSV	C G C T G G G A T A G C G C
SFFV	T C T G C G G T A A G C A G
SNV	T T C T C G G A A T C G G C
HSV tk	G G C G G G G T T T G T G T
Ig heavy chain	G C T G T G G T T T G A A G
Ig Kappa	C T C T C G G A A A G T C C

implications both for the bidirectional function of enhancer elements in a genomic environment and for the molecular mechanism of transcriptional enhancement.

From studies of naturally occurring viral variants (39) as well as *in vitro* recombinants (52), it was concluded that a single SV40 repeat was sufficient to retain enhancer function. Regions of the sequence essential for the enhancer element's effect on gene transcription were identified by deletions and point mutations in the repeat unit. Although the first two-thirds of a single repeat were sufficient to retain partial activity, deletion or mutation of a 10-bp sequence at the 5' end of the repeat unit greatly reduced the expression of a test gene (124). The critical nucleotides were referred to as the core of the enhancer element. Similar core sequences have been found in a number of enhancers (Table 2), a different nucleotide sequence has been described as the adenovirus E1A enhancer core sequence (54). It is not cleae how many different enhancer motifs with analogous function may exist.

Studies with several genes in tandem demonstrated that while the enhancer element could activate transcription from either end of a gene, activation of a distal gene could be blocked by an intervening promoter, or by certain sequences present in plasmid vectors (63). Thus, it has generally been concluded that the ability of an enhancer to work over distances is more a function of the nucleotide sequences between the enhancer and its promoter than the actual physical distance between them.

From the work summarized above, an operational definition of enhancer elements emerged. It was based primarily on the properties of the SV40 enhancer but has proved applicable to a number of similar eukaryotic transcriptional elements subsequently discovered:

- The enhancer sequence is usually a short set of nucleotides (50 to 100 bp) often repeated in tandem.

- Although it may contain an essential core (7 to 10 bp long), the nucleotide sequences of most known enhancers are dissimilar.
- Enhancers are cis-acting regulatory elements that are directly linked to the gene which is transcriptionally activated.
- The enhancer dramatically increases transcription from the nearest promoter(s) on either side of the enhancer in a manner relatively independent of position and orientation.
- This effect extends over as much as several kilobases if there is no other promoter (or "blocking" sequence) included in the intervening DNA.
- A number of enhancer elements appear to function with a tissue or species specificity (see below).

It is clear that our understanding of enhancer elements and their function is rudimentary. While the characteristics listed above represents a working definition of enhancers, it is obvious that each regulatory element will be in some ways unique. The precise function of these elements will no doubt be clarified by a better understanding of the molecular basis underlying transcriptional initiation.

Analysis and Detection of Enhancers

A number of different techniques have been employed to demonstrate that a particular set of nucleotides functions as an enhancer element. While the definition of enhancer elements implies an increase in the transcriptional activity of an associated gene, in certain circumstances it has proved easier to rely on indirect assays. Among the earliest assays of enhancer activity were tests that simply showed an increase in the number of colonies that would appear when a selectable gene was linked to a putative enhancer sequence and the resulting construct wss introduced into tissue culture cells under a selective pressure. Thus, as previously described (15), association of an enhancer element with a selectable marker such as the Herpes simplex virus thymidine kinase gene would lead to an increased number on colonies of cells which could grow in the presence of HAT (hypoxanthine, aminopterin, thymidine) medium. This type of assay suffers from a number of potential complications: (a) it is based on the detection of a protein product rather than specific RNA transcripts, (b) growth in selective media may require the production of a threshold level of the selectable protein which is not equivalent to a linear or quantitative assay of transcription, (c) the assay is based on long-term effects that may involve uncontrolled parameters such as the influence of variable integration sites of the selected gene in the host cell genome, and (d) the assay may detect an unexpected function of the test sequences, such as enhanced integration.

A second type of enhancer assay also relies on the linkage of a putative enhancer sequence with a "recorder" gene. However, the recombinant is introduced into cells and assayed for its gene product after only 24 to 72 hours. One of the most popular of these transient enhancer assays is the so-called

CAT ASSAY

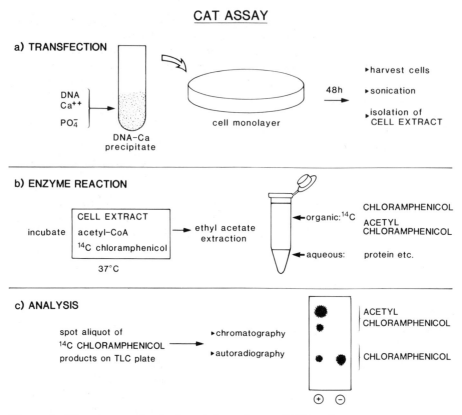

Figure 4 The choramphenicol acetyltransferase (CAT) enhancer assay.

"CAT assay" (47) (see Fig. 4). Putative enhancer sequences are linked to a plasmid construct with contains a classical set of promoter elements (such as the SV40 promoter) associated with the CAT gene which encodes the prokaryotic enzyme, chloramphenicol acetyltransferase. This construct is introduced into tissue culture cells by DNA transfection. After 48 hours a cell extract is made and the potency of the enhancer element under question is assayed as a function of CAT enzyme activity. The virtues of this technique are that it is extremely sensitive, quantitative, and reproducible. The disadvantage, as in the selectable marker assay, is that it indirectly measures the enhancer activity by assaying the levels of the induced protein product rather than the level of mRNA accumulation. While these two parameters are usually directly proportional to one another, complications can arise, for example, if the transcript starts at an unpredicted site and includes a proximal translational AUG initiator codon not in frame with the CAT protein.

Therefore, the preferred assay for an enhancer activity is the direct analysis of RNA using, for example, the S1 or primer extension method. These assays give an indication not only of the amount of RNA produced in a test situation, but the location of the 5′ ends of the transcripts relative to known promoter elements.

Enhancers in DNA Viruses

Since the discovery of the SV40 enhancer, similar elements have been found in analogous positions in a number of related papovaviruses (see Table 1). The genomic arrangement of the murine virus *polyoma* includes a noncoding region adjacent to the origin or replication, which, like the SV40 repeats, has been shown essential for early gene expression (65, 115). When linked to a rabbit β-globin gene, a 244-bp fragment from this noncoding region of the virus caused significant enhancement in the level of globin gene expression (27). More recently, two distinct nonoverlapping enhancer elements have been defined within the polyoma enhancer region, with different individual activities tested in mouse fibroblasts or embryonal carcinoma cells (56). Furthermore, it has been demonstrated that the replication of polyoma virus DNA requires a functional enhancer (28).

The two human papovaviruses, BKV and JCV, closely related to SV40, have been analyzed for the presence of enhancer elements. In BKV (Dunlop strain) three slightly shorter repeats (two of 68 bp flanking one of 50 bp) occur at the same genomic position as do the 72-bp repeats of SV40. The BKV repeats were shown to enhance the expression of a linked CAT test gene (see above), in either orientation at the 5′ or 3′ end of the gene (97). Similar studies were performed with the two 98-bp repeats of JCV, also located at the same relative position in the genome as the SV40 enhancer (67). The JCV enhancer appeared to function efficiently only in human fetal glial cells. This cell-specificity of the JCV enhancer may be an important determinant in the restricted host range of this papovavirus.

Although it is not clear what the relationship of the enhancer element is to such proximal promoter elements as the TATA box, CAAT box, and GC-rich motifs, it is interesting to compare their relative positions in the control regions of the papovaviruses (Fig. 5). As described above, the clear separation of these promoter elements in SV40 permitted their early identification and characterization through *in vitro* mutagenesis techniques. In BKV, although the TATA box is separate, the GC-rich sequences appear within each of the three tandem repeats. JCV has two 98-nucleotide tandem repeats, each of which contains its own promoter elements including a TATA box. Whether the variation in the organization of the promoter elements reflects a plasticity in the mechanism of transcriptional initiation, or rather significant differences in the early transcriptional programs of these three papovaviruses is unclear. Clarification of this question awaits a molecular characterization of the events involved in promoter activation.

Other DNA viruses shown to contain enhancers include *bovine papilloma virus* (BPV) which causes malignant epithial tumors in infected animals. (Although classified historically as papovaviruses, recent molecular character-

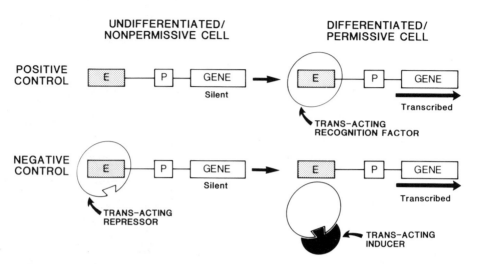

Figure 5 Comparison of regulatory sequence arrangements in the papovaviruses SV40, BKV, and JCV. Core nucleotides in the enhancers are indicated by shaded boxes, GC-rich hexamers by black boxes, and TATA boxes in AT-rich regions are marked.

ization has demonstrated that the papilloma viruses are members of a distinct group.) A BPV enhancer element was mapped to within a 59-bp fragment lying 3' (distal) to the early viral genes (82). Deletion of this fragment from the BPV genome eliminated the expression of early viral genes which cause neoplastic transformation of transfected cells. Further, when the BPV fragment was linked to a Herpes TK gene and the resulting recombinant molecule was introduced into TK⁻ tissue culture cells, the expression of the TK gene was substantially increased, producing large numbers of TK⁺ transformants (82). Like other papovaviral enhancers, the BPV enhancer element was active in either orientation relative to the TK gene, and functional at a distance from the TK promoter. It should be noted that in the virus, the BPV enhancer appears to activate early gene promoters several kilobases away. More recently, Howley and coworkers (personal communication) have located an additional enhancer element upstream from the early BPV transcriptional template. The relationship of this sequence to that described by Lusky and Botchan (83) awaits further analysis.

Human adenoviruses are DNA viruses that contain a genome nearly six times as large as the papovavirus SV40. Extensive studies have revealed a complex pattern of adenovirus gene expression in which an immediate early viral protein E1A is responsible for activation of the early and late adenovirus transcriptional units (5, 61). Several groups of researchers have recently identified enhancer elements located 200 to 300 bp upstream from the transcription start site of the E1A gene in adenovirus 2 (54, 55, 123). Each of these groups has described a different element which suggests a certain redundancy in enhancer functions within this region of the adenovirus genome. All of these enhancer elements manifest the characteristics observed for the SV40 enhancer element (see above). It is likely that the immediate early genes of the human adenovirus serotypes not yet analyzed have analogous enhancer elements. Whether similar enhancers are associated with delayed early, intermediate, and late adenovirus transcriptional units, and how such sequences may be activated by the E1A gene product is presently under investigation (see "Inducible genes," below). More recently, a very strong enhancer (several-fold stronger than the SV40 enhancer) has been identified in human cytomegalovirus (CMV), located upstream of the major immediate early gene transcription initiation site (8). Different subsets of sequences in this enhancer (including several repeated motifs) can substitute for the SV40 enhancer in an "enhancer trap" assay, showing no sequence or cell type preference. Which of these repeated sequences contributes to the unusually strong activity of the CMV enhancer remains to be determined.

Enhancers in Retroviruses

In the life cycle of retroviruses (RNA tumor viruses), the infectious viral RNA strand is converted by reverse transcriptase into double-stranded "proviral" DNA. Upon integration of the proviral DNA into the host chromosome, a terminal segment of the viral genome is duplicated. This segment, forming the LTRs that flank the provirus, is functionally divided into three regions, U3, R, and U5. In the LTRs of several retroviruses, the U3 regions (which are upstream of the transcription start site and include most of the transcriptional control sequences) have been shown to contain an element that acts as an enhancer of expression of the proviral genes.

The presence of an enhancer sequence in the LTR of the *murine sarcoma virus* (MSV) was demonstrated in a study where a fragment from the MSV LTR was introduced into an SV40 genome from which the 72-bp repeats had been deleted. Addition of the MSV LTR segment restored viral viability (78). In experiments similar to those designed to define enhancers in DNA viruses, several groups placed a portion of the MSV LTR, containing the 72-bp repeats, adjacent to either a CAT gene (72, 73), a Herpes virus TK gene or a bacterial neomycin resistance gene (71). In each case, the MSV repeats acted in a position- and orientation-independent manner to increase dramatically the level of expression in mouse cells.

Since the elucidation of the MSV enhancer, other retroviral LTRs have been tested for enhancer activity. Experiments with chimeric retroviral genomes demonstrated the presence of an enhancer in the U3 region of the *feline sarcoma virus* (FeSV) LTR (34). Using the TK assay, Luciw et al (81), demonstrated an enhancer activity associated with the 3' LTR and adjacent sequences in the Schmidt-Ruppen strain of *Rous sarcoma virus* (RSV), a retrovirus of chickens. The 3' LTR of RSV (Prague strain), contains an unusual arrangement of at least three distinct domains comprising two functional overlapping enhancer elements, as defined by a standard enhancer assay using a CAT test gene (23, 74). Other recent demonstrations of retroviral enhancers include a steroid hormone-inducible element in the LTR of the *mouse mammary tumor virus* (MMTV) (16), and tissue-specific enhancer sequences of several *T-cell specific murine leukemia viruses*, as discussed below (25, 76).

The regulatory properties of these retroviral LTRs suggest an interesting potential role played by *endogenous retroviruses*, represented in multiple copies throughout the genomes of birds and mammals. One might speculate that endogenous retroviral enhancers have the potential to regulate adjacent cellular genes. The possibility that these endogenous retroviral sequences may function as transposable regulatory elements is an exciting prospect. One such endogenous LTR, isolated from a monkey genome, does have enhancer activity as shown by an *in vitro* CAT assay (68). Whether endogenous LTRs have similar effects on adjacent genomic coding regions remains to be demonstrated.

In at least one well-described example, an exogenous avian retrovirus *avian leukosis virus* (ALV), carrying no transforming gene of its own, is capable of causing bursal lymphomas. It has been demonstrated that in these tumors, ALV integrates next to the cellular proto-oncogene myc, and activates its expression by a mechanism referred to as promoter insertion (113). Eventual tumor formation in the host is presumably related to inappropriate and/or enhanced levels of myc expression. The LTR of the ALV that is responsible for this myc activation can stimulate abnormal transcription of the adjacent gene through the action of its enhancer as well as its promoter. In fact, some of the ALV integration sites which have been mapped in avian bursal lymphomas are located downstream of myc (91). Thus, in some cases this mechanism may more appropriately be referred to as *enhancer insertion*. The ability of exogenous or endogenous retroviral enhancers to affect cellular gene expression suggests a potential symbiotic relationship between the retroviruses and the hosts they infect, and may hold important implications for both the normal (regulated) and abnormal (disease related) activation of cellular genes.

Host Range and Cell Specificity of Viral Enhancers

During initial tests of viral enhancers, it was noted that the level of activity of several enhancer sequences depended on the host cell type used in the study. For example, a chimeric SV40 genome in which the native enhancer was replaced by the MSV 72-bp repeats (see above) was significantly less active in monkey cells than was an intact SV40 virus; but the recombinant expressed higher levels of SV40 early proteins in mouse cells (87). These observations suggested that the host range of the MSV-SV40 recombinant was at least in part the result of the cell-specific function of its enhancer. Further studies on the isolated MSV and SV40 enhancers, comparing their relative effects on linked test genes (72, 73), supported this hypothesis. Whereas the MSV enhancer stimulated gene activity severalfold higher than did the SV40 enhancer in mouse L cells, the opposite was true in monkey cells, the natural host for SV40 infection.

TABLE 3 Tissue-specific viral enhancers and their role in viral-induced disease

Viral enhancer	Tissue specificity	Disease
DNA:		
JCV (human)	Fetal glial cells	PML
LPV (monkey)	B lymphocytes	?
Py-EC (mouse)	Differentiated F9 cells	?
BKV (human)	B pancreatic cells	?
RNA:		
RSV[TD] (chicken)	Fibroblasts	Sarcoma
M-MuLV (mouse)	T lymphocytes	Lymphoma
SLS-3 (mouse)	T lymphocytes	Lymphoma
HTLV (human)	T lymphocytes	T-cell leukemia
FrMuLV (mouse)	Erythroblasts	erythroleukemia

A host cell preference has also been observed for the polyoma enhancer which is optimally active in mouse cells (26). The multiple domains of the (RSV) chicken retrovirus enhancer are more active in chicken cells than in monkey cells, in contrast to the SV40 enhancer (74).

The cell preference of certain viral enhancers depends more on the particular host cell type than on the species of the infected cells (see Table 3). An especially interesting example of tissue-specific enhancer function is represented by a series of polyoma variants with sequence changes in the enhancer region. These variants differ from wild-type polyoma (Py) virus in their ability to productively infect undifferentiated murine embryonal carcinoma (EC) cells which do not support wild-type virus replication (40, 104, 111, 117). Some of these Py variants differ from the wild-type by a single base substitution in the enhancer region, suggesting a very specific interaction between potential recognition factors in the host cell and their target regulatory sequences in the virus.

Other examples of viral enhancers exhibiting tissue specificity include variants of the human papovavirus BKV. A series of small deletions and duplications in the enhancer region appears to be responsible for the ability of the variants to induce tumors in rodent pancreatic β cells, whereas the wild-type BK virus produces sarcomas (121). The enhancer region of the closely related human papovavirus JCV is active in human fetal brain cells, yet inactive in HeLa or CV-1 cells (67). This result is particularly intriguing in view of the fact that JCV appears to be closely associated with and perhaps is the etiological agent of progressive multifocal leukoencephalopathy, a degenerative neurological disease characterized by the presence and growth of the virus within brain lesions.

The possibility that the spectrum of diseases caused by viral infection is at least partially attributable to the action of tissue-specific enhancers is further supported by studies on a series of retroviruses. Even et al. (34) were able to demonstrate that the enhancer elements of certain sarcoma viruses were important elements in the transformation of tissue culture cells. Thus, by exchanging the transcriptional regulatory sequences from MSV for those of

FeSV, one could enhance the ability of the latter (which contains the fes oncogene) to transform murine cells 50- to 100-fold. The transcriptional signals in the LTRs of Friend leukemia virus, which induces erythroleukemia in mice and Moloney murine leukemia virus (M-MuLV) which induces thymic lymphomas, are probably responsible for the disease specificities of the two viruses, as shown by several groups. In one study, the 3′ LTR of the Friend viral genome, containing putative regulatory sequences, was replaced by a comparable fragment of M-MuLV. When introduced into mice, the recombinant virus produced almost exclusively thymic lymphomas (18). Conversely, the reverse recombinant in which the Friend viral regulatory sequences replaced the corresponding region of the Moloney virus induced primarily erythroleukemia (N. Hopkins, unpublished).

In other experiments designed to distinguish the disease specificities of MuLV variants, DNA recombinants were constructed between a strain that replicated efficiently in thymus tissue (T$^+$) and a strain that could not (T$^-$) (25). Since there appears to be a close correlation between the ability of M-MuLVs to replicate in thymocytes and their leukemogenicity, it was hoped that the behavior of the T$^+$/T$^-$ recombinants *in vivo* would help identify genomic segments responsible for causing disease. Infection of mice with several T$^+$/T$^-$ reciprocal recombinants showed that sequences in the T$^+$ strain required for replication in the thymus resided in the LTR. More recent data indicate that the tandem direct repeat in the T$^+$ strain LTR is the primary determinant of the disease (25).

A detailed analysis of a comparable pair of MuLV strains, Akv (non-leukemogenic) and SL3-3 (highly leukemogenic) not only narrowed the sequences in SL3-3 responsible for leukemogenicity to its LTR by construction of reciprocal recombinants between the two viruses, but also revealed that sequence differences between Akv and 3L3-3 LTRs were limited to a series of point mutations and internal rearrangements in the tandem direct repeats (76). More recent experiments from the same group established the repeats as tissue-specific enhancer elements. Linked to a CAT test gene, the SL3-3 repeats directed higher levels of CAT expression than did the Akv repeats in T cells, whereas the reverse was observed in non-T(B) cells (W. Haseltine, unpublished).

Differences in the enhancer regions of two closely related RSV viral variants, Prague B and Schmidt-Ruppin, may contribute to the differences in oncogenicity noted after infection of chickens. The 3′ terminal region of the Prague strain of RSV contains three distinct domains that comprise two functional enhancer elements. Two of these domains are found in the U3 region of the 3′ LTR and are almost identical to those sequences found in the related Schmidt-Ruppin strain. The third domain, which is located outside the LTR, is necessary for activation of downstream genes. These domains outside the LTR are quite different in the two viruses, suggesting a possible model for differential activation of cellular genes by either the Schmidt-Ruppin or Prague strains. In previous genetic studies, the determinants responsible for disease were localized to the region around the 3′ LTR (114). The td Schmidt-Ruppin strain induces bursal lymphomas, while the td Prague strain induces a low frequency of lymphomas. Thus, one possible explanation for the different disease spectrum of the two strains would be a differential activation of downstream genes by enhancer domains immediately preceding the 3' LTR (74).

A dramatic instance of tissue-specificity attributed to a viral LTR has been recently uncovered by studies of human T-cell leukemia viruses (HTLVs) associated with human T-cell malignancies. HTLV types 1 and 2 differ markedly in their ability to transform human lymphocytes, type 1 functioning much more efficiently than type 2. Testing LTRs from both strains for their ability to enhance expression of a linked CAT test gene revealed a striking result: while the HTLV-1 construct produced higher CAT levels in lymphocytes than in fibroblasts, the highest CAT activity was observed in lymphocytes that had been preinfected or transformed with HTLV (107). Further, HTLV-2 preinfected cultures were the only cells in which the HTLV-2 construct was active. Not only did these experiments establish differential enhancer strengths between the HTLV-1 and HTLV-2 LTRs, but they also suggested that the presence of a viral induced transacting factor is necessary for optimal activity of the viral LTR (see "Inducible Enhancers," below). These studies have now been extended to other members of this T-lymphotrophic retrovirus group including bovine leukemia virus (BLV) (24, 96) and autoimmune deficiency syndrome (AIDS) virus (HTLV-3) (108). The remarkable feature of the trans-activation of these viral LTRs is the specific activation of expression in cells infected or transformed by the homologous agent. Thus, BLV transformed cells provide for high levels of activity from a transfected BLV-CAT construct but not from HTLV-CAT (96). This observation has suggested that the induction may be related to a viral encoded trans-activating protein. A good candidate is the product of the pX or LOR regions, transcribed downstream from envelope genes in these viruses (106).

The studies discussed above demonstrate that viral enhancers not only regulate the transcription of the associated viral genes in a variety of cell types, but also modulate the expression of linked cellular genes in transfection assays. Furthermore, proviruses, the DNA forms of retroviruses, are capable of activating adjacent cellular genes following proviral insertion into cellular DNA (6, 88, 91). Thus, the numerous endogenous viral genomes present in cellular DNA may constitute a large reservoir of regulatory sequences that retain their ability to act as enhancers for resident cellular genes. These examples underscore a close correspondence between the modes of transcriptional regulation in viruses and the cells they infect.

Cellular Enhancers

Once the prevalence of enhancers in viral genomes had been established, many investigators turned their research efforts toward locating and characterizing equivalent sequences in cellular DNA. It seemed logical that some of the recognition factors involved in viral enhancer activity would be cellular in origin (the size of many viruses limits the numbers of genes they carry), and therefore might have homologous target sites to the analogous regulatory regions in cellular DNA. Furthermore, the host and tissue specificities of certain viral enhancers suggested that their action was modulated by cell-encoded recognition molecules.

The problems associated with this ongoing line of research are not trivial and deserve mention. First, it was not clear that enhancers controlling cellular genes would be as strong as the previously identified viral enhancers. In fact, for some genes (e.g., housekeeping genes) a very strong enhancer might be inappropriate. In the case of cellular protooncogenes, activation to high levels by a viral enhancer has been shown to be associated with malignant transformation (6, 88, 91). Yet in tests designed to uncover the potential enhancer activity of a given cellular sequence, a low level of enhancer activity might not be detected. Second, one characteristic of viral enhancers is their sequence disparity; thus many potential cellular enhancer elements could be overlooked in a sequence homology scan based on the known viral enhancers. Finally, because enhancers can exert an effect on the transcription of genes thousands of base pairs away, a search for enhancers directly adjacent to known genes could prove futile.

Initially, two avenues of investigation were pursued. One set of experiments relied on the hypothesis that certain viruses had recently acquired their enhancers through the transduction of segments of cellular DNA which presumably performed a similar function for cellular genes. These transduced regulatory elements would then allow the virus to grow on that particular cell type. In view of the fact that closely related viruses, such as the members of the papovavirus family (SV40, JCV, BKV, and Py) share as much as 80 percent nucleotide sequence homology, yet have distinct enhancer elements, this seemed a reasonable model.

To find segments of chromosomal DNA that included sequences homologous to a viral regulatory region, a number of investigators screened libraries of monkey (95) and human (21) DNA, using as a probe the HindIII C fragment from the SV40 genome containing the enhancer sequences, the early gene promoter elements and origin of viral DNA replication. Three clones isolated by screening a monkey-phage λ library with the SV40 fragment contained regions of high GC content that had multiple and disconnected homologies with sequences directly surrounding the SV40 replication origin. Comparing the positive clones to each other and to total genomic monkey DNA by hybridization revealed that the GC-rich regions belonged to a family of moderately repetitive monkey sequences with approximately 80 members (95), some of which were transcribed in monkey CV-1 cells (79). One of the members studied in more detail resembled the SV40 regulatory region; it was associated with a DNase I hypersensitive site in monkey chromatin (79) and was transcribed in both directions, although the sequence did not contain a TATA box and did not have the ability to enhance the expression of linked genes (99).

The human clones isolated by their homology to the same SV40 probe were also shown to be partially homologous to each other, by restriction enzyme and hybridization analysis. Determination of the sequences of two SV40 hybridizing segments showed that similar to the monkey clones described above, the human SV40-like sequences contained repetitive regions of high GC content with multiple homologies to sequences in the SV40 origin of replication, the 21-bp and the 72-bp repeats. In experiments designed to test one of the human clones for enhancing effect on a linked Herpes TK gene, the level of TK transformation of TK⁻ human cells was raised about 20-fold by the human sequences when they were present in either orientation at the 3' end of the transfected TK gene. This suggested that these human sequences could function like viral enhancers, although their native position relative to coding regions in the human genome was not determined (21).

Another unrelated human locus was isolated by hybridizing a human λ genomic library to a probe made from the isolated enhancer element of BKV (97). The homologous locus was about 1.5 kb long, appeared only once in the human genome, and was conserved in monkey DNA but not mouse DNA. The sequence of the homologous region consists of tandem direct repeats, either 20 or 21 bp long. Unlike the repetitive clones isolated by other groups, these repeats were not GC-rich, although each repeat contained a BKV-like GC-rich hexamer, or a variation thereof. Each repeat also included a sequence homologous to a portion of the BKV enhancer element tentatively identified as a core sequence by virtue of its similarity to the SV40 enhancer core (124) (see Table 2). In tests for potential enhancer activity of the human sequences, the entire repeat locus, or as few as four repeat units, were placed at 5' and 3' positions relative to a test gene CAT and the resulting constructions were introduced into human, monkey, or mouse cells by DNA transfections. Like viral enhancers, the human repeats increase CAT expression in a position- and orientation-independent fashion irrespective of the number of repeats; but the human constructs were only 20 percent as active as the BKV enhancer element in analogous tests (97). A similar enhancing ability has been associated with a 58-bp fragment of mouse DNA, isolated by its ability to restore the transformation potential of a polyoma virus whose own enhancer had been deleted. The mouse sequences do not appear to be homologous to other viral or cellular enhancer elements, although they restored the transformation efficiency of the virus to 20 to 40 percent of that of wild-type polyoma (37).

Whether these mouse, monkey, and human loci function as enhancers for resident cellular genes remains to be determined. Without knowing what loci are associated with these putative enhancer sequences, it is difficult to assess the potential properties, such as tissue-specificity, that they might confer on adjacent gene expression.

A second approach to the identification of cellular enhancer elements involved testing genomic sequences near known cellular genes for enhancer function. The earliest successful experiments using this approach involved functional analyses of DNA sequences associated with the sea urchin H2A gene. By injecting the gene and varying lengths of upstream genomic sequences into frog oocytes and measuring the levels of H2A transcription from the transfected gene, Grosschedl et al. (50) identified a DNA segment lying between 165 bp and 711 bp away from the H2A capsite that imparted a maximal rate of H2A transcription. Nucleotide transitions of C to T in this segment resulted in dramatic reductions in H2A transcription levels. This region contained a 14-bp out of 17-bp homology to the sequences in the LTRs of several retroviruses (the simian sarcoma and Friend murine leukemia viruses) related to the Moloney murine sarcoma virus.

Analysis of cloned mammalian genetic loci for enhancer sequences has also been successful in a number of laboratories. In studies of the immunoglobulin multigene family whose members are transcribed only in mature B-lymphocytes, it was noted that in cells where immunoglobulin genes were active, a nucleosome-free area sensitive to DNase I developed in the intron between the joining (J) region and the constant (C) region coding sequences (19). Since activation of an immunoglobulin gene involves DNA rearrangement to join an upstream variable (V) segment to the D region and/or the J region, the intron containing the DNase-sensitive sequences would not be lost during the rearrangement and thus was a likely location for an enhancer element controlling the expression of the rearranged gene.

Using cloned immunoglobulin genes, several groups linked intron segments from a mouse heavy-chain gene (2, 44) and a mouse light-chain gene (94) to various test genes. The resulting constructs were introduced via DNA transfection into either mouse fibroblasts, where the native immunoglobulin genes were unrearranged and silent, or into murine myeloma cells, derived from B plasma cells, where the resident immunoglobulin genes were active. For each gene, the results were similar. Independent of their orientation or position, sequences in the introns dramatically increased the expression of linked test genes in myeloma cells but had no effect on the transcription of the test genes in fibroblasts. In segments of both heavy- and light-chain gene introns, sequences similar to the core elements of viral enhancers have been noted (Table 2).

Other well-characterized genes that are expressed optimally in a specific cell type have been subjected to similar tests to define sequences in and around the coding regions that might act as tissue-specific enhancers. The genes coding for mammalian insulins are expressed at high levels only in the endocrine beta-cells of the pancreas. By linking DNA sequences from the 5' flanking regions of rat [(119) and L. Laimins, unpublished] and human (119) insulin genes to a CAT test gene, researchers were able to measure the relative levels of expression induced by the mammalian sequences either in pancreatic beta-cells or fibroblasts transfected with the recombinant constructions. The levels of CAT transcription were significantly higher in beta-cells than in fibroblasts when sequences 200 to 300 bp upstream of the insulin cap site were included in the constructs. Similar studies of sequences upstream of the rat chymotrypsin gene, expressed preferentially in the exocrine alpha cells of the pancreas, revealed tissue-specific regulatory sequences at a similar position that enhanced CAT gene expression preferentially in the appropriate cell type (119).

The control of specialized gene expression by the action of tissue-specific enhancers provides an attractive and testable model for the regulation of transcription in differentiated cell types. Although the number of cellular genes shown to be associated with a tissue-specific enhancer is still small, experiments to test sequences near many other cloned genes for enhancer activity are currently underway. In many cases these experiments are limited only by the availability of appropriate cell types in tissue culture.

Mechanism of Enhancer Action

From the studies described above, it is evident that enhancer elements from diverse sources share functional properties, implying that there may be a common mechanism by which they increase the transcription of associated genes. Structurally, the enhancer elements identified thus far have very little sequence homology. A short consensus element of about 8 bp (Table 2) appears to be essential for the activity of at least the SV40 enhancer (124). Another core sequence is common to the adenovirus E1A enhancer, the polyoma virus enhancer and several others (54). Yet the role of these sequences in activation of transcription (e.g., whether they alter DNA structure and/or confer on it the ability to bind proteins) is not yet understood.

One striking structural similarity of many enhancer elements is the motif of tandemly repeated sequences. Since at least several viral enhancers are active in monomeric form, the significance of the repeated structure is not entirely clear. Interestingly, attempts by two groups to "trap" cellular enhancers—by ligating random fragments of genomic DNA into a virus whose enhancer was deleted and selecting for virus viability among the recombinants—instead produced variants in which viral sequences near the point of the enhancer deletion were duplicated (110, 112). A similar conclusion has been reached based on studies that specifically mutagenize and/or reconstruct enhancers from oligonucleotide subunits (58). These results suggest that, in at least some instances, a repeated structure may be the critical element responsible for enhancer function.

Several models for enhancer function have been proposed, based in part on the observations that enhancer regions alter chromatin conformations. These models are not mutually exclusive but rather emphasize different aspects of the role enhancers may play in transcription. One model suggests that enhancer sequences provide an active site for altering chromatin structure perhaps through binding of a protein which excludes nucleosomes. In several studies, a nucleosome-free region observed in SV40 minichromosomes was mapped to the vicinity of the enhancer, although additional sequences (the TATA box and 21-bp repeats) are important in forming this structure (38, 62, 100). Experiments probing chromatin configurations by DNase I sensitivity have revealed changes associated with the presence of both viral (62, 80, 100) and cellular (19, 79, 127) enhancer sequences.

In other studies, changes in the configuration of the DNA helix from the B to the Z form have been observed with sequences present in several enhancers (90). It has been suggested that the binding proteins that preferentially interact with enhancer regions contribute to the transition between B and Z configurations (1). In either case, the configurational changes of enhancer elements caused by nucleosome alteration and/or conformational shifts may act to expose these sequences to the entry of transcriptional complexes. Alternatively, the presence of an enhancer sequence near a gene could target that DNA segment to associate with a cellular compartment, such as the nuclear matrix, where transcription is activated.

Another model not necessarily exclusive from those described above proposes that enhancer sequences serve directly as recognition sites for molecules such as polymerase and polymerase-associated (σ-like) factors actively involved in transcription. A corollary to this hypothesis is that upon recognition, the RNA polymerase (perhaps a symmetrical oligomer or subunit) should be able to move in either direction to engage with appropriate nucleoprotein structures for the initiation of transcription. This would explain the flexibility in position and orientation associated with enhancer function. Recently, data demonstrating the interaction between cellular DNA binding proteins and enhancer regions have in part confirmed this premise (see below). Future studies on the identity and cell-specific nature of these proteins should provide a better understanding of the role of enhancers in the transcriptional process.

Ideally, an *in vitro* assay would allow identification of cellular fractions that increase transcription of enhancer-associated test genes. Subsequent enrichment by further protein fractionation of factors responsible for enhancer activity would provide the basis for subsequent biochemical analyses. Recent efforts by several laboratories (101, 105, 125) have provided convincing evidence that some aspects of enhancer function can be mimicked *in vitro*. In nuclear extracts from HeLa cells, a 5- to 10-fold stimulation of transcription is observed when the SV40 enhancer was present upstream of either the SV40 early promoter or the rabbit β-globin promoter. The SV40 enhancer activated transcription independent of orientation; and point mutations known to reduce enhancer activity *in vivo* also reduced the activity *in vitro*. When the SV40 enhancer was inserted upstream of the β-globin promoter, the stimulation of transcription decreased as distance from the promoter was increased; however, stimulation was still observed to be significant over a distance of 400 bp. The development of *in vitro* systems should allow for the purification and identification of enhancer-stimulating factors which will be a critical step in understanding the mechanisms of enhancer function.

In addition to establishing an *in vitro* system, investigators have developed experimental strategies with which one can demonstrate competition for enhancer binding factors in tissue culture using either identical or alternative enhancer sequences as competitors (86, 102). If such factors are present in limited amounts, these experiments could potentially provide information on the number and specificity of enhancer-regulatory molecules in the cell nucleus under a given set of conditions. Whether different enhancer elements could be activated by the same trans-acting cellular factors would be of particular interest, since most enhancers, such as those of SV40 and BKV, are active in the same cell types but do not share extensive stretches of sequence homology.

While enhancer-binding molecules have not yet been isolated, the presence of factors involved in tissue-specific gene expression has been suggested in several studies. The expression of the chick β-globin gene, which occurs as early as day 8 of chick development, seems to coincide with the appearance of a DNase I hypersensitive region approximately 70 nucleotides upstream from the transcriptional start site of the gene. The upstream sequences corresponding to the DNase-sensitive region were shown to retain a protein fraction from chick erythrocytes in a filter binding assay. By combining plasmids carrying the upstream β-globin DNA sequences with histones and extracts from 9-day chick embryo erythrocytes, Emerson and Felsenfeld (31) were able to reconstitute *in vitro* the DNase I hypersensitivity. This result was not obtained when they used extracts from embryos at earlier stages of development when the β-globin gene was inactive, suggesting the presence of a stage-specific factor involved in β-globin gene expression.

More recently, the same group has identified two distinct sites upstream of the β-globin gene which are protected in footprinting studies by specific binding of a cytoplasmic factor(s) *in vitro*. When physically separated, the two domains interact independently with the factor(s), suggesting that two or more complexes are formed (32). Since the footprint patterns obtained *in vitro* are similar to patterns observed in intact chicken erythrocyte nuclei, it is likely that, at least in the case of the β-globin gene, the isolation of cellular factors involved in trans-regulation of transcription may be imminent.

While the activity of enhancers has yet to be fully reconstituted in an *in vitro* assay, another associated viral control element, the GC-rich 21-bp repeats of SV40, also appears to be necessary for efficient *in vitro* transcription of both early and late genes (9, 13, 35, 38, 53). Recent studies by Tjian and Dynan (29) have demonstrated that the GC-rich hexamers associated with the 21-bp repeats of both SV40 and BKV bind a protein, SP1, which is critical for the transcriptional efficiency of both the SV40 early and late promoters *in vitro*. It is noteworthy that this same factor does not appear to be critical to transcription from the promoters of certain other genes such as the adenovirus major late gene. Although fragments from the SV40 genome containing the 21-bp repeats do not inhibit SV40 enhancer-driven activity in cell culture competition assays described above (102), they act as effective inhibitors of SV40 late transcription *in vitro*, presumably by competing for trans-acting factors, such as Sp1, which are involved in SV40 promoter activation in the *in vitro* system.

The discovery of enhancer sequences associated with genes transcribed in a tissue-specific manner suggests that the specificity may lie in a cell's production of enhancer-regulating molecules which can act in trans to activate transcription. A direct interaction between cellular factors and enhancer sequences has been suggested by a study in which the enhancer region of an endogenous immunoglobulin heavy-chain gene was modified by dimethyl sulfate (DMS) treatment of intact myeloma cells and the extracted DNA was subsequently examined with genomic sequencing techniques (20, 33). Only specific nucleotides near, but not in, the core sequences were protected or enhanced by DMS treatment, and only in the B cells but not in fibroblasts. If cellular factors are involved in recognizing a tissue-specific enhancer sequence, it will be crucial to identify the genes coding for these factors in order to further elucidate the mechanisms by which enhancers operate.

These putative enhancer binding proteins may act as positive regulators,

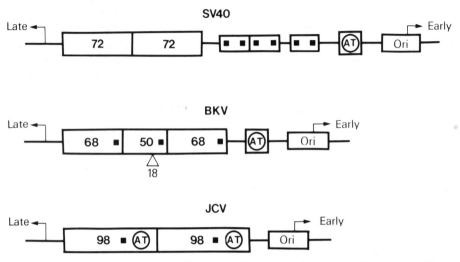

Figure 6 Models of tissue-specific transcription regulation by enhancers. In the positive regulation scheme, trans-acting factors, absent in the undifferentiated or nonpermissive cell, drive transcription in the differentiated or permissive cell by recognizing the enhancer sequence. In the negative regulation scheme, trans-acting suppressor factors in the nonpermissive cell recognize the enhancer sequence and prevent an active chromatin configuration by binding to upstream DNA. Inducers present in the permissive cell compete away the negative factors allowing transcription to proceed. In a combination regulation scheme, negative suppressor factors in the nonpermissive cell would block the action of positive factors on the transcription unit. Removal or absence of the suppressor factors in the permissive cell type would allow positive factors to drive transcription.

for example, by facilitating the interactions of genes adjacent to enhancers with RNA polymerase or by altering local chromatin structure to make genes more accessible to transcription complexes. Alternatively, in some cases an inducer molecule may complete for enhancer binding against suppressor molecules (e.g., repressor-like proteins present in "nonpermissive" cells) which limit transcription of certain genes to specific tissues (Fig. 6).

Inducible Enhancers

Although trans-acting factors that interact with the enhancer elements discussed above have not yet been identified, there is a group of well-characterized genes whose expression is stimulated by known effector molecules. In many cases, the regulatory sequences involved in induction of these genes can also activate transcription of heterologous test genes in the presence of the appropriate inducer, and as such they may be defined by their mode of action as a potential subset of classical enhancers.

Inducible genes in viruses include the adenovirus early genes E2, E3, and E4, and late genes controlled by the major late promoter (MLP), all of which are induced by the E1A protein (the first viral product synthesized after infection of permissive cells). By linking the E2, E3, or E4 upstream sequences to test genes and either cotransfecting these constructions with E1A-producing expression plasmids (59, 123) or coinjecting them into *Xenopus* oocytes (60), or somatic cells (36), several groups have recorded the induction of test gene expression in the culture cells. In the case of the E2 promoter, a region between −79 and −28 bp upstream of the cap site will stimulate gene expression in response to the presence of E1A from either a 5′ or 3′ position relative to the test gene (59).

The adenovirus E1A protein can also induce gene expression in trans from heterologous viral and cellular promoters, including the SV40 early promoter (lacking the enhancer) [(5, 61) and M. Loeken, unpublished], human β-globin (48), rat preproinsulin (41), and Drosophila heat-shock gene 70 (89) promoters. Interestingly, E1A protein induces the expression of these genes only when they are transfected, as the corresponding endogenous genes are not stimulated. This suggests that although these genes are potentially susceptible to activation by E1A induction, a permissive chromosomal configuration is an additional prerequisite for the induction to take place.

A similar mechanism of induction by viral gene products operates in the Herpes virus life cycle. Expression of immediate early genes such as ICP4 may require a late viral protein which is part of the inoculated virion (70). In turn ICP4 is required for the transcription of intermediate genes, such as the TK (70, 77, 93). Sequences between −109 and −20 upstream of the TK cap site have been shown to be responsible for the induction effect in the presence of immediate early genes, as well as representing integral elements in the TK gene promoter (84). Finally, it appears that HSV early and intermediate gene products induce expression of a number of other viral and cellular genes, in many ways analogous to the action of E1A (14, 48, 59, 92).

The late genes of papovaviruses SV40 and polyoma, which are induced specifically by the early viral gene product T antigen, represent other well-documented examples of inducible gene expression (10, 66). The property of T antigen as a replication origin-specific DNA binding protein leads one to consider this DNA-protein interaction as the direct stimulus for late gene activation, although rigorous proof for this hypothesis has not yet been obtained. The concept of trans-activation of the late viral DNA transcription units by the early and intermediate viral gene products provides a potential model for the activation of enhancers by trans-acting proteins. In the case of the viral systems, the sequential expression of early, intermediate, and late viral functions is essential to the efficiency of viral reproduction. These patterns of viral gene regulation may reflect analogous mechanisms which presumably control tissue-specific and cell cycle-specific expression of higher eukaryotic genes.

While the nature of these putative cellular trans-activating genes remains obscure, an insight is obtained from the analysis of transforming proteins which share properties with the early viral functions. Both the papovaviral T antigens and adenoviral E1A protein can immortalize primary cells (98), an activity they share the viral and cellular myc genes (75). Since v-myc or c-myc proteins are also capable of inducing the transcription of various transfected genes (69), it is possible that all three categories of inducers (E1A, T antigen, and myc) act in a similar fashion. If this is the case, c-myc would represent the first example of a cellular gene that functions as a transcriptional trans-activator. Its role in transformation is conceivably related to its ability to stimulate the expression of cellular genes that govern cell growth and division. A number of laboratories have recently devised schemes for isolating other cellular genes that play a role in transcriptional activation or manifest the ability to complement the ras group of oncogenes. Thus, we should soon learn considerably more about their role in transcription as well as their contribution to cell transformation.

Finally, a number of other eukaryotic genes that respond to specific inducing agents by an increase in the level of transcription initiation have been examined in detail. The transcription of genes coding for metallothioneins is induced by both heavy metal ions and by hormones, each interacting with the 5' flanking region at distinct locations (64). The induction of transfected β-globin genes in mouse erythroleukemia cells by dimethyl sulfoxide (DMSO) seems to be controlled by at least two regions, one of which is in the body of the coding sequences (126, 17). This emphasizes the importance of examining, when possible, intact eukaryotic genes rather than chimeras when searching for transcriptional regulation elements. Whether these sequences are analogous to enhancer elements, however, will require further study. Repeated sequences upstream of the β-interferon gene have the characteristics of an inducible enhancer element. In response to viral or poly(I)-poly(C) activation, they can act upstream or downstream of the gene, in either orientation from as far as 1 kb from their normal location (46). In the above cases, the nucleotide sequences that respond either directly or indirectly to the inducing stimulus are being examined in detail by mutagenesis and transposition.

Enhancers in Animal Systems

In recent experiments, tissue-specific gene expression has been dramatically documented in transgenic mice, produced by microinjecting specific genes into the male pronucleus of fertilized eggs. For example, the expression of immunoglobulin in the appropriate tissues of these transgenic mice and their progeny has been observed following injection of the cloned genes (12, 51, 109). Although both of these genes are known to be controlled by tissue-specific enhancers, it remains to be seen whether the defined enhancer sequences are specifically responsible for tissue-specific expression *in vivo*. A intriguing example of what may prove to be tissue-specific enhancer activity in transgenic mice was provided by experiments in which the SV40 early region, including the 72-bp repeats, was introduced into the germlines of mice (11). A high percentage of the mice, which carry the SV40 sequences at numerous integration sites in most tissues, eventually developed tumors only in the choroid plexus. SV40 T antigen was expressed only in tumor tissue, but not in unaffected tissues or in pretumorous tissues, suggesting that tumorigenesis depended on the activation of the early SV40 genes. Further evidence that the SV40 enhancer is in some way responsible for choroid plexus-specific T-antigen expression comes from the experiments using enhancer-deleted constructs. Mice carrying these sequences tended to develop peripheral neuropathy, but no choroid plexus adenomas (Brinster et al., unpublished). Thus, it appears that the SV40 enhancer may be active only in selected tissues of the brain and stimulate transcription of the SV40 early genes independent of its site of integration.

Transgenic mice represent a powerful tool for studying the tissue-specific action of enhancers *in vivo*. As in the case of SV40, the function of an enhancer in tissue culture may not reflect its behavior in the environment of a developing animal. The reasons for this discrepancy may include the presence of other promoter elements in the SV40 that are differentially regulated in developing tissues. Although the production of transgenic mice to characterize the behavior of genes with enhancer sequences that have been deleted or mutagenized is relatively impractical, the action of regulatory elements *in vivo* will continue to be the focus of intense investigation.

Cell Specificity of Other Promoter Elements

Recent studies have suggested that the cell specificity of certain transcriptional elements may not be a property only of enhancer elements but of other promoter sequences as well. Although each of these sequences has not yet been clearly segregated and defined, it is clear in the case of immunoglobulin genes that the enhancer element that resides in an intron is separate and distinct from upstream cap site-proximal sequences that dictate the start of transcription. Chimeric constructs in which the immunoglobulin (2, 44, 94) gene enhancer was substituted by another enhancer element from MSV (73) with broad host range, nevertheless still functioned with B-cell specificity (49). The most likely interpretation of these results is that the upstream promoter elements also activate transcription in a tissue-specific manner. Since both elements are essential for transcription initiation, the tissue specificity of either one or both would dictate gene activity only in the specific tissue. One hypothesis is that some promoter elements such as the immunoglobulin promoter or enhancer bind regulatory factors that are present only in a specific cell (e.g., B lymphocytes) while other promoter elements, like the MSV enhancer or promoter, interact with analogous factors that are present in many types of cells.

Antienhancers

Based on prokaryotic models, it would seem that transcriptionally active DNA binding proteins could function either positively (inducers) or negatively (repressors). Perhaps the best characterized of these DNA binding proteins is the SV40 T antigen. Several studies have conclusively demonstrated the ability of T antigen to repress transcription from the early SV40 promoter. Since this promoter controls the transcription of the T-antigen coding sequences, this phenomenon has been referred to as autoregulation.

Recent evidence has suggested that through its role in binding to DNA, T antigen also activates the late SV40 transcriptional unit (9, 66). Although the mechanism of this activation is still unclear, the 72-bp repeated enhancer element and its role in late transcription have been implicated (10).

Although the 289 amino acid gene product of the adenovirus E1A region does not appear to bind DNA directly, there is speculation that it may do so through interactions with other DNA binding proteins (M. Rosenberg and N. Jones, personal communication). In any event, its role as a transcriptional regulator is clear (see above). Recently, several groups have made the important observation that this E1A 13S gene product can function as an "antienhancer" in blocking the function of a transcriptional control element (i.e., the enhancer) in SV40 (7) or polyoma virus (118) from activating the early SV40 or the Ad 2 E1A promoters. Furthermore, the level at which this inhibition occurs has been shown by nuclear "run-off" experiments to be at the initiation of transcription. The details of the antienhancer effect are just now being elucidated, but there is no reason to believe that this mechanism will be substantially different from prokaryotic repression, in which a DNA-binding protein associates with DNA at a critical region and blocks a step essential to the initiation of the transcriptional complex.

It seems reasonable to speculate that a mechanism akin to this antienhancer affect may be responsible for the recent phenomenon described for adenovirus-transformed cells by Van der Eb and his colleagues. It appears that two variants of adenovirus, Ad 5 and Ad 12, both transform murine cells in culture, but only the former is tumorigenic in histocompatible syngeneic animals (103, 112). The most likely explanation of this phenomenon relates to the observation that while a stimulation of class I genes (histocompatibility antigens) occurs in Ad 5 transformed cells, a repression of the transcription of these same class I genes is found in Ad 12 transformed cells. This phenomenon has been traced to the E1A 13S mRNA gene products, which differ between the two virus serotypes. Thus, one possibility is that Ad5 E1A acts as an inducer while Ad12 E1A acts as a repressor on class I genes, perhaps by antienhancer mechanism postulated above.

Enhancers and DNA Replication

The potential role of enhancers in DNA replication deserves some comment. It is clear that for replication and maintenance of episomal sequences BPV requires in addition to a plasmid maintenance sequence (PMS), an enhancer element included in the plasmid in cis (83). Likewise, efficient replication of papovaviruses has been shown to be dependent on enhancer elements (28). Whether the contribution of these sequences to replication occurs at the level of transcription is not at all clear. One reason to suspect a direct coupling of transcription in DNA replication comes from prokaryotic model systems. In lamda, for example, it has been shown that transcription across the origin for

DNA replication serves to activate this sequence for replication. An analogous role of transcription in cellular DNA replication is yet to be demonstrated.

Conclusions

The discovery of enhancer elements represents a significant advance in defining the regulatory steps involved in eukaryotic gene transcription. It remains to be seen how prevalent enhancers will be in the genes of higher organisms. They may be relatively rare, associated only with highly specialized genes involved in luxury functions limited to specific cell types. Since both mouse immunoglobulin and rat insulin enhancer sequences do not cross hybridize with other mammalian genomic sequences, enhancers may be unique elements that evolved with the genes they control.

Alternatively, enhancers may belong to families of related sequences which have diverged far enough to perform disparate functions in specific tissues, while retaining in common certain elements, such as the core sequence. Examples of such a potential evolutionary relationship among enhancer elements are the LTRs of the murine retrovirus AKV and a leukemogenic 3L3-3 discussed above. The enhancers of these two viruses share essentially the same sequences, but the scrambled arrangement in the 3L3-3 enhancer allows a T-cell tropism directly related to the tissue specificity of this virus (76). The clear evolutionary relationship between known cellular enhancer elements will first require identification of the transcriptional regulatory elements controlling functionally related genes, such as the members of a multigene family. It now seems possible that sequences like enhancers will be found associated with many transcriptional units, differing substantially from one another in strengths, host cell specificity, and inducibility (i.e., in response to a protein or hormone produced only transiently during a particular phase of the cell cycle).

The extent to which enhancers facilitate tissue-specific transcription remains unclear. First, cellular enhancers associated with genes such as immunoglobulin may not be the sole determinants of tissue-specific expression; the relationship between a given enhancer and the gene it activates may play a role in determining the timing and tissue specificity of transcription. In other genes such as α- and β-globin, transcriptional regulatory elements also appear to reside within the body of the coding sequences (126, 17), in which case the structural portions of genes themselves would be recognized by tissue-specific trans-acting factors. In light of these recent findings, experiments designed to locate tissue-specific regulatory elements associated with structural genes will require testing both noncoding and coding sequences for enhancing activity. This is a consideration that has been neglected in many experiments designed to measure gene activity through sensitive assays (e.g., CAT, gpt, TK), or to mark genes bypartial deletion of coding sequences so they could be distinguished from endogenous counterparts.

Finally, enhancers that activate transcription of associated genes eclusively in limited tissue types may be valuable tools in gene therapy experiments. Introduction into animals of an enhancer-gene combination that is appropriately expressed in specific tissues could allow the correction of a genetic dysfunction due to mutations or deletions in endogenous genes. The same experimental plan would also facilitate the study of normal gene expression by *in vitro* mutation of the enhancer-gene construct and subsequent introduction into an animal model. The successful production of transgenic mice expressing exogenously introduced genes with the tissue specificity of their enhancer/promoter elements is an exciting first step in this direction. As new cellular and viral enhancer elements are identified, the repertoire of target tissues for the transcription and expression of introduced genes should expand to include a variety of cell types. This may make available to researchers a library of sequences each of which regulates transcription of associated genes in a different group of cells. In some cases, tissue-specific expression of enhancer-recognition molecules may be directly responsible for the establishment of the cell's differentiated state. The identification of these regulatory factors will be necessary for the molecular dissection of the events leading to differentiation. In this research direction, enhancers promise to play a crucial part.

General References

CHAMBON, P., DIERICH, A., GAUB, M. P., JAKOWLEV, S., JONGSTRA, J., KRUST, A., LE PENNEC, J. P., OUDET, P., and REUDELHUBER, T. 1984. Promoter elements of genes coding for proteins and modulation of transcription by estrogens and progesterone, *Recent Prog. Hormone Res.* 40:1–42.

FELSENFELD, G. 1985. DNA, *Sci Amer.* 253:58–67.

References

1. Azorin, F., and Rich, A. 1985. Isolation of Z-DNA binding proteins from SV40 minichromosomes: Evidence for binding to the viral control region, *Cell* 41:365–374.

2. Banerji, J., Olson, L., and Schaffner, W. 1983. A lymphocyte-specific cellular enhancer is located downstream of the joining region in immunoglobulin heavy chain gene, *Cell* 33:729–740.

3. Banerji, J., Rusconi, S., and Schaffner, W. 1981. Expression of a β-globin gene is enhanced by remote SV40 DNA sequences, *Cell* 27:299–308.

4. Benoist, C., and Chambon, P. 1981. *In vivo* sequence requirements of the SV40 early promoter region, *Nature (London)* 290:304–310.

5. Berk, A. J., Lee, F., Harrison, T., Williams, J., and Sharp, P. A. 1979. Pre-early adenovirus 5 gene products regulates synthesis of early viral messenger RNAs, *Cell* 17:935–944.

6. Blair, D. G., Oskarsson, M., Wood, T. G., McClements, W. L., Fischinger, P. J., and Vande Woude, G. F. 1981. Activation of the transforming potential of a normal cell sequence: A molecular model for oncogenesis, *Science* 212:941–943.

7. Borelli, E., Hen, R., and Chambon, P. 1984. Adenovirus-2 E1a products repress enhancer-induced stimulation of transcription, *Nature (London)* 312:608–612.

8. Boshart, M., Weber, F., Jahn, G., Dorsch-Häsler, K., Fleckenstein, B., and Schaffner, W. 1985. A very strong enhancer is located upstream of an immediate early gene of human cytomegalovirus, *Cell* 41:521–530.

9. Brady, J. N., Bolen, J. B., Radonovich, M., Salzman, N. P., and Khoury, G. 1984. Stimulation of simian virus 40 late expression by simian virus 40 tumor antigen, *Proc. Nat. Acad. Sci. U.S.A.* 81:2040–2044.

10. Brady, J. N., and Khoury, G. 1985. *Trans*-activation of the SV40 late transcription unit by T-antigen, *Mol. Cell. Biol.* 5:1391–1399.

11. Brinster, R., Chen, H., Messing, A., Van Dyke, T., Levine, A., and Palmiter, R. 1984. Transgenic mice harboring SV40 T-antigen genes develop characteristic tumors, *Cell* 37:367–379.

12. Brinster, R. L., Ritchie, K. A., Hammer, R. E., O'Brien, R. L., Arp, B., and Storb, U. 1983. Expression of a microinjected immunoglobulin gene in the spleen of transgenic mice, *Nature (London)* 306:332–336.

13. Byrne, B. J., Davis, M. S., Yamaguchi, J., Bergsma, D., and Subramanian, K. N. 1983. Definition of the simian virus 40 early promoter region and demonstration of a host range bias in the enhancement effect of the simian virus 40 72-base-pair repeat, *Proc. Nat. Acad. Sci. U.S.A.* 80:721–725.

14. Campbell, M. E. M., Palfreyman, J. W., and Preston, C. M. 1984. Identification of HSV DNA sequences which encode a *trans*-acting polypeptide responsible for stimulation of immediate early transcription, *J. Mol. Biol.* 180:1–19.

15. Capecchi, M. R. 1980. High efficiency transformation by direct microinjection of DNA into cultured mammalian cells, *Cell* 22:479–488.

16. Chandler, V. L., Maler, B. A., and Yamamoto, K. R. 1983. DNA sequences bound specifically by glucocorticoid receptor in vitro render a heterologous promoter hormone responsive in vivo, *Cell* 33:489–499.

17. Charnay, P., Treisman, R., Mellon, P., Chao, M., Axel, R., and Maniatis, T. 1984. Differences in human α and β globin gene expression in mouse erythroleukemia cells: The role of intragenic sequences, *Cell* 38:251–263.

18. Chatis, P., Holland, C., Hartley, J., Rowe, W., and Hopkins, N. 1983. Role of the 3' end of the genome in determining the disease specificity of Friend and Moloney murine leukemia viruses, *Proc. Nat. Acad. Sci. U.S.A.* 80:4408–4411.

19. Chung, S.-Y. and Wooley, J. 1983. DNase I-hypersensitive sites in chromatin of immunoglobulin κ light chain genes, *Proc. Nat. Acad. Sci. U. S. A.* 80:2427–2431.

20. Church, G., Ephrussi, A., Gilbert, W., and Tonegawa, S. 1985. Cell-type specific contacts to immunoglobulin enhancers in nuclei, *Nature (London)* 313:798–800.

21. Conrad, S. E., and Botchan, M. 1982. Isolation and characterization of human DNA fragments with nucleotide sequence homologies with SV40 regulatory region, *Mol. Cell. Biol.* 2:949–965.

22. Cordon, J., Wasylyk, B., Buchwalder, A., Sassone-Corsi, P., Kedinger, C., and Chambon, P. 1980. Promoter sequences of eukaryotic promoter-coding genes, *Science* 209:1406–1414.

23. Cullen, B. R., Raymond, K., and Ju, G. 1985. Transcriptional activity of avian retroviral long terminal repeats directly correlates with enhancer activity, *J. Virol.* 53:515–521.

24. Derse, D., Caradonna, S., and Casey, J. 1985. Bovine leukemia virus long terminal repeat: A cell-type specific promoter, *Science* 227:317–320.

25. DesGroseillers, L., Rassart, E., and Jolicoeur, P. 1983. Thymotropism of murine leukemia virus is conferred by its long terminal repeat, *Proc. Nat. Acad. Sci. U.S.A.* 80:4203–4207.

26. de Villiers, J., Olson, L., Tyndall, C., and Schaffner, W. 1982. Transcriptional "enhancers" from SV40 and polyoma virus show a cell type preference, *Nucleic Acids Res.* 10:7965-7976.

27. de Villiers, J., and Schaffner, W. 1981. A segment of polyoma virus DNA enhances the expression of a cloned β-globin gene over a distance of 1400 base pairs, *Nucleic Acids Res.* 9:6251–6264.

28. de Villiers, J., Schaffner, W., Tyndall, C., Lupton, S., and Kamen, R. 1984. Polyoma virus DNA replication requires an enhancer, *Nature (London)* 312:242–246.

29. Dynan, W. S., and Tjian, R. 1983. Isolation of transcription factors that discriminate between different promoters recognized by RNA polymerase II, *Cell* 32:669–680.

30. Efstratiadis, A., Posakony, J. W., Maniatis, T., Lawn, R. M., O'Connell, C., Spritz, R. A., DeRiel, J. K., Forget, B. G., Weissman, S. M., Slightom, J. L., Blechl, A. E., Smithies, O., Baralle, F. E., Shoulders, C. C., and Proudfoot, N. J. 1980. The structure and evolution of the human β-globin gene family, *Cell* 21:653–668.

31. Emerson, B. M., and Felsenfeld, G. 1984. Specific factor conferring nuclease hypersensitivity at the 5′ end of the chicken adult β-globin gene, *Proc. Nat. Acad. Sci. U.S.A.* 81:95–99.

32. Emerson, B. M., Lewis, C. D., and Felsenfeld, G. 1985. Interaction of specific nuclear factors with the nuclease-hypersensitive region of the chicken adult β-globin gene: Nature of the binding domains, *Cell* 41:21–30.

33. Ephrussi, A., Church, G. M., Tonegawa, S., and Gilbert, W. 1985. B-lineage-specific interactions of immunoglobulin enhancer with cellular factors *in vivo*, *Science* 227:134–140.

34. Even, J., Anderson, S. J., Hampe, A., Galibert, F., Lowy, D., Khoury, G., and Sherr, C. J. 1983. Mutant feline sarcoma proviruses containing the viral oncogene (v-*fes*) and either feline or murine control elements, *J. Virol.* 45:1004–1016.

35. Everett, R. D., Baty, D., and Chambon, P. 1983. The repeated G-C rich motifs upstream from the TATA box are important elements of the SV40 early promoter, *Nucleic Acids Res.* 11:2447–2464.

36. Ferguson, B., Krippl, B., Andrisani, O., Jones, N., Westphal, H., and Rosenberg, M. 1985. E1A 13S and 12S mRNA products made in *E. coli* both function as nucleus-localized transcription activators but do not directly bind DNA, *Mol. Cell. Biol.* 5:2653–2661.

37. Fried, M., Griffiths, M., Davies, B., Bjursell, G., LaMantia, G., and Lania, L. 1983. Isolation of cellular DNA sequences that allow expression of adjacent genes, *Proc. Nat. Acad. Sci. U.S.A.* 80:2117–2121.

38. Fromm, M., and Berg, P. 1982. Deletion mapping of DNA regions required for SV40 early region promoter function *in vivo*, *J. Mol. Appl. Genet.* 1:457–481.

39. Fromm, M., and Berg, P. 1983. Simian virus 40 early- and late-region promoter functions are enhanced by the 72-bp repeat inserted at distant locations and inverted orientations, *Mol. Cell. Biol.* 3:991–999.

40. Fujimura, F. K., Deininger, P. L., Friedmann, T., and Linney, E. 1981. Mutation

near the polyoma DNA replication origin permits productive infection of F9 embryonal carcinoma cells, *Cell* 23:809–814.

41. Gaynor, R. B., Hillman, D., and Berk, A. J. 1984. Adenovirus early region 1A protein activates transcription of a nonviral gene introduced into mammalian cells by infection or transfection, *Proc. Nat. Acad. Sci. U.S.A.* 81:1193–1197.

42. Gerard, R. D., Woodworth-Contai, M., and Scott, W. A. 1982. Deletion mutants which affect the nuclease-sensitive site in simian virus 40 chromatin, *Mol. Cell. Biol.* 2:782–788.

43. Ghosh, P. K., Lebowitz, P., Frisque, R. J., and Gluzman, Y. 1981. Identification of a promoter component involved in positioning the 5′ termini of simian virus 40 early mRNAs, *Proc. Nat. Acad. Sci. U.S.A.* 78:100–104.

44. Gillies, S. D., Morrison, S. L., Oi, V. T., and Tonegawa, S. 1983. A tissue-specific transcription enhancer element is located in the major intron of a rearranged immunoglobulin heavy chain gene, *Cell* 33:717–728.

45. Gluzman, Y., Sambrook, J. F., and Frisque, R. J. 1980. Expression of early genes of origin defective mutants of simian virus 40, *Proc. Nat. Acad. Sci. U.S.A.* 77:3898–3902.

46. Goodbourn, S., Zinn, K., and Maniatis, T. 1985. Human β-interferon gene expression is regulated by an incredible enhancer element. *Cell* 41:509–520.

47. Gorman, C. M., Moffat, L. F., and Howard, B. H. 1982. Recombinant genomes which express chloramphenicol acetyltransferase in mammalian cells, *Mol. Cell. Biol.* 2:1044–1051.

48. Green, M., Treisman, R., and Maniatis, T. 1983. Transcriptional activation and cloned human β-globin genes by viral immediate early gene products, *Cell* 35:137–148.

49. Grosschedl, R., and Baltimore, D. 1985. Cell-type specificity of immunoglobulin gene expression is regulated by at least three DNA sequence elements, *Cell* 41:885–897.

50. Grosschedl, R., Machler, M., Rohrer, U., and Birnstiel, M. L. 1983. A functional component of the sea urchin H2A gene modulator contains an extended sequence homology to a viral enhancer, *Nucleic Acids Res.* 11:8123–8136.

51. Grosschedl, R., Weaver, D., Baltimore, D., and Constantini, F. 1984. Introduction of a μ immunoglobulin gene into the mouse germ line: Specific expression in lymphoid cells and synthesis of functional antibody, *Cell* 38:647–658.

52. Gruss, P., Dhar, R., and Khoury, G. 1981. Simian virus 40 tandem repeated sequences as an element of the early promoter, *Proc. Nat. Acad. Sci. U.S.A.* 78:943–947.

53. Hartzell, S., Yamaguchi, J., and Subramanian, K. 1983. SV40 deletion mutants lacking the 21-bp repeated sequences are viable, but have uncomplementable deficiencies, *Nucleic Acids Res.* 11:1601–1616.

54. Hearing, P., and Shenk, T. 1983. The adenovirus type 5 E1A transcriptional control region contains a duplicated enhancer element, *Cell* 33:695–703.

55. Hen, R., Borrelli, E., Sassone-Corsi, P., and Chambon, P. 1983. An enhancer element is located 340-base pairs upstream from the adenovirus-2 E1A cap site, *Nucleic Acids Res.* 11:8747–8760.

56. Herbomel, P., Bourachot, B., and Yaniv, M. 1984. Two distinct enhancers with different cell specificities coexist in the regulatory region of polyoma, *Cell* 39:653–662.

57. Herbomel, P., Saragosti, S., Blangy, D., and Yaniv, M. 1981. Fine structure of the origin-proximal DNAase I-hypersensitive region in wild-type and EC mutant polyoma, *Cell* 25:651–658.

58. Herr, W., and Gluzman, Y. 1985. Duplications of a mutated simian virus 40 enhancer restore its activity, *Nature (London)* 313:711–713.

59. Imperiale, M. J., Feldman, L. T., and Nevins, J. R. 1983. Activation of gene expression by adenovirus and herpes virus regulatory genes acting in *trans* and by a *cis*-acting adenovirus enhancer element, *Cell* 35:127–136.

60. Jones, N. C., Richter, J. D., Weeks, D. L., and Smith, L. D. 1983. Regulation of adenovirus transcription by an E1A gene in microinjected *Xenopus laevis* oocytes, *Mol. Cell. Biol.* 3:2131–2142.

61. Jones, N., and Shenk, T. 1979. An adenovirus 5 early gene function regulates expression of other early viral genes, *Proc. Nat. Acad. Sci. U.S.A.* 76:3665–3669.

62. Jongstra, J., Reudelhuber, T. L., Oudet, P., Benoist, C., Chae, C.-B., Jeltsch, J.-M., Mathis, D. J., and Chambon, P. 1984. Induction of altered chromatin structures by simian virus 40 enhancer and promoter elements, *Nature (London)* 307:708–714.

63. Kadesh, T., and Berg, P. 1983. Effects of the position of the 72-bp enhancer segment on transcription from the SV40 early region promoter, in: *Current Communications in Molecular Biology* (Y. Gluzman and T. Shenk, eds.), pp. 21–29, Cold Spring Harbor Laboratory, Cold Spring Harbor, N. Y.

64. Karin, M., Haslinger, A., Holtgreve, H., Cathala, G., Slater, E., and Baxter, J. 1984. Activation of a heterologous promoter in response to dexamethasone and cadmium by metallothionein gene 5'-flanking DNA, *Cell* 36:371–379.

65. Katinka, M., Yaniv, M., Vasseur M., and Blangy, D. 1980. Expression of polyoma early functions in mouse embryonal carcinoma cells depends on sequence rearrangements in the beginning of the late region, *Cell* 20:393–399.

66. Keller, J. M., and Alwine, J. C. 1984. Activation of the SV40 late promoter: Direct affects in the absence of viral DNA replication, *Cell* 36:381–389.

67. Kenney, S., Natarajan, V., Strike, D., Khoury, G., and Salzman, N. P. 1984. JC virus enhancer-promoter active in the brain cells, *Science* 226:1337–1339.

68. Kessel, M., and Khan, A. 1985. Nucleotide sequence analysis and enhancer function of long terminal repeats associated with an endogenous African green monkey retroviral DNA, *Mol. Cell. Biol.* 5:1335–1342.

69. Kingston, R., E., Baldwin, A. S., Jr., and Sharp, P. A. 1984. Regulation of heat shock protein 70 gene expression by c-*myc*, *Nature (London)* 312:280–282.

70. Kit, S., Dubbs, D. R., and Schaffer, P. A. 1978. Thymidine kinase activity of biochemically transformed mouse cells after superinfection by thymidine kinase negative, temperature-sensitive herpes simplex virus mutants, *Virology* 85:456–463.

71. Kriegler, M., and Botchan, M. 1983. Enhanced transformation by a simian virus 40 recombinant virus containing a Harvey murine sarcoma virus long terminal repeat, *Mol. Cell. Biol.* 3:325–339.

72. Laimins, L., Gruss, P., Pozzatti, R., and Khoury, G. 1984. Characterization of enhancer elements in the long terminal repeat of Moloney murine sarcoma virus, *J. Virol.* 49:183–189.

73. Laimins, L. A., Khoury, G., Gorman, C., Howard, B., and Gruss, P. 1982. Host-specific activation of gene expression by 72 base pair repeats of simian virus 40 and Moloney murine leukemia virus, *Proc. Nat. Acad. Sci. U.S.A.* 79:6453–6457.

74. Laimins, L. A., Tsichlis, P., and Khoury, G. 1984. Multiple enhancer domains in the 3' terminus of the Prague strain of Rous sarcoma virus, *Nucleic Acids Res.* 12:6427–6442.

75. Land, H., Parada, L. F., and Weinberg, R. A. 1983. Tumorigenic conversion of primary embryo fibroblasts requires at least two cooperating oncogenes, *Nature (London)* 304:596–602.

76. Lenz, J., Celander, D., Crowther, R., Patarca, R., Perkins, D., and Haseltine, W. 1984. Determination of the leukaemogenicity of a murine retrovirus by sequences within the long terminal repeat, *Nature (London)* 308:467–470.

77. Leung, W. C. 1978. Evidence for a herpes simplex virus-specific factor controlling the transcription of deoxypyrimidine kinase, *J. Virol.* 27:269–274.

78. Levinson, B., Khoury, G., Vande Woude, G., and Gruss, P. 1982. Activation of SV40 genome by 72–base pair tandem repeats of Moloney sarcoma virus, *Nature (London)* 295:568–572.

79. Lord, S., and Singer, M. 1984. A transcriptionally active monkey genomic segment homologous to the regulatory region of simian virus 40 is associated with DNase I-hypersensitive sites, *Mol. Cell. Biol.* 4:1635–1637.

80. Luchnik, A. N., Bakayev, V. V., Zbarsky, I. B., and Georgiev, G. P. 1982. Elastic torsional strain in DNA within a fraction of SV40 minichromosomes: Relation to transcriptionally active chromatin, *EMBO J.* 1:1353–1358.

81. Luciw, P. A., Bishop, J. M., Varmus, H. E., and Capecchi, M. R. 1983. Location and function of retroviral and SV40 sequences that enhance biochemical transformation after microinjection of DNA, *Cell* 33:705–716.

82. Lusky, M., Berg, L., Weiher, H., and Botchan, M. 1983. Bovine papilloma virus contains an activator of gene expression at the distal end of the early transcription unit, *Mol. Cell. Biol.* 3:1108–1122.

83. Lusky, M., and Botchan, M. 1984. Characterization of the bovine papilloma virus plasmid maintenance sequences, *Cell* 26:391–401.

84. McKnight, S. 1982. Functional relationships between transcriptional control signals of the thymidine kinase gene of herpes simplex virus, *Cell* 31:355–365.

85. McKnight, S. L., and Kingsbury, R. 1982. Transcriptional control signals of a eukaryotic protein coding gene, *Science* 217:316–324.

86. Mercola, M., Goverman, J., Mirell, C., and Calame, K. 1985. Immunoglobulin heavy-chain enhancer requires one or more tissue-specific factors, *Science* 227:266–270.

87. Moreau, P., Hen, R., Wasylyk, B., Everett, R., Gaub, M. P., and Chambon, P. 1981. The SV40 72 base pair repeat has a striking effect on gene expression both in SV40 and other chimeric recombinants, *Nucleic Acids Res.* 9:6047–6068.

88. Neel, B. G., Hayward, W. S., Robinson, H. L., Fang, J., and Astrin, S. 1981. Avian leukosis virus-induced tumors have common proviral integration sites and synthesize discrete new RNAs oncogenesis by promoter insertion, *Cell* 23:323–334.

89. Nevins, J. R. 1982. Induction of the synthesis of a 70,000 dalton mammalian heat shock protein by the adenovirus E1A gene product, *Cell* 29:913–919.

90. Nordheim, A., and Rich A. 1983. Negatively supercoiled simian virus 40 DNA contains Z-DNA segments within transcriptional enhancer sequences, *Nature (London)* 303:674–679.

91. Payne, G. S., Bishop, J. M., and Varmus, H. E. 1982. Multiple arrangements of viral DNA and an activated host oncogene in bursal lymphomas, *Nature (London)* 295:209–214.

92. Pellet, P., McKnight, J., Jenkins, F., and Roizman, B. 1985. Nucleotide sequence and predicted amino acid sequence of a protein encoded in a small HSV DNA fragment capable of *trans*-inducing genes, *Proc. Nat. Acad. Sci. U.S.A.* 82:5870–5874.

93. Post, L. E., Mackem, S., and Roizman, B. 1981. Regulation of α genes of herpes simplex virus: Expression of chimeric genes produced by fusion of thymidine kinase with α gene promoters, *Cell* 24:555–565.

94. Queen, C., and Baltimore, D. 1983. Immunoglobulin gene transcription is activated by downstream sequence elements, *Cell* 33:741–748.

95. Queen, C., Lord, S. T. McCutchan, T. F., and Singer, M. F. 1981. Three segments from the monkey genome that hybridize to simian virus 40 have common structural elements, *Mol. Cell. Biol.* 1:1061–1068.

96. Rosen, C., Sodorski, J., Kettman, R., Burny, A., Haseltine, W. 1985. Trans-activation of the bovine leukemia virus long terminal repeat in BLV-infected cells, *Science* 227:320–323.

97. Rosenthal, N., Kress, M., Gruss, P., and Khoury, G. 1983. BK viral enhancer element and a human cellular homolog, *Science* 222:749–755.

98. Ruley, H. E. 1983. Adenovirus early region 1A enables viral and cellular transforming genes to transform primary cells in culture, *Nature (London)* 304:602–606.

99. Saffer, J., and Singer, M. 1984. Transcription from SV40-like monkey DNA sequences, *Nucleic Acids Res.* 12:4769–4788.

100. Saragosti, S., Moyne, G., and Yaniv, M. 1980. Absence of nucleosomes in a fraction of SV40 chromatin between the origin of replication and the region coding for the late leader RNA, *Cell* 20:65–73.

101. Sassone-Corsi, P., Dougherty, J. P., Wasylyk, B., and Chambon, P. 1984. Stimulation of *in vitro* transcription from heterologous promoters by the simian virus 40 enhancer, *Proc. Nat. Acad. Sci. U.S.A.* 81:308–312.

102. Scholer, H. R., and Gruss, P. 1984. Specific interaction between enhancer-containing molecules and cellular components, *Cell* 36:403–411.

103. Schrier, P., Bernards, R., Vassen, R., Houweling, A., Van der Eb, A. 1983. Expression of class I major histocompatibility antigens switched off by highly oncogenic adenovirus 12 in transformed cells, *Nature (London)* 305:771–775.

104. Sekikawa, K. and Levine, A. 1981. Isolation and characterization of polyoma host range mutants that replicate in nullipotential embryonal carcinoma cells, *Proc. Nat. Acad. Sci. U.S.A.* 78:1100–1104.

105. Sergeant, A., Bohmann, D., Zentgraf, H., Weiher, H., and Keller, W. 1984. A transcription enhancer acts *in vitro* over distances of hundreds of base-pairs on both circular and linear templates but not in chromatin-reconstituted DNA, *J. Mol. Biol.* 180:577–600.

106. Sodroski, J., Rosen, C., Goh, W. C., and Haseltine, W. 1985. A transcriptional activator protein encoded by the X-LOR region of the human T-cell leukemia virus. *Science* 228:1430–1434.

107. Sodroski, J. G., Rosen, C. A., Haseltine, W. A. 1984. Trans-acting transcriptional activation of the long terminal repeat of human T lymphotrophic viruses in infected cells, *Science* 225:381–385.

108. Sodroski, J., Rosen, C., Wong-Staal, F., Salahuddin, S., Popovic, M., Arya, S., Gallo, R., and Haseltine, W. 1985. Trans-acting transcriptional regulation of human T-cell leukemia virus type III long terminal repeat, *Science* 227:171–173.

109. Storb, U., O'Brien, R., McCullen, M., Gollahan, K., and Brinster, R. 1984. High expression of cloned immunoglobulin K gene in transgenic mice is restricted to B lymphocytes, *Nature (London)* 310:238–241.

110. Swimmer, C. and Shenk, T. 1984. A viable simian virus 40 variant that carries a newly generated sequence reiteration in place of the normal duplicated enhancer element, *Proc. Nat. Acad. Sci. U.S.A.* 81:6652–6656.

111. Tanaka, K., Chowdhury, K., Chang, K. S., Israel, M., and Ito, Y. 1982. Isolation and characterization of polyoma virus mutants which grow in murine embryonal carcinoma and trophoblast cells, *EMBO J.* 1:1521–1527.

112. Tanaka, K., Isselbacher, K. J., Khoury, G., and Jay, G. 1985. Reversal of oncogenesis by the expression of a major histocompatibility complex class I gene, *Science* 228:26–30.

113. Temin, H. M. 1982. Function of the retrovirus long terminal repeat, *Cell* 28:3–5.

114. Tsichlis, R. N., Donehower, L., Hager, G., Zeller, N., Malavarca, R., Astrin, S., and Skalka, A. M. 1982. Sequence comparison in the crossover region of an oncogenic avian retrovirus recombinant and its nononcogenic parent: Genetic regions that control growth rate and oncogenic potential, *Mol. Cell. Biol.* 2:1331–1338.

115. Tyndall, C., La Mantia, G., Thacker, C. M., Favaloro, J., and Kamen, R. 1981. A region of the polyoma virus genome between the replication origin and late protein coding sequences is required in *cis* for both early gene expression and viral DNA replication, *Nucleic Acids Res.* 9:6231–6250.

116. Van Heuverswyn, H., and Fiers, W. 1979. Nucleotide sequence of Hind C fragment of simian virus 40 DNA, *Eur. J. Biochem.* 100:51–63.

117. Vasseur, M., Kress, C., Montreau, N., and Blangy, D. 1980. Isolation and characterization of polyoma virus mutants able to develop in embryonal carcinoma cells, *Proc. Nat. Acad. Sci. U.S.A.* 77:1068–1072.

118. Velcich, A., and Ziff, E. 1985. Adenovirus E1A proteins repress transcription from the SV40 early promoter, *Cell* 40:705–716.

119. Walker, M. D., Edlund, T., Boulet, A. M., and Rutter, W. S. 1983. Cell-specific expression controlled by the 5′ flanking region of insulin and chymotrypsin genes, *Science* 306:557–561.

120. Wasylyk, B., Wasylyk, C., Augereau, P., and Chambon, P. 1983. The SV40 72 bp repeat preferentially potentiates transcription starting from proximal natural or substitute promoter elements, *Cell* 32:503–514.

121. Watanabe, S., Yoshiike, K., Nozawa, A., Yuasa, Y., and Yoshida, S. 1979. Viable deletion mutant of human papovavirus BK that induces insulinoma F1 in hamster, *J. Virol.* 32:934–942.

122. Weber, F., de Villiers, J., and Shaffner, W. 1984. An SV40 "enhancer trap" incorporates exogenous enhancer or generates enhancers from its own sequences, *Cell* 36:983–992.

123. Weeks, D. L., and Jones, N. C. 1983. E1A control of gene expression is mediated by sequences 5′ to the transcriptional starts of the early viral genes, *Mol. Cell. Biol.* 3:1222–1234.

124. Weiher, H., König, M., and Gruss, P. 1983. Multiple point mutations affecting the simian virus 40 enhancer, *Science* 219:626–631.

125. Wildeman, A., Sassone-Corsi, P., Grundstrom, T., Zenke, M., and Chambon, P. 1984. Stimulation of *in vitro* transcription from the SV40 early promoter by the enhancer involves a specific *trans*-acting factor, *EMBO J.* 3:3129–3133.

126. Wright, S., Rosenthal, A., Flavell, R., and Grosveld, F. 1984. DNA sequences

required for regulated expression of β-globin genes in murine erythroleukemia cells, *Cell* 38:265–273.

127. Wu, C., and Gilbert, W. 1981. Tissue-specific exposure of the 5′ terminus of the rat preproinsulin II gene, *Proc. Nat. Acad. Sci. U.S.A.* 78:1577–1581.

Questions for Discussion with the Editors

1. *Did the discovery of "enhancers" come as a surprise?*

The discovery of enhancers was a surprise, in part because there is no known prokaryotic counterpart. Specifically, the ability of enhancer elements to stimulate transcription of a gene from a position before, behind, or within the transcribed sequence is unique. Since the molecular basis of enhancer actions remains to be described, we do not know what the mechanisms of enhancers and other known transcriptional regulatory elements, such as the UAS sequences in yeast, have in common. Furthermore, the recent discovery that some eukaryotic promoters are tissue-specific suggests that enhancers are probably only one of many ways to regulate genes in higher organisms.

2. *Since the existence of enhancer sequences is so well documented, a logic is easily developed that predicts that "diminisher sequencers" which function to oppose enhancer effects may also exist. What sort of experimental strategy would you suggest to detect the presence of "diminisher sequences"?*

The same experimental approach that was used extensively to characterize enhancers (measuring the effect of a potential enhancer sequence on the transcription of a heterologous test gene) can be used to identify negative control elements, as in a paper by Brand et al.[1] demonstrating the presence of silencer elements in yeast and in a study recently completed in our laboratory defining similar elements in the upstream regions of the rat insulin 1 gene.[2]

3. *How large is the increment in level of in vivo transcription when an enhancer sequence is linked to a heterologous gene?*

The strength of the enhancer effect varies between different elements, in some cases related to the species or tissue type of the cell in which they are tested. For example, an enhancer in the CMV (cytomegalovirus) genome has been shown to be 3 to 5 times stronger than the SV40 enhancer, which generally increases the transcription of heterologous test genes by at least two orders of magnitude. Another potent enhancer in the immunoglobulin heavy chain gene is a strong or stronger than the SV40 enhancer when the activities of the two elements are compared in lymphoid cells. Yet, the immunoglobulin enhancer is virtually silent in fibroblasts, in which the SV40 enhancer is very active. The strength of a given enhancer element is undoubtedly related to its function *in vivo*, as a regulator of a specific cellular or viral gene.

[1]Brand, A. H., Breeden, L., Abraham, J., Sternglanz, R., and Nasmyth, K. 1985. Characterization of a "silencer" in years: A DNA sequence with properties opposite to those of a transcriptional enhancer, *Cell* 41:41-48.
[2]Laimins, L., et al. 1986. *Proc. Nat. Acad. Sci. (U.S.A.)* (in press).

CHAPTER 4

Housekeeping Genes

Joo-Hung Park, Howard V. Hershey,
and Milton W. Taylor

SINCE THE ELUCIDATION of the three-dimensional structure of
DNA in 1953, a tremendous amount of information has accumulated on the
molecular basis of gene regulation. Most of this information has come from
studies of prokaryotes, due to their rapid growth rate and the ease of manipulating large populations. Studies on mammalian gene regulation were very
limited because of the long generation time of the mammalian cell in culture
and difficulty of isolating mutants. These problems were partially solved with
the development of certain somatic cell lines in 1958 by Puck et al (37) who
promoted the use of somatic cell lines as the equivalent of microorganisms.

Very few established cell lines remain truly diploid. Rather, in cell culture they become pseudodiploid, pseudotetraploid or heteroploid. However,
some cell lines such as Chinese hamster ovary (CHO) have a relatively stable
karyotype, and have thus been useful for the molecular analysis of mutagenesis and gene expression.

Because most cell lines capable of long-term growth are fibroblastic and
do not express highly differentiated phenotypes, most of the genetic work
with such cells has involved mainly the class of genes termed "housekeeping
genes," that is, genes that are constitutively expressed in essentially all
replicating cells *in vivo*. Housekeeping genes are often, but not always, vital
for cell function and account for up to 90 percent of the expressed eukaryotic
genome, even in highly differentiated tissues. Beyond this, however, housekeeping genes are highly diverse, including genes expressed at very high
levels (e.g., actin, rRNA, tRNA, and histone genes) and genes expressed at
moderate to low levels [e.g., thymidine kinase (TK), dihydrofolate reductase
(DHFR), and adenine phosphoribosyltransferase (APRT) genes.] Housekeeping genes include genes that are highly regulated by environmental condi-

79

tions [e.g., the ribosomal protein genes, and the 3-hydroxy-3 methylglutaryl coenzyme A (HMG CoA) reductase gene] or by cell cycle [e.g., TK, DNA polymerase, histone (16, 31)], and genes that appear to be constitutively produced (e.g., hypoxanthine guanine phosphoribosyltranferase (HGPRT). APRT).

Until recently, much of the molecular analysis of mammalian genes has been of those which are either expressed at high levels or present in multiple copies (e.g., histones, rRNA, actin), present on viruses (e.g., oncogenes, Herpes TK gene), or are tissue-specific (e.g., immunoglobulins, hemoglobins). Because genes expressed under these conditions produce high levels of mRNA, it has been relatively easy to isolate appropriate cDNA probes. In contrast, much of the work in mutagenesis in tissue culture has involved genes expressed in moderate or low levels that are not vital for cell survival in an appropriate environment (usually media supplemented with the product of the gene). This includes many of the genes involved in purine and pyrimidine metabolism since the alternate pathways of synthesis and salvage allow cell survival in the absence of one or the other pathway.

This chapter concentrates on those housekeeping genes that lend themselves to analysis by mutagenesis in tissue culture and that are dispensable and selectable. They are usually expressed only to a moderate extent and often constitutively expressed. Since there are several good reviews on mutagenesis in somatic cell culture (4, 8), this chapter concentrates on a molecular analysis of the mechanism of regulation of housekeeping genes, rather than on mechanisms of mutation.

Expression of Housekeeping Genes

Although there is a huge literature on the molecular mechanism of transcriptional regulation in prokaryotes via feedback and allosteric modification of regulatory proteins, similar studies are much less complete for mammalian cells in culture. Perhaps the best analyzed of the housekeeping genes is the HMG CoA reductase, a transmembrane glycoprotein that catalyzes the conversion of HMG CoA to mevalonate. It is negatively regulated at the level of transcription by cholesterol and other metabolic end products. This makes the HMG CoA reductase system a good model for studying the molecular mechanism of feedback regulation in mammalian cells (27).

Analysis of the mechanism of temporal regulation of genes associated with the cell cycle is also at a rudimentary level in mammalian cell culture, with the best evidence for cell cycle regulation involving the histone genes. It has been shown that when certain toxic analogs are used for selection, mammalian cells are able to respond via amplification of a number of housekeeping genes (e.g., the DHFR, HGPRT, HMG CoA reductase, and adenosine deaminase (ADA) genes) [for review, see (39) and Chaps. 5 and 6 in this volume]. This selective amplification may be an important mechanism in the evolution of normally amplified genes (e.g., histone and rRNA genes).

However, amplification has not been shown to contribute to the normal physiological mechanism of gene regulation (at least for single-copy housekeeping genes). A notable exception, however, is the rRNA gene amplification in amphibian oocytes.

Considering the characteristics associated with housekeeping genes, it might be expected that these genes would have some regulatory signals different from those of tissue-specific genes. In this chapter several housekeeping genes will be discussed in terms of gene organization, 5'- and 3'-end cis-acting structures, methylation, chromatin-structure, DNase hypersensitivity, trans-acting factors involved in transcriptional and posttranscriptional regulation, and temporal expression.

Organization of Single-Copy Housekeeping Genes

Housekeeping genes can be divided into two classes on the basis of the size of genomic DNA (see Table 1). One class of genes includes the glyceraldehyde-3-phosphate dehydrogenase (GAPDH) gene, the chicken TK gene, and the human and mouse APRT genes. These genes are relatively small relative to their mRNA size. The other class of genes is very large compared to its mRNA size and includes the human superoxide dismutase, human ADA, human DHFR, mouse HGPRT, and hamster HMG CoA reductase genes.

Intron-exon boundaries for a number of housekeeping genes have been sequenced and compared with the consensus sequence (5'-GT-intron-AC-3') in other eukaryotic genes. Most of the genes listed in Table 1 have intron-

TABLE 1   Organization of housekeeping genes

Gene	Size of functional genomic DNA (kb)	Size of mRNA (kb)	Number of introns	Reference
Chicken glyceraldehyde-3-phosphate dehydrogenase	4.65	1.3	11	(45)
Chicken TK	3.0	2.0	6	(22)
Mouse APRT	3.1	~0.85	4	(9)
Human superoxide dimutase	11	0.7, 0.9	4	(23)
Human adenosine deaminase	32	1.5	11	(47)
Human DHFR	30	0.8, 1.0, 3.8	5	(5)
Mouse HGPRT	33	1.3	8	(30)
Hamster HMG CoA reductase	25	4.5	19	(38)

exon boundaries that are in reasonable accord with the consensus sequence. There are, however, a few exceptions: a GT→ TG change at intron 4 in the GAPDH gene (45); a GT → GC change at intron 1 in the superoxide dismutase gene (23); and a GT → GC change at intron 2 in the mouse APRT gene (9). However, it is clear that these transcripts are adequately processed to produce functional mRNAs, perhaps by a variant splicing system using SnRNPs other than U1 SnRNP (20, 34). Although *these* 5' splice site variants are expressed, several cases have been reported in which changes of the GT sequence abolish 5' splice site activity with consequent production of abnormally spliced mRNAs: a G→A change in human β-globin resulting in β^{0}-thalasemia, lack of nucleotides 1 to 5 of intervening sequence (IVS) 1 in a mutant human α_2-globin gene, and inactivation of the 5' splice site in a rabbit β-globin gene IVS2 (46).

Because the functions that housekeeping genes perform are often present across the entire evolutionary spectrum, it may well be possible to analyze the position and content of introns for a particular gene across a wide phylogenetic spectrum. Such studies may shed light on questions on the antiquity of introns and whether the absence of introns in prokaryotes is a primary condition or secondary specialization. Indeed the presence of introns in prokaryotic tRNAs provides a hint of these possibilities.

Prokaryotic, yeast, and chicken GAPDH proteins are very similar, which indicates a strong evolutionary relationship. However, there are no introns in two yeast genes, yet chicken GAPDH has 11 introns interspersed along a 4.65-kb stretch of DNA (94 to 621 bp in length). Examining the GAPDH genes in other eukaryotic species should allow one to follow the fate of the various introns.

Pseudogenes, which are structural homologs to the counterpart functional genes but are functionally silent, have been reported for a variety of eukaryotic genes (e.g., the small nuclear RNA, human immunoglobulin γ-light chain, β-tubulin, and mouse globulin genes) (6). Pseudogenes are nonfunctional because of a variety of reasons: many have acquired frameshift mutations and internal stop codons, some lack introns, and some are missing coding sequences at the 5' or 3' end. For example, there are three human DHFR pseudogenes, one of which is missing the 5' end coding region and the other two are lacking introns (6). The 5' and 3' ends of the two pseudogenes are flanked by a short DNA sequence, 60 nucleotides in length, which is a characteristic of transposable elements. There are four pseudogenes of the human HGPRT gene which map to chromosomes 11, 3, and 5 (36). There also exists one mouse APRT pseudogene (43). The origin of pseudogenes is not clear, but observations suggest that many intronless (processed) pseudogenes are derived from mRNA that is transcribed into DNA by reverse transcriptase and then inserted into the genome elsewhere (6). However, there are other kinds of pseudogenes that have introns but lack some 5'- or 3'-end sequence or coding sequences. These suggest alternate mechanisms for the evolution of pseudogenes, such as genomic rearrangement by translocation or retroviruses.

The 5'-End Sequence

The promoter sequence of a typical eukaryotic protein-coding gene consists of three domains: initiator, selector, and modulator. The initiator, the mRNA start site, seems to consist of an "A" residue surrounded by pyrimidines and to correspond to cap sites in most genes studied. The selector sequence, the TATA sequence, which is found 26 to 34 nucleotides upstream from the cap site of many eukaryotic genes and most viral genes transcribed by RNA polymerase II, acts as a selector that determines the transcription initiation site rather than being the direct functional equivalent of a prokaryotic Pribnow box. The modulator sequence, which is usually located 50 to approximately 200 nucleotides upstream from the cap site, plays a role in controlling transcription quantitatively, in contrast to the TATA sequence which controls transcription qualitatively. The model described above is based on data for a limited subset of tissue-specific and viral genes. In this section, the 5'-end sequence of several housekeeping genes will be discussed in conjunction with function and compared with that of tissue-specific genes.

The eukaryotic TATA sequence acts as a selector that determines the transcription initiation site. Heterogeneity of the 5' end of mRNA transcripts was observed when the TATA box was deleted in mutants of the SV40 early genes, the H2A histone gene, and the HSV-1 TK gene, and also was found in genes that are naturally lacking the TATA sequence, although there was some difference in the extent of the heterogeneity. Most housekeeping genes analyzed so far (except for the GAPDH and superoxide dismutase genes) lack the consensus TATA or CAAT sequence in the 5'-end region, resulting in multiple transcription initiation sites in agreement with the assigned function for the TATA sequence (Table 2).

One common feature among the 5'-end sequences from different housekeeping genes (Table 2) is the prevalence of the GC-rich sequences, implying some functional role for these. The 5'-CCGCCC-GGGCGG-3' sequence upstream of the cap site in the HSV-1 TK gene is necessary for the expression of the TK gene, possibly by forming a stem-and-loop structure (28). The 5'-CCGCCC-3' sequence, which is repeated six times within the three 21-bp repeats of the SV40 promoter, is essential for expression of the SV40 early and late genes. This sequence is recognized by the trans-acting factor Sp_1 (11). Furthermore, a 440 bp fragment of the African green monkey cell genomic DNA which shares homology with the transcriptional regulatory sequence of the SV40 gene contains the sequence GGGCGG, which also is recognized by Sp_1, implying that the "GC box" is generally involved in Sp_1 mediated transcription (10).

How is the GC sequence involved in gene expression? One possible explanation is that the GC sequence and the flanking sequences may form a stem-loop structure, which could be recognized by trans-acting factors such as Sp_1, thus facilitating RNA polymerase binding to the DNA. Alternatively, the GC sequence may play a role mediated by methylation. In eukaryotic cells, the C residue of the CpG sequence is readily methylated, and methylation of

TABLE 2 5′-terminal sequences of housekeeping genes

Gene	−25 region	−60 region	Possible upstream modulating sequences[a]	Heterogeneity of transcription initiation site[b]	G + C content (%)[c]	Reference
Chicken glyceraldehyde-3-phosphate dehydrogenase	TATAAA	CCATT	GGGGCGGGGC at −58, GCCCGCCCC at −93	1	80	(45)
Human superoxide dismutase	TATAAA	CCATT	GGCGGG at −90, CCGCCC at −137, GGCGGG at −173, GGCGGG at −244	1	70	(23)
Human DHFR	AAATAG	GCGCC	GCGCGG at −40, GGCGGG at −49, CCGCCC at −190	4 at one cluster	80	(5)
Chicken TK	ATAACT	CTCCC	CCGCCC at −55	3 at one cluster	76	(22)
Mouse HGPRT	GGAGCC	AGCCT	CGGAGCCTGG at −21, CGGAGCCTGG at −57, GGCGCGG at −45	3 at one cluster	67	(30)
Human adenosine deaminase	CGTTAA	GGCGG	GGGCGGGG at −50, −73, −93 or −99	7 at one cluster	82	(47)
Hamster HMG CoA reductase	No TATA-like sequence	No CAAT-like sequence	3 CCGCCC hexamers interspersed through −240	9 at 5 clusters	65	(38)
Mouse DHFR	No TATA-like sequence	No CAAT-like sequence	4 tandem 48 bp repeats	Multiple at 3 clusters	ND[d]	(7)

[a]Determined by nucleotide sequencing.
[b]Determined by S1 mapping or primer extension analysis. A cluster denotes a group of protected bands that are different in length by one or two nucleotides. A single cluster is more than five nucleotides distant from the next cluster.
[c]Calculated over 200 nucleotides upstream from the transcription initiation site.
[d]ND, not determined.

the CpG sequences has been shown to regulate gene expression. Thus the GC sequence in housekeeping genes could be involved in gene regulation via alterations in the methylation pattern.

The HMG CoA reductase gene is subject to feedback inhibition of transcription by cholesterol (35). When the 5′ sequences from the hamster HMG CoA reductase gene were fused to the coding region of chloramphenicol acetyltransferase (CAT) gene and assayed for CAT activity, deletion within the 500 bp from +200 to −300 from the AUG site abolished not only transcription but also feedback inhibition. This 500-bp region has five CCGCCC or GGCGGG, suggesting a role for the hexanucelotide repeat in transcription or feedback inhibition of the HMG CoA reductase gene. The HMG CoA reductase gene has a 640- to 729-nucleotide leader, which is the longest leader sequence found in a eukaryotic gene (38). In addition, there is an 3.5 kb long intron in this leader, raising the possibility that this intron may be involved in regulation of the HMG CoA reductase gene.

The 3′–End Sequence

In eukaryotes, mRNA precursors are processed to mature mRNAs through 3′ processing and intron splicing. As for poly (A)mRNAs, 3′ processing of mRNA precursors occurs at two steps either sequentially or in concert. These are cleavage of the 3′ sequence 10 to 30 nucleotides downstream of the polyadenylation site and addition of the poly (A) tail (260 to 300 nucleotides in length) to the polyadenylation site. Nonpolyadenylated mRNAs, such as mammalian histone mRNAs, are cleaved only at the authentic 3′ end which consists of a potential exonuclease-resistant stem-and-loop structure [for review, see (2)].

The function of the poly(A) tail is still controversial, but it seems to be involved in stabilizing mRNAs after processing. The AAUAAA sequence located 10 to 30 nucleotides upstream of the polyadenylation site has been shown to be an element directing cleavage and polyadenylation of pre-mRNA. In addition to the consensus hexamer AAUAAA, other related sequences, such as AAUUAA, AAUACA, AAUAAC, CAUAAA, AUUAAA, AGUAAA, UAUUAG, and AAUAUA, have been seen in different genes and appear to be fully functional in the appropriate context.

Polyadenylation signals in housekeeping genes are generally in accord with one of the above hexamers. Two exceptions are the human DHFR gene which uses AAAAGG and the mouse DHFR gene which uses AUAA (Table 3). The most abundant polyadenylation signal seen in housekeeping genes is the consensus AAUAAA hexamer. It is not known which hexamer is most efficient in 3′ processing, but their relative efficiencies may be an additional aspect of gene regulation. For example, three mRNAs of 0.8, 1.0, and 3.8 kb in length are produced from the human DHFR gene, the 3.8-kb mRNA being the most abundant form. The polyadenylation signal for the 3.8-kb mRNA has the consensus hexamer sequence, AAUAAA, whereas the other two forms have AAAAGG or AUUAAA as their polyadenylation signals, suggesting different

TABLE 3 3'-terminal sequence of housekeeping genes

Gene	Polyadenylation signal	Distance from polyadenylation site (bp)	Poly(A) addition site[a]	Length in mRNA (kb)	Reference
Mouse DHFR	AUAA	24	→TAGATCTATGAGTAATTATTTCTTTA	0.75	(41)
	AUAAC	53	CAAGTTATGTCATCTGCATAAGGTCT	1.0	
	AUAA	25	TAGAGACTGGATTATTGTATGCACAT	1.2	
	AAUAAA	13	TAAAATGGCTTTCCTCATCTCAGCAG	1.6	
Human DHFR	AAAAGG	16	TAGATCTATAATTATTTCTAAGCAAC	0.8	(5)
	AUUAAA	22	CAACTGAGCAGTTTCACTAGTGGAAA	1.0	
	AAUAAA	18	AAGTGCTGGCCCTCCTGTTGTTCTTTG	3.8	
Hamster	AAUAAA	19	TATTCTATCTAAGCTTCGAGTTCAG	4.2	(38)
HMG CoA reductase	AAUUAU	15	CATTAACAGGTGTAAGCGGTGGCTTT	4.7	
	AAUAAA	ND[b]	N.D.	4.9	
Human superoxide dismutase	AUUAAA	16	AAGTGTAATTGTGTGACTTTTTCAGA	0.7	(23)
	AAUAAA	14	CACTTATTATGAGGCTATTAAAAGAA	0.9	
Mouse HGPRT	AAUAAA	12	TCGCCACTTTTTGAATTC	1.3	(30)
Human adenosine deaminase	AAUAAA	21	ND	1.5	(47)
Chicken TK	AAUUAA	ND	ND	2.0	(22)
Mouse APRT	AAUAAA	17	ND	0.85	(9)

[a]TCT nucleotide sequences that are found in a region about 30 bp downstream of the polyadenylation signal are underlined.
[b]ND, not determined.

processing efficiencies. However, the hypothesis that the most efficient hexamer in 3′ processing is the AAUAAA sequence does not fit for two other genes, the human superoxide dismutase and HMG CoA reductase. In the superoxide dismutase gene, the 0.7-kb mRNA which is 4 times more abundant than the 0.9-kb mRNA has the AUUAAA sequence while the latter has the AAUAAA sequence (Table 3). The least abundant form among three mRNAs produced in the HMG CoA reductase gene also uses an AAUAAA sequence as a polyadenylation signal. Considering the above data, it is difficult to conclude which hexamer is the most efficient in 3′ processing, but it's not un-likely that the polyadenylation hexamer and downstream sequences could interact with each other to form a complex recognized by the processing machinery, i.e., the context in which the signal is embedded is important.

In the housekeeping genes the sequences 20 to 50 nucleotides downstream from the polyadenylation signal do not fit to the consensus sequence, YGTGTTYY (29), but contain one or two more interspersed GT sequences per gene. It has yet to be tested whether the various downstream sequences in housekeeping genes presented in Table 3 are involved in processing of the polyadenylation site.

Several housekeeping genes produce more than one mRNA which differ in length in the 3′-untranslated region (as discussed above) raising the question of why different length mRNAs are transcribed from the same gene. Is there any function assigned to the 3′-heterogeneity of mRNAs? Many viruses use overlapping mRNAs to produce different proteins from the same transcript. However, the heterogeneity at the 3′ end of transcripts of housekeeping genes does not result in different proteins. Rather the 3′ ends appear to be involved in regulation. There are two human α-globin genes (coding for α_1 and α_2 protein) and the corresponding mRNAs (α_1 and α_2 mRNAs) have similar nucleotide sequences except for their 3′ untranslated region, but are translated differently, suggesting that the 3′ untranslated region is involved in the translational regulation of this gene (25). The 3′ untranslated region of the c-pos mRNA inhibits the translation of its own mRNA (32). Expression of the mouse histone H4, which is regulated cell-cycle dependently is controlled by the 3′ untranslated sequence of the gene (26). Considering these data, the heterogeneity of the 3′ end of mRNAs in housekeeping genes could play a role in the regulation of gene expression at a posttranscriptional level. There is some evidence for posttranscriptional regulation of several housekeeping genes.

Methylation and DNase Hypersensitivity

The presence of 5-methylcytosine in DNA was first described for calf thymus DNA about three decades ago. This observation has led to the idea that gene expression is proportional to the extent of DNA methylation. In addition to the regulatory role of DNA methylation in gene expression, there may be several other roles for DNA methylation such as protection of DNA from endonuclease, DNA recombination, DNA replication, and viral latency (see Chap. 1).

In vertebrates, the only modified base seen in DNA is 5-methylcytosine, present primarily (if not exclusively) in the relatively rare CpG dinucleotide. Most of the CpG dinucleotides in vertebrates are methylated (approximately 70 percent), although the specific methylated sites vary from tissue to tissue. CpG dinucleotides are clustered in the 5' region of certain genes, and treatment of cells with 5-azacytidine (which inhibits DNA methylation) can activate previously silent genes. Thus, methylation may play a significant role in the regulation of eukaryotic genes.

Hypersensitivity to endonucleases such as DNase I has been shown to accompany the switching of inactive genes to the active state. Activation of eukaryotic genes, such as the *Drosophila* heat shock genes, histone genes, globin genes and preproinsulin genes, is accompanied by DNase hypersensitivity with the most hypersensitive sites being clustered around the 5' promoter region (12). Although there is a substantial amount of evidence for the role of methylation and hypersensitivity in regulation of gene expression, there are many problems still not resolved (e.g., which 5-methycytosine residues are important; what functional changes are induced by methylation; whether there are any trans-acting factors that interact with methylated or unmethylated sequences, and whether there is a relationship between methylation and DNA hypersensitivity).

As mentioned earlier, several housekeeping genes have unusual 5'-end sequences in which the GC content is extremely high with a significant number of GC clusters. The 5' region of the CHO APRT gene is unmethylated relative to potential methylation sites within the gene (Fig. 1). Moreover, when unmethylated sites are methylated *in vitro*, the methylated APRT DNA is unable to transform APRT mouse cells implying that at least some of the potential methylation sites must remain unmethylated for gene expression to occur (21). Similarly, the 5' end of the human DHFR gene is hypomethylated, while the remaining segment is heavily methylated. The 5' end of the HGPRT gene present in the active X chromosome is completely unmethylated whereas the 3' end is methylated. In contrast, in the inactive chromosome there is no distinctive methylation pattern except for a general hypermethylation of the 5'-end region. Treatment of "silent" genes with 5-azacytidine often results in gene activation, perhaps by an effect on methylation. In 5-azacytidine-treated cells, the 5' end of the HGPRT gene on reactivated X chromosome is unmethylated in contrast to methylation of the parental counterpart, although the 3' end is still unmethylated.

Hypomethylation at the 5' end and hypermethylation at the 3' end may be prerequisites for activation of some housekeeping genes (Fig. 1). There are, however, several exceptions to this rule. The human GAPDH gene, an X-linked gene, has CpG clusters around the 3' region. These 3' clusters are undermethylated in the active X and more methylated in the inactive X.

In the case of the human HGPRT gene, the 5-azacytidine reactivated gene showed a different methylation pattern at the 3' region from that of the

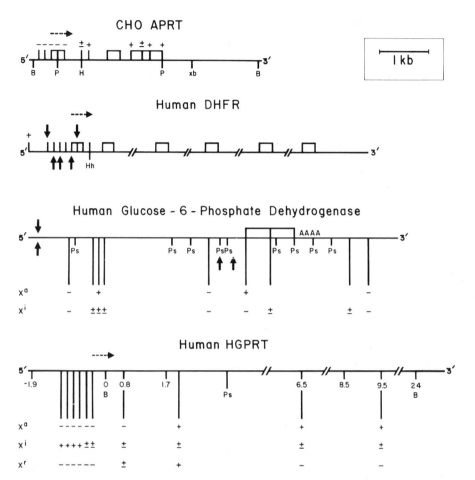

Figure 1 Map showing methylation and nuclease hypersensitivity pattern in selected housekeeping genes. B, BamHI; P, PvuII; H, HincII; Xb, XbaI; Ps, PstI; Hh, HhaI. Open boxes denote exons. The symbols indicate the methylation status of cytosine in a number of CpG sites. Minus (−), unmethylated; plus (+), fully methylated; (≠), partially methylated. Horizontal arrows indicate direction of transcription. Transcription starts at the left of each arrow, except for human GPDH (*) for which only the 3′ end of the gene is given. AAAA represents the polyadenylation signal for GPDH. Downward arrows indicate DNase I hypersensitive sites. Upward arrows indicate S1 nuclease hypersensitive sites. X_i inactive X chromosome; X-, active chromosome; X′, reactivated X chromosome. Based on data of Stein et al. (44) for CHO APRT; Shimada et al. (42) for human DHFR; Wolf et al. (49) and Battistuzzi et al. (1) for human GPDH; Wolf et al. (48, 49) for human HGPRT.

HGPRT gene on the active X chromosome. There are also differences in the methylation pattern around the 3′ end of the human GAPDH gene in different tissues; some CpG clusters around the 3′ end of the human GAPDH gene are unmethylated in the active X chromosome. In addition, the Herpes

TK gene when methylated at either the 5'- or 3'-end region loses its transforming ability (21). Thus, the overall pattern of methylation may facilitate interactions between the correct initiation site and the transcriptional machinery, while inhibiting nonfunctional associations (i.e., unmethylated 5' clusters could be essential for initiation of transcription while the 3'-end methylation or unmethylation may be involved in maintaining or enhancing the transcriptional activity).

It has been shown that the appearance of the 5'-end DNase I hypersensitive sites is a prerequisite for gene activity in such genes as the *Drosophila* heat shock genes. In most cases, the DNase hypersensitivity is clustered around the 5'-promoter region. In this context, it is interesting that the DNase I or S1 nuclease hypersensitive sites were found in the 5' CpG clusters of the human DHFR, GAPDH, and HGPRT genes. Since gene activity generally requires 5' hypomethylation, there is a correlation between hypomethylation of 5' CpG clusters and the presence of the DNase hypersensitive sites.

The mechanism by which methylation exerts its effect is unknown, although it may affect the conformation of DNA and activity of the transcriptional complex. This could occur either indirectly by facilitation of a B-to-Z transition, or directly by affecting the binding of transacting factors which recognize methylcytosine in DNA (18).

Temporal Expression

A number of housekeeping genes are regulated during the cell cycle; for example, there is a class of genes that exhibit higher activity during S phase than in other phases of the cell cycle. These include TK, deoxycytidine kinase, thymidylate synthetase, ribonucleotide reductase, DHFR, ornithine decarboxylase, and DNA polymerase. The tissue-specific regulation of gene activity has been found to be influenced by cis-acting DNA sequences found within both the promoter and intron regions (14). Thus one can propose that at least some regulating signals include gene-specific sequences which affect cell cycle-regulated genes. It has been shown that the chicken TK gene (but not the Herpes virus TK gene) is temporally regulated when transfected into a mouse cell line, implying that at least some signals necessary for temporal regulation are present in the chicken TK gene itself (40).

In addition to transcriptional regulation, there are several possible post-transcriptional levels at which regulatory signals may exert their effect: posttranscriptional processing, transcript stability, translation rates, and posttranslational modification. Indeed, in the case of the mouse DHFR gene, the rate of transcription of primary transcript and the half-life of cytoplasmic mRNAs are the same in both dividing and nondividing cells, but the primary transcripts are processed and transported to the cytoplasm in dividing cells, whereas in resting cells the majority of transcripts are rapidly degraded in the nucleus (24). Similarly, there is no difference in the methylation pattern, chromatin structure, or the rate of transcription in the chicken TK gene in either resting or dividing cells, but there is a several-fold increase in stability of the mRNAs in the dividing cells (15). The steady-state levels of histone H4 mRNA varied 10-50-fold over the cell cycle while transcriptional activity varies only 2- 5-fold (26). The α-tubulin gene is highly expressed in stem cells, but not in differentiated cells. Here, the steady-state levels also are quite different, but the transcription and translation rates are the same, implying posttranscriptional regulation of α-tubulin mRNAs during differentiation (17).

However, the human histone H4 gene seems to be regulated at both the transcriptional and posttranscriptional level (3). When *in vitro* transcriptional activity of the gene was measured using nuclear extracts from cells at different stages of the cell cycle, the nuclear extracts from S-phase cells gave the highest activity. In addition, when mouse L cells were transfected with the human histone H4 gene (60 copies per haploid genome), expression of the human histone H4 gene varied more than 10-fold during the cell cycle, but the endogeneous mouse histone H4 gene was not regulated, and the mouse H2A and H3 genes were expressed normally. In cell revertants that lost the majority of the human H4 gene copies, the expression of the mouse H4 gene was restored, implying that there is a trans-acting transcription factor (3) involved in cell-cycle-dependent regulation that is specific for the histone H4 gene and for which the transfected human H4 gene can compete with the endogeneous mouse H4 gene. There are subtype-specific conserved sequences in the 5'-flanking region of histone genes from a variety of organisms. These conserved sequences may be recognized by the trans-acting factors.

How does posttranscription regulation occur? In the mouse histone H4 gene a 3'-terminal segment of the gene appears to control cell-cycle dependent regulation (26). When SV40 early promoter was fused to a 463-bp fragment containing the 3'-terminal half of the mouse H4 gene and a 230-bp spacer fragment, the fused gene showed regulated expression. But transcripts that read through the 3' end of the histone gene and the downstream *E. coli* galactokinase gene to a SV40 polyadenylation signal were not expressed cell-cycle dependently, indicating the position of the 3' end of histone gene relative to the end of transcripts could be important. The 3' end of the mouse histone H4 gene has a potential stem-and-loop structure which may contribute to the stability of mRNA when it is positioned at an appropriate distance from the end of the transcript, protecting it from nucleases. Another example of the role of cis-acting sequences in cell-cycle-dependent regulation is to be found in the chicken TK gene. This gene is expressed in differentiating muscle cells but not in differentiated nondividing cells. When Merrill et al. (31) fused the chicken TK promoter to the coding sequence of the HSV TK gene, expression of the HSV TK gene was not dependent on differentiation, that is, the fused gene was constitutively expressed in both differentiating and differentiated cells. However, when the HSV TK gene fused to the 3'-end coding sequence of the chicken TK gene was examined for expression in different cells, the fused gene was expressed in differentiating cells, but not in differentiated cells, implying that the 3'-end coding sequence of the chicken TK gene is involved in cell cycle regulation. Intragenic sequences have been shown to be involved positively in expression of 5S rRNA and tRNA genes, and negatively in tissue-specific genes. It will be intriguing to examine how the intragenic sequence in the chicken TK gene is involved in gene regulation.

Overall then, a number of the cell-cycle-regulated housekeeping genes exhibit posttranscriptional regulation (although transcriptional regulation does occur) and in some cases, the 3' end of the transcript appears to play a significant role.

Conclusions

In contrast to tissue-specific or cell-type-specific genes which are expressed at high levels in some cell types but not at all in other cell types, housekeeping genes are expressed in all cells at some time during the cell cycle. Thus, some of the regulatory mechanisms controlling the expression of housekeeping genes may differ from those controlling the expression of tissue specific genes.

One prominent feature of the DNA structure of the housekeeping genes is the large number of GC clusters at the 5' end of the gene and, typically, an absence of the eukaryotic consensus TATA or CAAT sequence. In general, the TATA sequence in eukaryotes has been shown to be a selector sequence determining precise initiation of transcription, and in some cases is necessary for efficient transcription. It is possible that the lack of a TATA sequence in some housekeeping genes may contribute to the low level of gene expression, although this hypothesis has not yet been tested. It can be tested using *in vitro* mutagenesis to insert a canonical TATA sequence into an appropriate region of the 5' end. However, the insertion of the TATA sequence may disturb the potential secondary structure of the 5' region and interfere with transcription. Indeed in some housekeeping genes there is a 5' potential stem-loop structure which is presumably associated with promoter activity (Fig. 2).

One possible function of the GC cluster in the 5' region is the formation of a potential stem-and-loop structure which may facilitate entry of RNA polymerase II onto template DNA (Fig. 2). A second, and possibly related, hypothesis is that the GC-rich sequence may be recognized by some transcription initiation factors such as Sp_1 which is known to specifically bind to the GC-rich sequence of the SV40 21-bp repeat (10, 11). Also the GC-rich sequence contains several CpG sequences that are capable of being methylated in eukaryotes.

Experimental evidence for the role of the 3' end in the regulation of gene expression has been reported for some cell-specific eukaryotic and viral genes: e.g., two mRNAs with different 3' ends encode membrane-bound and secreted forms of immunoglobulin mu heavy chain, respectively, and different polyadenylation sites are important for maturation of transcripts from adenovirus. Among housekeeping genes, cell cycle-dependent expression of the mouse histone H4 gene depends on the 3'-untranslated region (19). Moreover, the mRNAs for some housekeeping genes are heterogeneous at the 3' end (Table 3). All the mRNAs for the mouse DHFR and human superoxide dismutase genes were shown to be functional, raising the ques-

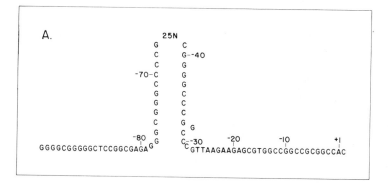

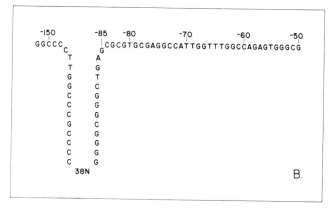

Figure 2 Potential secondary structures in the promoter region of (A) human adenosine deaminase and (B) superoxide dismutase genes. Numbers indicate the distance from the cap site in base pairs. The full energy per strand is (A) −42.4 and (B) −33 cal/mol. Data of Valerio et al. (47) for the human adenosine deaminase gene and of Levanon et al. (23) for the human superoxide dismutase gene.

tion of why there are multiple polyadenylation sites for each gene? One possible explanation is that the presence of multiple polyadenylation sites could result in weak activity in 3′ processing. When the mouse DHFR cDNA was fused to the SV40 late polyadenylation signal and examined for the usage of each polyadenylation signal, the most prominent mRNA signals, suggesting that DHFR polyadenylation signals were relatively weak. The usage of different polyadenylation signals varied over the cell cycle, implying that the metabolic state of cells determined the efficiency of the polyadenylation signals at different sites (19). These different sequences may affect the stability of mRNAs or affect processing of precursor mRNAs, thus contributing to the regulation of gene expression.

A surprising, recent discovery has been the finding of an oppositely transcribed RNA utilizing the major promoter region of the mouse DHFR gene (about 200 bp upstream of the AUG). Although there is disagreement concerning whether this oppositely transcribed RNA is part of a large (14,000 bp) mRNA (7) or only present as an abundant, small (180 to 240 nucleotide), nuclear RNA (13), it is interesting that this transcript does overlap DHFR transcripts which initiate from a more distal promoter (500 to 560 bp from the AUG) and that these transcripts are only found in the nucleus (7). Whether this represents a feature similar to prokaryotic antisense RNA or unique to the DHFR gene, or represents a more general phenomenon remains to be examined.

General References

REYNOLDS, G. A., BASU, S. K., OSBORNE, T. F., CHIN, D. J., GOL, G., BROWN, M. S., GOLDSTEIN, J. L., AND LUSKEY, K. L. 1984. HMG CoA reductase: a negatively regulated gene with unusual promoter and 5′ untranslated regions. *Cell* 38:275–285.

WOLF, S. F., and MIGEON, B. R. 1985. Clusters of CpG dinucleotides implicated by nuclease hypersensitivity as control elements of housekeeping genes, *Nature* 314:467–469.

References

1. Battistuzzi, G., D'Urso, M., Toniolo, D., Persico, G. M., and Luzzatto, L. 1985. Tissue-specific levels of human glucose-6-phosphate dehydrogenase correlate with methylation of specific sites at the 3′ end of the gene, *Proc. Nat. Acad. Sci. U.S.A.* 82:1465–1469.

2. Birnstiel, M. L., Busslinger, M., and Strub., K. 1985. Transcription termination and 3′ processing: The end is in site!, *Cell* 41:349–359.

3. Capasso, O., and Heintz, N. 1985. Regulated expression of mammalian histone H4 genes *in vivo* requires a trans-acting transcription factor, *Proc. Nat. Acad. Sci. U.S.A.* 82:5622–5626.

4. Caskey, C. T., and Robbins, D. C. 1982. *Somatic Cell Genetics*, Plenum Press, New York.

5. Chen, M. J., Shimada, T., Moulton, A. D., Cline, A., Humphries, R. K., Maizel, J., and Nienhuis, A. W. 1984. The functional human dihydrofolate reductase gene, *J. Biol. Chem.* 259:3933–3943.

6. Chen, M. J., Shimada, T., Moulton, A. D., Harrison, M., and Nienhuis, A. W. 1982. Intronless human dihydrofolate reductase genes are derived from processed RNA molecules, *Proc. Nat. Acad. Sci. U.S.A.* 79:7435–7439.

7. Crouse, G. F., Leys, E. J., McEwan, R. N., Frayne, E. G., and Kellems, R. E. 1985. Analysis of the mouse dhfr promoter regions: Existence of a divergently transcribed gene, *Mol. Cell. Biol.* 5:1847–1858.

8. Davidson, R. L. 1984. *Somatic Cell Genetics*, Hutchinson Ross Publishing Company, Stroudsburg.

9. Dush, M. K., Sikela, J. M., Khan, S. A. Tischfield, J. A., and Stambrook, P. J. 1985. Nucleotide sequence and organization of the mouse adenine phosphoribosyltransferase gene: Presence of a coding region common to animal and bacterial

phosphoribosyltransferases that has a variable intron/exon arrangement, *Proc. Nat. Acad. Sci. U.S.A.* 82:2731–2735.

10. Dynan, W. S., Saffer, J. D., Lee, W. S., and Tjian, R. 1985. Transcription factor Sp1 recognizes promoter sequences from the monkey genome that are similar to the Simian virus 40 promoter, *Proc. Nat. Acad. Sci. U.S.A.* 82:4915–4919.

11. Dynan, W. S., and Tjian, R. 1983. The promoter-specific transcription factor Sp1 binds to upstream sequences in the SV40 early promoter, *Cell* 35:79–87.

12. Elgin, S. C. 1981. DNase I-hypersensitive sites of chromatin, *Cell* 27:413–415.

13. Farnham, P. J., Abrams, J. M., and Schimke, R. T. 1985. Opposite-strand RNAs from the 5′ flanking region of the mouse dihydrofolate reductase gene, *Proc. Nat. Acad. Sci. U.S.A.* 82:3978–3982.

14. Grosschedl, R., and Baltimore, D. 1985. Cell type specificity of immunoglobulin gene expression is regulated by at least three DNA sequence elements, *Cell* 41:885–897.

15. Groudine, M., and Casimir, C. 1984. Post-transcriptional regulation of the chicken thymidine kinase gene, *Nucleic Acids Res.* 12:1427–1446.

16. Hentschel, C. C., and Birnstiel, M. L. 1981. The organization and expression of histone gene families, *Cell* 25:301–313.

17. Howe, C. C., Lugg, D. K., and Overton, G. C. 1984. Post-transcriptional regulation of the abundance of mRNAs encoding α-tubulin and a 94,000 dalton protein in teratocarcinoma-derived stem cells versus differentiated cells, *Mol. Cell. Biol.* 4:2428–2436.

18. Huang, L. -H., Wang, R., Gama-Sosa, M. A., Shenoy, S., and Ehrlich, M. 1984. A protein from human placenta nuclei binds preferentially to 5-methylcytosine-rich DNA, *Nature (London)* 308:293–295.

19. Kaufman, R. J., and Sharp, P. A. 1983. Growth-dependent expression of dihydrofolate reductase mRNA from modular cDNA genes, *Mol. Cell. Biol.* 3:1598–1608.

20. Keller, E. B., and Noon, W. A. 1984. Intron splicing: A conserved internal signal in introns of animal pre-mRNAs, *Proc. Nat. Acad. Sci. U.S.A.* 81:7417–7420.

21. Keshet, I., Yisraeli, J., and Cedar, H. 1985. Effect of regional DNA methylation on gene expression, *Proc. Nat. Acad. Sci. U.S.A.* 82:2560–2564.

22. Kwoh, T. J., and Engler, J. A. 1984. The nucleotide sequence of the chicken thymidine kinase gene and the relationship of its predicted polypeptide to that of the vaccinia virus thymidine kinase, *Nucleic Acids Res.* 12:3959–3971.

23. Levanon, D., Lieman-Hurwitz, J., Dafni, N., Wigderson, M., Sherman, L., Bernstein, Y., Lalver-Rudich, Z., Danciger, E., Stein, O., and Groner, Y. 1985. Architecture and anatomy of the chromosomal locus in human chromosome 21 encoding the Cu Zn superoxide dismutase, *EMBO J.* 4:77–84.

24. Leys, E. J., Crouse, G. F., and Kellems, R. E. 1984. Dihydrofolate reductase gene expression in cultured mouse cells is regulated by transcript stabilization in the nucleus, *J. Cell Biol.* 99:180–187.

25. Liebhaber, S. A., and Kan, Y. 1982. Different rates of mRNA translation balance the expression of the two human α-globin loci, *J. Biol. Chem.* 257:11852–11855.

26. Lüscher, B., Stauber, C., Schindler, R., and Schümperli, D. 1985. Faithful cell-cycle regulation of a recombinant mouse histone H4 gene is controlled by sequences in the 3′-terminal part of the gene, *Proc. Nat. Acad. Sci. U.S.A.* 82:4389–4393.

27. Luskey, K. L., Faust, J. R., Chin, D. J., Brown, M. S., and Goldstein, J. L. 1983.

Amplification of the gene for 3-hydroxy-3-methylglutaryl coenzyme A reductase, but not for the 53-K Da protein, in UT-1 cells, *J. Biol. Chem.* 258:8462–8469.

28. McKnight, S. L., Kingsbury, R. C., Spence, A., and Smith, M. 1984. The distal transcription signals of the herpes virus tk gene share a common hexanucleotide control sequence, *Cell* 37:253–262.

29. McLauchlan, J., Gaffney, D., Whitton, J. L., and Clements, J. B. 1985. The consensus sequence YGTGYYTT located downstream from the AATAAA signal is required for efficient formation of mRNA 3′ termini, *Nucleic Acids Res.* 13:1347–1368.

30. Melton, D. W., Konecki, D. S., Brennand, J., and Caskey, C. T. 1984. Structure, expression, and mutation of the hypoxanthine phosphoribosyltransferase gene, *Proc. Nat. Acad. Sci. U.S.A.* 81:2147–2151.

31. Merrill, G. F., Hauschka, S. D., and McKnight, S. L. 1984. tk Enzyme expression in differentiating muscle cells is regulated through an internal segment of the cellular tk gene, *Mol. Cell. Biol.* 4:1777–1784.

32. Miller, A. D., Curran, T., and Verma, M. 1984. c-fos protein can induce cellular transformation: A novel mechanism of activation of a cellular oncogene, *Cell* 36:51–60.

33. Mizuno, T., Chou, M. -Y., and Inouye, M. 1984. A unique mechanism regulating gene expression: Translational inhibition by a complementary RNA transcript (micRNA), *Proc. Nat. Acad. Sci. U.S.A.* 81:1966–1970.

34. Oshima, Y., Itoh, M., Okada, N., and Miyata, T. 1981. Novel models for RNA splicing that involve a small nuclear RNA, *Proc. Nat. Acad. Sci. U.S.A.* 78:4471–4474.

35. Osborne, T. F., Goldstein, J. L., and Brown, M. S. 1985. 5′-end of HMG CoA reductase gene contains sequences responsible for cholesterol-mediated inhibition of transcription, *Cell* 42:203–212.

36. Patel, P. I., Nussbaun, R. L., Framson, P. E., Ledbetter, D. H., Caskey, C. T., and Chinault, A. C. 1984. Organization of the human hgprt gene and pseudogenes, *Somat. Cell Mol. Genet.* 10:483–493.

37. Puck, T. T., Cieciura, S. J., and Robinson, A. 1958. Genetics of somatic mammalian cells III. Long term cultivation of euploid cells from human and animal subjects, *J. Exp. Med.* 108:945–956.

38. Reynolds, G. A., Basu, S. K., Osborne, T. F., Chin, D. J., Gol, G., Brown, M. S., Goldstein, J. L., and Luskey, K. L. 1984. HMG CoA reductase: A negatively regulated gene with unusual promoter and 5′ untranslated regions, *Cell* 38:275–285.

39. Schimke, R. T. 1984. Gene amplification in cultured animal cells, *Cell* 37:705–713.

40. Schlosser, G., Steglich, J., de Wet, J., and Scheffler, I. 1981. Cell cycle dependent regulation of thymidine kinase activity introduced into mouse LMTK⁻, cells by DNA and chromatin-mediated gene transfer, *Proc. Nat. Acad. Sci. U.S.A.* 78:1119–1123.

41. Setzer, D. R., McGrogan, M., and Schimke, R. T. 1982. Nucleotide sequence surrounding multiple polyadenylation sites in the mouse dihydrofolate reductase gene, *J. Biol. Chem.* 257:5143–5147.

42. Shimada, T., and Nienhuis, A. W. 1985. Only the promoter region of the constitutively expressed normal and amplified human dihydrofolate reductase gene is DNase I hyper-sensitive and unmethylated, *J. Biol. Chem.* 260:2468–2474.

43. Sikela, J. M., Khan, S. A., Feliciano, E., Trill, J., Tischfield, J. A., and Stambrook, P. J. 1983. Cloning and expression of a mouse adenine phosphoribosyl transferase gene, *Gene* 22:219–228.

44. Stein, R., Sciaky-Gallili, N., Razin, A., and Cedar, H. 1983. Pattern of methylation of two genes coding for housekeeping functions. *Proc. Nat. Acad. Sci. U.S.A.* 80:2422–2426.

45. Stone, E. M., Rothblum, K. N., Alery, M. C., Kuo, T. M., and Schwartz, R. J. 1985. Complete sequence of the chicken glyceraldehyde-3-phosphate dehydrogenase gene, *Proc. Nat. Acad. Sci. U.S.A.* 82:1628–1632.

46. Treisman, R., Proudfoot, N.M., Shander, M., and Maniatis, T. 1982. A single-base change at a splice site in a $\beta°$-thalassemic gene causes abnormal RNA splicing, *Cell* 29:903–911.

47. Velerio, D., Duyvesteyn, M. G. C., Dekker, B. M. M., Weeda, G., Berkvens, T. M. van der Voorn, L., van Ormondt, H., and van der Eb, A. J. 1985. Adenosine deaminase: Characterization and expression of a gene with a remarkable promoter, *EMBO J.* 4:437–443.

48. Wolf, S. F., Jolly, D. J., Lunnen, K. D., Friedmann, T., and Migeon, B. R. 1984. Methylation of the hypoxanthine phosphoribosyltransferase locus on the human X chromosome: Implications for X-chromosome inactivation, *Proc. Nat. Acad. Sci. U.S.A.* 81:2806–2810.

49. Wolf, S. F., and Migeon, B. R. 1985. Clusters of CpG dinucleotides implicated by nuclease hypersensitivity as control elements of housekeeping genes, *Nature (London)* 314:467–469.

Questions for Discussion with the Editors

1. *By how much does the concentration of a typical housekeeping gene vary among different cell types in any one organism? Does that range of differences suggest that the same housekeeping gene might be subject to alternative regulation mechanisms in different cells? For example, could the extent of posttranscriptional regulation vary among different cells?*

There is probably no such thing as a "typical" housekeeping gene. Some of these genes are clearly regulated in a tissue-specific manner, depending upon the metabolic and structural requirements of that tissue. Other housekeeping genes, if measured on a per-cell basis, appear to be essentially constant.

In those housekeeping genes that exhibit tissue-specific variation, the levels of protein can vary as much as 1000-fold (e.g., adenosine deaminase). This variation can be due to differential transcription (perhaps related to differential methylation) or to posttranscriptional processes (such as mRNA stability) or even posttranslational processes (such as protein stability).

2. *What sort of speculations have been offered to account for the functional significance of the apparent size categories shown in Table 1? Has the size of the genomic DNA for any of the individual genes been compared between species in order to determine whether the two sizes of categories are really legitimate?*

There are a few cross-species comparisons of introns involving widely divergent species. However, the number (but not usually the position) of introns in the few studies done have been seen to vary. Some species, e.g., yeast, appear to have a generally reduced number of introns. In mammalian species, however, the size, number, and position of introns appear to be generally conserved, e.g., the HGPRT and DHFR genes. The significance of introns, generally, is not known except for those instances where differential splicing is used to produce alternate, related proteins. For example, HGPRT and APRT are quite similar small proteins yet have dramatically different gene constructions (30 kb and 9 introns for HGPRT and 3 to 4 kb and 4 introns for APRT). One can suspect that this is related to the fact that HGPRT appears to be more highly regulated than APRT. And, indeed, some introns (but not, as yet, those in these housekeeping genes) have been shown to contain enhancer elements, Alu and Alu-like sequences, and other repetitive sequences. The immunoglobin's intron enhancer sequence has been shown to play a role in immunoglobin regulation. However, in many cases, intronless constructs appear to work as well as those with introns, although this may reflect a somewhat artifactual situation. In short, the question is unresolved, but one would suspect that large introns are more likely to contain regulatory elements.

CHAPTER 5

Drug-Resistant Dihydrofolate Reductases

Christian C. Simonsen

DIHYDROFOLATE REDUCTASE (DHFR) catalyzes the NADPH-dependent reduction of dihydrofolate to tetrahydrofolate, a cofactor in the biosynthesis of purines, pyrimidines, and glycine (15). It is an essential enzyme in both eukaryotes and prokaryotes. In eukaryotes DHFR mediates both steps of the conversion of folate to tetrahydrofolate, whereas in prokaryotes dihydrofolate is formed directly from paraaminobenzoate; thus prokaryotic DHFRs are not required to reduce folate to dihydrofolate (6, 15, 41). Over the last 30 years a variety of folate analogs have been described (81–83). Several of these agents, for example, trimethoprim and methotrexate (MTX), have become widely used as antibacterial or antineoplastic agents. However, the use of these inhibitors as drugs is strongly compromised by the emergence of cells resistant to the effects of the selective agent.

Three mechanisms have been shown to generate folate analog resistance in mammalian and bacterial cells. Since many of the folate analogs utilize an active transport system, resistance can be generated by mutations affecting the cells' ability to internalize the drug (73, 85, 91). Resistant cells also can be isolated that possess drug resistant DHFRs (5, 7, 8, 14, 30, 31, 38, 39, 69, 85, 88, 90, 92, 94, 98, 100). Finally, cells expressing increased levels of DHFR, which render the cells resistant to the substrate analog, have been observed after stepwise selection in the presence of the antifolate (2, 3, 16, 31, 50, 62, 63).

Schimke and his colleagues have shown that mammalian cells have increased their levels of DHFR by amplifying the DHFR gene, thus allowing the increased synthesis of the enzyme (see Chap. 6). This latter finding has resulted in a new appreciation of the plasticity of the mammalian chromosome. Gene amplification is a general phenomenon that may be considered both as a mechanism of gene regulation as witnessed by the increase in

Drosophila chorion and *Xenopus* ribosomal genes during development (81) and as a systemic response to stress caused by drug selection (4, 81–83, 105, 106). Despite the general nature of gene amplification, the DHFR system has continued to be the most intensively studied of the amplified genes. This is due in part to the ease with which the DHFR gene can be amplified, the plethora of folate analogs available, and the development of drug-resistant tumor cells when MTX is employed as a human antineoplastic drug. But with the emergence of recombinant DNA technologies and the development of somatic cell techniques to transfer genetic information to mammalian tissue-culture cells, DHFR has become a widely used selectable marker as well. The availability of Chinese hamster ovary (CHO) cell mutants lacking a functional DHFR gene (103) has greatly facilitated the use of DHFR as a selectable marker. Moreover, treatment of cells expressing a transfected DHFR gene with MTX results in the amplification of the transfected gene (11, 18, 33, 45–48, 53, 77). It is interesting to note that the amplified heterologous DHFR genes display structural characteristics similar to the amplified endogenous genes of the host cell. Murine 3T3 cells made resistant to high levels of MTX by amplification of a modular DHFR gene have the amplified DHFR genes situated within acentric extrachromosomal elements called "double minute chromosomes" (12). In contrast, transfected CHO cells containing amplified DHFR genes possess chromosomes having expanded homogeneously staining regions (HSRs) (48), similar to the HSRs observed in MTX resistant CHO cells containing the amplified endogenous hamster gene (22, 44, 71).

Gene amplification in mammalian tissue-culture cells does not appear to require specific DNA sequences (see Chap. 6). Studies by Hamlin and her colleagues have suggested that the amplified unit is determined at least in part by the physical proximity of an origin of DNA replication near the DHFR gene (61–63). This has provided a rationale for the observation that sequences that lie a considerable distance from the DHFR gene can also be amplified during MTX selection (16). In a similar manner, it is possible to amplify genes that are cotransfected with modular DHFR cDNA genes after stepwise MTX selection (77). This has allowed the high-level expression of a variety of large, glycosylated polypeptides such as hepatitis B and Herpes simplex viral surface antigens (11, 18, 74), truncated (34) viral surface antigens for a potential Herpes simplex vaccine (53), tissue-type plasminogen activator (49), various interferons (40, 57, 79), and clotting factor VIII (102, 109).

Although in principle it should be possible to use MTX to directly select for any cell that has taken up and expressed the vectorborne DHFR (68), the use of the murine DHFR cDNA as a dominant selectable marker has proved difficult due to the relatively high level of expression of the endogenous DHFR gene. The DHFR⁻ CHO cell lines isolated by Urlaub and Chasin (103) have permitted the use of DHFR as a selectable marker. These are the only mutant cell lines available that lack the DHFR gene. Yet the study of gene expression cannot be limited to a particular mutant cell line, nor can it be limited by the often low-level expression of one or a few copies of the transfected sequences obtained when dominant selectable markers such as the *Escherichia coli* xanthine-guanine phosphoribosyltranferase (Ecogpt) (66, 67) or aminoglycoside 3'-phosphotransferase (neo) (20, 97) are used. In this chapter efforts are described that are aimed toward the characterization and expression of drug-resistant DHFRs isolated from mammalian cells and bacteria.

Characterization of a Murine MTX-Resistant DHFR

Isolation of a cDNA Encoding an Altered DHFR

In order to develop a dominant amplifiable vector we reasoned that an MTX-resistant DHFR might confer on recipient cells a level of MTX-resistance sufficient to permit the use of MTX as the selective agent. Previously, Wigler et al. (108) demonstrated that it was in fact genomic DNA from an MTX-resistant hamster cell line that transformed an MTX-sensitive murine cell line to MTX-resistance.

We chose to utilize a murine cell line expressing a DHFR having a reduced affinity for MTX. This cell line, designated 3T6-R400 (38, 39), had been extensively characterized by Schimke and his coworkers. It was shown to contain approximately 30 copies of a DHFR gene (39) and had been used to introduce DNA sequences into bone marrow cells [(19) and Chap. 15]. The properties of the enzyme included an impaired ability to reduce folate to tetrahydrofolate, a much lower affinity for MTX and a more basic pI. A further consideration was the fact that molecular clones of the murine DHFR gene and cDNA were available (23, 70) and the sequence of the wild-type DHFR cDNA was known (23, 59, 70, 87).

A cDNA library cloned into plasmid pBR322 was prepared using total cytoplasmic polyadenylated RNA isolated from the 3T6-R400 cell line (90). Clones that hybridized to the 5' end of the wild-type murine DHFR cDNA were isolated and two independent clones were sequenced. Compared to the previously published wild-type murine DHFR cDNA (23, 70), only a single nucleotide difference in the coding region of the R400 DHFR cDNA was observed. Most significantly, the nucleotide sequence predicted the substitution of an arginine residue for a leucine residue at position 22 of the wild-type enzyme. That change is consistent with the basic shift in the migration of the altered enzyme in two-dimensional gels observed by Haber et al. (38) To directly confirm that result, the R400 and wild-type DHFR cDNAs were sequenced in parallel. Figure 1 clearly shows that where a CTA (leu) codon is present in the wild-type DHFR cDNA a CGA (Arg) codon is located in the R400 cDNA.

Expression of R400 DHFR in Transfected Cells

In order to unequivocally establish that the single nucleotide change observed in the DHFR cDNA clones isolated from the 3T6-R400 cell line was indeed responsible for the MTX resistance exhibited by the R400 cell line, the altered cDNA was inserted into a mammalian expression plasmid vector.

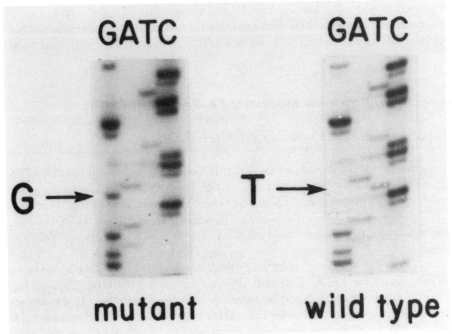

Figure 1 Comparison of mutant and wild-type DHFR cDNA sequence auto-radiographs. Plasmids pDHFR-11 (11) and pR400-12 were digested with Fnu4HI and BglII and the sequence of the fragment spanning the DHFR coding region was determined as described (60, 74). A region of the sequencing gel illustrating the T-to-G nucleotide substitution at position +68 is shown.

It was placed between a simian virus 40 (SV40) early promoter (9) and the polyadenylation signal (76) of the hepatitis B virus surface antigen gene. As depicted in Fig. 2, the wild-type and altered DHFR cDNA clones were used to construct parallel expression plasmids which differed only by the alteration in the twenty-third codon of the DHFR gene. Any difference in MTX resistance conferred by the R400 expression vector as compared to the wild-type DHFR expression vector could be directly attributable to the nucleotide change at codon 22.

We first determined if these vectors would convert DHFR⁻ CHO cells to a DHFR⁺ phenotype. Although both vectors were capable of converting cells from DHFR⁻ to DHFR⁺ efficiently, we noticed that the wild-type vector consistently produced higher numbers of transfected cells. However, when individual colonies were isolated and grown in various levels of MTX we observed that the cells expressing the R400 DHFR were capable of growing in considerably higher levels of the drug (Fig. 3). The information in Fig. 3 also demonstrates why we had been unable to utilize the wild-type DHFR as a dominant marker: the level of MTX that inhibited cells expressing the plasmid-borne wild-type DHFR was less than one-twentieth the level that

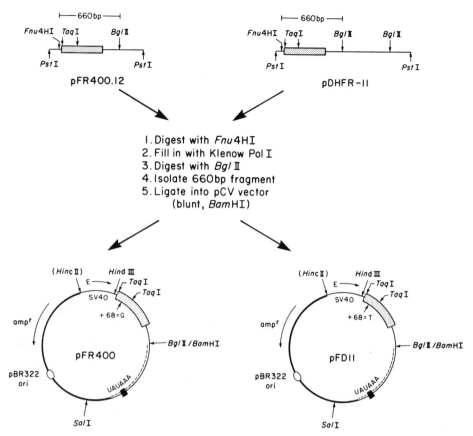

Figure 2 Construction of DHFR expression plasmids. The DHFR cDNA plasmids pR400-12 and pDHFR-11 were digested with Fnu4HI and the single-stranded fragment ends were filled in by *E. coli* DNA polymerase I large fragment. After cleavage with BglII, the 660-bp fragment was isolated by electroelution from polyacrylamide gels and inserted into the expression plasmid pCV342E (54). This plasmid is a derivative of p342E [containing the early promoter of SV40 (9), and the late 548-bp BamHI-BglII fragment containing the polyadenylation site of the surface antigen gene from hepatitis B virus] (88). The heavy line represents pBR322-derived sequences (ori, origin; amp^r, ampicillin resistance) (55); parallel broken and solid lines represent hepatitis B virus-derived sequences; and the solid line, SV40 or DHFR sequences (E shows the direction of transcription of SV40 early mRNA); the stippled bar represents the DHFR protein-encoding sequence. Plasmid manipulations were as described (13, 24, 37).

inhibited the growth of wild-type CHO-K1 cells and but one-thousandth the level that inhibited growth of the cells expressing the R400 DHFR.

In order to examine the MTX resistance of a greater number of colonies, we assayed the ability of the vectors to directly generate MTX-resistant colonies by introducing the vectors into the DHFR⁻ CHO cell line using the cal-

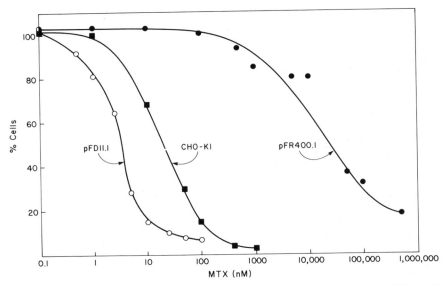

Figure 3 Inhibition of growth of HGT-independent cells by MTX. DHFR⁻ cells were separately transfected using the calcium-phosphate coprecipitation method (32, 35) with pFD11 and pFR400 DNA, and DHFR⁺ colonies were selected in medium lacking HGT. After subcloning, individual colonies were passaged into larger culture dishes. The cells were treated with trypsin and plated onto 60-mm dishes at approximately 10⁵ cells per dish. MTX was added to the growth medium at the concentrations indicated. The cells were incubated for 3 days and counted. The data from a typical subclone after transfection with pFD11 and pFR400 DNA are presented, along with the MTX sensitivity of the CHO wild-type parent. The data are expressed as the percentage of cells growing at the indicated level of MTX relative to growth in the absence of MTX.

cium-phosphate coprecipitation technique (32, 35, 36, 107) and selecting the transfected cells in various levels of MTX (Table 1). Only the cells containing the MTX-resistant DHFR cDNA were capable of forming colonies in levels of MTX greater than 25 nM. Moreover, the colony-forming ability of the R400 DHFR was unaffected by levels of MTX known to inhibit the growth of virtually all cells; this suggested that it would be possible to use the MTX-resistant DHFR vector in wild-type cell lines other than the DHFR⁻ CHO cell line. Accordingly, the R400 expression vector was introduced into murine, hamster, and human cells by calcium phosphate mediated gene transfer and the recipient cells selected by growth in the presence of MTX. Table 2 shows this result. MTX-resistant colonies were efficiently formed when wild-type CHO cells HeLa cells, and a murine cell line (Ltk⁻) were transfected with the pFR400 vector; in contrast, no MTX-resistant colonies appeared when either the wild-type DHFR expression plasmid (pFD11) or a control plasmid were used.

TABLE 1 Frequency of colony formation by DHFR⁻ cells transfected with mutant and wild-type DHFR expression plasmids

MTX (nM)	Transfection efficiency, colonies/μg DNA per 10^6 cells		
	No plasmid	pFD11	pFR400
0	<0.3	>10^3	330
1	—	330	310
5	<0.3	260	230
10	—	140	240
25	—	30	250
50	<0.3	<0.3	240
100	<0.3	<0.3	200
250	—	<0.3	170
500	—	<0.3	85
1,000	—	<0.3	75
10,000	—	—	45
100,000	—	—	10

Amplification of the Transfected R400 DHFR Sequences

The data shown in Tables 1 and 2 indicate that the DHFR encoded by the R400 cell line is resistant to intermediate levels of MTX. Fewer colonies were formed at the higher levels of MTX. That observation suggested that increased concentrations of MTX would inhibit the enzyme. When cells arising

TABLE 2 Frequency of colony formation by CHO-K1 HeLa, and Ltk⁻ cells transfected with DHFR expression plasmids

MTX (nM)	Cell line	Transfection efficiency, colonies/μg DNA per 10^6 cells		
		No plasmid	pFD11	pFR400
250	CHO	<1	5[a]	270
500	CHO	<1	5[a]	90
100	Ltk⁻	<2	<2	80
250	Ltk⁻	<2	<2	44
500	Ltk⁻	<2	<2	14
250	HeLa	<2	<2	23
500	HeLa	<2	<2	12

[a]Nonviable colonies that did not propagate upon subcloning

after selection at 250 nM MTX were grown in increasing levels stepwise of the drug, it was possible to isolate subclones capable of growing in levels of MTX as high as 1 mM (58), suggesting that the cells had amplified the transfected MTX^r DHFR cDNA. Southern blot analysis of these cells demonstrated the presence of increased copies of the transfected sequences, thus directly proving the amplifiability of the R400 vector. More recently, cells expressing a retroviral vector based on the altered DHFR cDNA have been shown to contain amplified copies of the R400 cDNA (64). These amplified cell lines have proved useful in providing high titers of helper-free retroviral vector stocks.

Expression of in vitro Mutagenized DHFRs

The change in the codon specifying amino acid 22 of the altered DHFR encoded by the R400 cell line is provocative. An identical leucine-to-arginine substitution at an analogous position has been identified in a bacterial DHFR that also has a reduced affinity for the antifolate drug (7). Leu-22 has been implicated by x-ray crystallographic studies to be one of several residues that form the binding pocket for folate, and by inference, MTX (56, 104). The replacement of the hydrophobic side chain of leucine with the positively charged guanidinium group of arginine could easily perturb this interaction. The alteration of the binding pocket could also account for the decrease in enzyme activity that has been observed for the R400 enzyme (38).

We attempted to create by *in vitro* mutagenesis (110) a murine DHFR with a decreased affinity for MTX and a relative increase in enzyme activity. Two different mutations were inserted into the DHFR cDNA at position 22; the CTA codon of Leu-22 was converted to AAA (lysine) and to GAA (glutamic acid). When inserted into equivalent mammalian expression vectors, the variant DHFRs were capable of converting cells to MTX resistance; as seen in Table 3, both pSVDHFR-Lys and pSVDHFR-Glu were capable of converting

TABLE 3   Frequency of colony formation by CHO-K1 DHFR ⁻ cells transfected with variant DHFR expression plasmids

MTX (nM)	pFD11	pFR400	Transfection efficiency, colonies/μg DNA per 10^6 cells	
			pSVDHFR-lys	pSVDHFR-glu
0	>1000	450	135	35
50	2	465	128	42
100	0	435	151	37
1000	0	225	53	41

cells to a MTX-resistant phenotype, but at a frequency less than the R400 DHFR containing an arginine at position 22.

Even though the colony-forming ability of the two DHFRs created by *in vitro* mutagenesis were decreased, it was of interest to determine the MTX resistance of the transformed cells, since colony-formation in the presence of MTX reflected both enzyme activity as well as affinity for the substrate inhibitor. A decrease in the MTX resistance of the enzyme would be manifested in an increased ability to be amplified by higher levels of MTX. A growth curve of cells expressing the variant DHFRs was performed in the presence of MTX. Figure 4 shows this result; cells expressing pSVDHFR-lys were consistently more sensitive to MTX than the R400 DHFR. Conversely, cells expressing the negatively charged glutamic acid variant were significantly more resistant to the effects of MTX.

Isolation of the Type 1 Bacterial DHFR

As we previously noted, the MTX resistant DHFR contained in the murine 3T6-R400 cell line has a bacterial counterpart in the Arg-28 variant bacterial DHFR found in a trimethoprim-resistant *E. coli* (7). During our studies on the murine-altered DHFR, we began a project to identify and characterize the gene encoding a plasmidborne folate analog-resistant DHFR. There exists in *E. coli* two plasmidborne DHFRs that confer resistance to trimethoprim (a

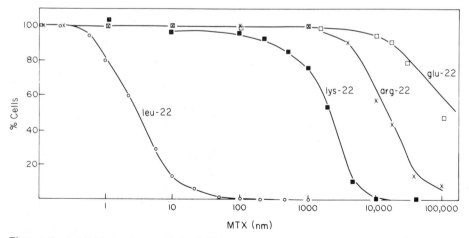

Figure 4 Inhibition of growth by MTX of cells containing position 22 variant DHFR expression vectors. CHO DHFR⁻ cell lines generated by expression of Leu-22 (wild-type), Lys-22, Arg-22 (R400), or Glu-22 enzymes were passaged into the indicated levels of MTX. Conditions are as described for Fig. 3.

functional analog of MTX). The type 1 DHFR is contained within transposon 7, as part of the multidrug resistance episome R483 (5, 8, 28, 29, 92). Although the amino acid sequence of the type 1 DHFR was unknown, it had previously been shown to be a homodimeric enzyme of 36,000 MW (69). In contrast to the R483 DHFR, which is resistant to intermediate levels of trimethoprim, the type 2 DHFR is resistant to extremely high levels of trimethoprim *in vivo* (and MTX *in vitro*) (94). It is also a 36,000-MW enzyme, but it is composed of four identical 9000 subunits (69, 94). The amino acid (98) and nucleotide (101) sequence of the type 2 DHFR shows no homology to other bacterial DHFRs (7, 93, 100).

Despite the lack of homology between the mammalian DHFRs and the type 2 DHFR, attempts to express the MTX-resistant type 2 DHFR in mammalian cells have been reported (72), although with limited success. Murine L cells made resistant to MTX by virtue of expression of the bacterial type 2 DHFR arise at a low frequency after gene transfer (72). However, the extreme drug resistance exhibited by the type II enzyme precludes amplification of the transfected DNA at reasonable levels of MTX. It was our intent to utilize the intermediately trimethoprim-resistant DHFR encoded by Tn7 as a selectable marker for tissue culture cells.

The drug-resistance factor contained within *E. coli* K-12 R483 is a large plasmid having a streptomycin-resistance marker in addition to the trimethoprim-resistant DHFR gene (8). The approximately 100-kb plasmid was isolated from the trimethoprim-resistant *E. coli* and the DNA was digested with EcoRI and BamHI. The resulting fragments were inserted into plasmid pBR322, which had been similarly cleaved. When *E. coli* harboring the recombinant plasmids were selected with trimethoprim, only a single class of plasmid containing a 36-kb region of R483 was isolated that was capable of conferring trimethoprim resistance. This plasmid, designated pR483, was restriction mapped. As shown in Fig. 5, pR483 contains two fragments from the R483 resistance factor: an approximately 30-kb EcoRI fragment and a 5.4-kb EcoRI-BamHI fragment. This plasmid was digested with PvuII under conditions that produced incomplete digestion of the DNA.

Upon religation and transformation of competent *E. coli*, trimethoprim-resistant colonies were again isolated. The plasmid contained in one of the trimethoprim-resistant colonies was grown up for further analysis. The structure of this plasmid, designated pTM4 is also shown in Fig. 5. Plasmid pTM4 evidently arose from pR483 by the deletion of an approximately 25-kb region of the R483 resistance factor. The 6-kb EcoRI fragment and the 5.4-kb EcoRI-BamHI fragments from pTM4 were cloned into pBR322; only the plasmid containing the 6-kb EcoRI fragment of pTM4 was capable of conferring trimethoprim-resistance to recipient bacteria. To further localize the region within the 6-kb EcoRI fragment responsible for trimethoprim resistance, it was next partially digested with TaqI, a restriction enzyme having the four-base recognition sequence TCGA. Trimethoprim-resistant bacteria were again isolated. As shown in Fig. 5c, this plasmid, designated pTM10, has a 1600-bp insert spanning two internal TaqI sites.

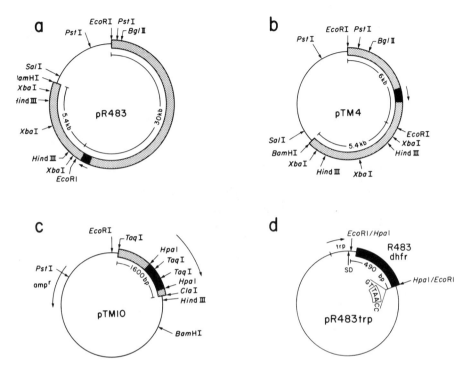

Figure 5 Plasmid constructions. The derivations of the pBR322-R483 recombinant plasmids are described in the text. DNA derived from R483 is shaded; the type 1 DHFR gene is shown in black. SD, Shine-Dalgarno sequence of trp promoter fragment. (a) pR483; (b) pTM4; (c) pTM10; (d) pR483trp.

Sequence of the Type 1 Bacterial DHFR

The three TaqI fragments comprising the 1600-bp insert were cloned separately into phage M13 and sequenced using the dideoxynucleotide chain termination method of Sanger et al. (60, 78). The nucleotide sequence of the entire 1600-bp sequence is shown in Fig. 6. Two open reading frames, arrayed in opposite orientations, were observed. Each was of a size capable of encoding the 18,000-MW type 1 DHFR monomer.

One of these open reading frames spanned both TaqI sites contained in the 1600-bp insert. This was most likely the type 1 DHFR since the other open reading frame should have been isolated as a 900-bp TaqI fragment were it the type 1 DHFR. Of greater significance is the observation that the deduced amino acid sequence of the open reading frame exhibits homology to bacterial and mammalian DHFRs (30, 41, 88, 93, 99). Although the amino acid homology shown in Fig. 7 ranges only from 22 to 28 percent, the homology is evident particularly in the regions of the enzyme that form the active site (56, 104).

```
HpaI
GTTAACCTCTGAGGAAGAATT

              1                                              10
              met lys leu ser leu met val ala ile  ser lys asn gly val ile gly asn gly
              GTG AAA CTA TCA CTA ATG GTA GCT ATA  TCG AAG AAT GGA GTT ATC GGG AAT GGC
              ATG
                                      20
              asp ile pro trp ser
              GAT ATT CCA TGG AGT

              30
              ala lys gly glu gln leu phe lys ala ile thr tyr asn gln trp
              GCC AAA GGT GAA CAG CTC TTT AAA GCT ATT ACC TAT AAC CAA TGG
                                          40                              50
              leu leu val gly arg lys thr phe glu  ser met gly ala leu
              CTG TTG GTT GGA CGC AAG ACT TTT GAA  TCA ATG GGA GCA TTA

              60
              pro asn arg lys tyr ala val val thr arg ser ser phe thr
              CCC AAC CGA AAG TAT GCG GTC GTA ACA CGT TCA AGT ACA TCT
                                       70                              80
              ser asp asn glu asn val leu ile leu phe  ile lys asp ala leu
              GAC AAT GAG AAC GTA TTG ATC TTG TTT     ATC AAA GAT GCT TTA

              90
              thr asn leu lys ile asp his val ile
              ACC AAC CTA AAA ATA GAT CAT GTC ATT
                                       100                             110
              ser gly gly glu ile tyr lys ser leu ile  asp gln val asp thr leu his
              TCA GGT GGT GAG ATA TAC AAA AGC CTG ATC  GAT CAA GTA GTA GAT ACA CTA CAT

              120
              ile ser thr ile asp ile glu pro gly glu ile
              ATA TCT ACA ATC GAC ATC GAG CCG GGA GAA ATC
                                       130                             140
              pro ser asn phe arg pro val phe  thr gln asp phe ala
              CCC AGC AAT TTT AGG CCA GTT TTT  ACC CAA GAC TTC GCC

              150
              ser asn ile asn tyr ser tyr gln ile trp gln lys gly
              TCT AAC ATA AAT TAT AGT TAC CAA ATC TGG CAA AAG GGT
                                                     157
                                                     gly  OC
                                                     GGT  TAA C
                                                          HpaI
```

Translated Mol. Weight = 17545.91

Figure 6 Nucleotide sequence and deduced amino acid sequence of the type 1 plasmid-encoded DHFR. The HpaI fragment from plasmid pTM10 was inserted into M13 mp10 and sequenced using the dideoxynucleotide chain termination method. A synthetic oligonucleotide primer was designed to specifically alter the initiation codon from a GTG to an ATG. The primer was used in a Klenow repair reaction, and the resulting mutant phage were identified by hybridization with the primer at elevated temperatures. The sequence of the wild-type and altered initiation codons are shown.

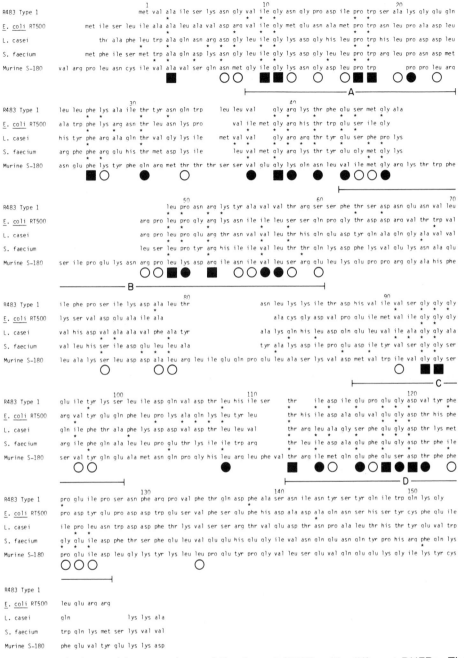

Figure 7 Amino acid comparison of the type 1 DHFR with different DHFRs. The sequences were aligned, using the Smith-Waterman algorithm (95), and were manually adjusted to give the best fit relative to the type 1 enzyme (26). The sequence of the mouse S-180 cell line enzyme has been deduced from its nucleotide sequence (23). The sequence of the *E. coli*, *S. faecium*, and *L. casei* enzymes were obtained by direct sequence analysis of the purified polypeptides or genes (14, 41, 85, 93, 98, 100). Symbols: *, residues homologous to the R483 enzyme; ■, residues which are common to all five enzymes; ●, homologous residues which are common to four of the DHFRs; ○, homologous residues found in three-fifths of the DHFRs. The regions of the enzyme forming the hydrophobic binding pocket (104) are underlined.

111

In order to prove that this open reading frame was the type 1 DHFR gene, a 490-bp HpaI fragment was isolated from pTM10. This fragment contains only 18 bp of untranslated sequence at the 5′ end. The HpaI recognition sequence (GTTAAC) interrupts the translational termination sequence of the open reading frame. This fragment was inserted into a plasmid vector so that the gene was under the transcriptional control of the trp promoter (25). In addition, the 3′ HpaI site was fused to a filled-in BamHI site so that the translational termination codon (GTTGATCC) was restored.

Introduction of this plasmid (pR483trp, Fig. 5d) into bacteria permitted colony formation in the presence of trimethoprim. As can be seen in Table 4, the presence of pR483trp allowed the growth of bacteria in high levels of trimethoprim. When extracts were prepared from cells containing pR483trp, trimethoprim-resistant DHFR activity could be easily observed (Fig. 8). Proteins from these same extracts were analyzed on sodium dodecyl sulfate (SDS)-acrylamide gels. A 17,000-MW band could be observed in the extracts prepared from the pR483trp containing cells (data not shown).

Expression of the Bacterial Type 1 DHFR in Mammalian Cells

Since the goal of this work was to express the type 1 bacterial gene in mammalian cells, it was desirable to place the R483 DHFR gene under the appropriate mammalian regulatory signals (9, 17, 46, 47, 65, 89). Initial inspection of the nucleotide sequence spanning the trimethoprim-resistance marker did not allow the assignment of the 5′ end of the coding sequence. The open reading frame contains an ATG codon at the 5′ end. It lies 25 bp downstream of the promoter and Shine-Dalgarno ribosome binding-site sequence (75, 86),

TABLE 4 Growth of *E. coli* 294 transformed with plasmids containing type 1 plasmid-encoded DHFR[a]

Trimethoprim (mg/ml)	Growth[b] with			
	pRB483	pR483trp	pTaq900[c]	No plasmid
0	+	+	+	+
0.0005	+	+	+	+
0.005	+	+	−	−
0.05	+	+	−	−
0.5	+	+	−	−
5.0	−	−	−	−

[a]Little or no effect on the number of colonies was observed until levels of trimethoprim were sufficient to totally inhibit growth.

[b]+, Colony formation observed; −, no colonies observed.

[c]pTaq900 contains the 900 bp TaqI fragment spanning the second open reading frame contained within pTM10.

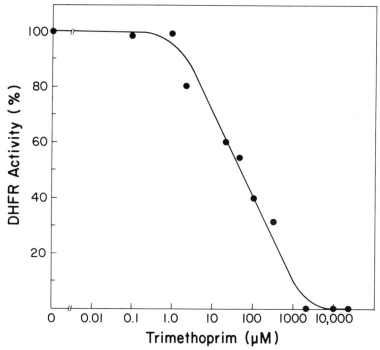

Figure 8 Inhibition of DHFR activity by trimethoprim. Cells containing pR483trp were induced by growing in minimal M-9 medium to an optical density at 550 nm of 1.0. Extracts (0.5 ml each) were prepared from 10 ml of induced cells as described previously (6). The reduction of dihydrofolate to tetrahydrofolate at 340 nm was measured in increasing levels of trimethoprim and plotted as shown.

a distance that is greater than exhibited by most bacterial genes (25). More closely situated to the Shine-Dalgarno sequence is the triplet GTG, located 15 bp upstream of the ATG codon. The assignment of the 5′ end of the coding sequence was provided by Stone and his colleagues, who were able to determine the amino acid sequence of the amino terminus of the enzyme (69). The protein sequence was in agreement with the assignment of the GTG codon as the initiation codon. Alternative initiation codons are common in certain bacteria, but GTG and TTG codons are incapable of being utilized by mammalian cells. This is partly attributable to the observation that mammalian cells, unlike bacteria, do not have a stringent requirement for a ribosome-binding sequence. Rather, the first ATG codon at the 5′ end of a mammalian gene is utilized as the translational start codon (51, 52). Since the R483 DHFR possessed a GTG initiation codon, it was incapable of being translated in mammalian cells.

We therefore had to engineer the bacterial gene to a form capable of being expressed in mammalian cells by converting the GTG initiation codon to an ATG by *in vitro* mutagenesis. A side-by-side comparison of the normal and *in vitro* altered genes is shown in Fig. 9. Both the GTG and the ATG

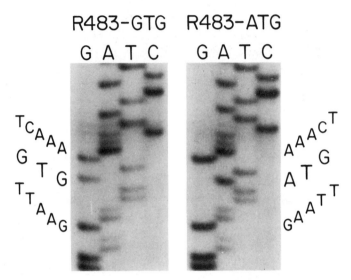

Figure 9 Nucleotide sequence of the wild-type (*left*) and altered (*right*) initiation codons. The M13 mp10 template used for the conversion of the GTG initiation codon to ATG and a mutant phage which hybridized to the ATG primer were sequenced using the dideoxynucleotide chain termination method (60, 74). A portion of the sequencing gel corresponding to the 5' end of the DHFR gene is shown. The sequence of a region including the initiation codon is listed beside each sample.

forms were inserted into a plasmid vector, pR483-ATG and pR483-GTG (Fig. 10). pR483-ATG contains the SV40 early promoter directing the transcription of the heterologous gene. Donor and acceptor splice sites (1, 17, 65) and an efficiently utilized polyadenylation signal (17, 76), derived from the hepatitis B virus surface antigen gene (89), flank the DHFR sequence at the 3' end.

pR483-ATG and pR483-GTG were introduced into the DHFR⁻ CHO cell line (103) by the calcium phosphate coprecipitation technique (32, 35), selecting for cells capable of utilizing exogenously added dihydrofolate. As a positive control, cells were transfected with pFD11 and pFR400, expressing the murine DHFR cDNA (90). As shown in Table 5, DHFR⁺ colonies arose when transfected with pR483-ATG but not with pR483-GTG. Addition of MTX prevented the formation of colonies by the wild-type murine DHFR vector pFD11. However, it did not affect the number of colonies formed by expression of the type 1 DHFR gene. That the type 1 DHFR was responsible for the ability of the recipient cells to grow in the selective medium was further evidenced by the observation that DHFR⁺ cells arose only when the cells transfected with pR483-ATG were supplemented with dihydrofolate, consistent with the observation that bacterial DHFRs, unlike the mammalian enzymes, are not able to reduce folate to dihydrofolate (6, 15).

The results shown in Table 5 suggest that the ability of the type 1 DHFR to transform DHFR⁻ cells to a DHFR⁺ phenotype was virtually unaffected by the addition of intermediate levels of MTX. A more quantitative

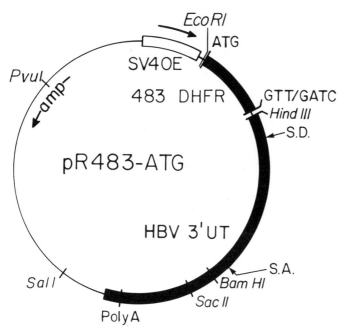

Figure 10 Construction of type 1 DHFR expression plasmid pR483-ATG. The 510-bp EcoRI-HindIII fragment spanning the R483 DHFR was isolated from the M13 replicative forms containing either the altered (ATG) or the wild-type (GTG) initiation codons. These were inserted into EcoRI-HindIII cleaved pEHBa114 between the SV40 early promoter and the hepatitis B virus sequences. The R483 DHFR encoding sequences are indicated by the filled in region. The sequences derived from the hepatitis B virus surface antigen gene are stippled. The site of polyadenylation of the hepatitis B virus surface antigen transcript is indicated. The 342-bp PvuII-HindIII fragment spanning the SV40 early promoter and modified to be bounded by an EcoRI site at the 3' boundary is shown by the open box; pR483-GTG contains the GTG triplet. The sequence of the Hpa-BamHI fusion at the 3' end of the bacterial gene is shown. S.D., spice donor site; S.A., splice acceptor site of chimeric SV40-HBsAG mRNA (88).

TABLE 5 Frequency of colony formation by CHO-K1 DHFR⁻ cells transfected with R483 expression plasmids

| MTX (nM) | Transfection efficiency, colonies/μg DNA per 10^6 cells | | | |
	pFD11	pFR400	pR483-GTG	pR483ATG
0	875	355	0	112
100	0	285	0	145
1000	0	75	0	85

assay of MTX-resistance was obtained by selecting several clones of CHO
cells expressing pR483-ATG and growing them in varying levels of the drug.
A comparison of the MTX growth inhibition of CHO cells expressing the
type 1 DHFR gene is shown in Fig. 11. Although the type 1 enzyme is only
intermediately resistant to MTX *in vitro,* under the selective conditions
used, the cells are extremely resistant to the drug, thus effectively preclud-
ing the use of MTX to amplify the gene.

Supertransfection of R483 DHFR Expressing Mammalian Cells

The apparent greater resistance of the type 1 DHFR *in vivo* as opposed to *in
vitro* is due in part to the presence of the dihydrofolate supplement which was

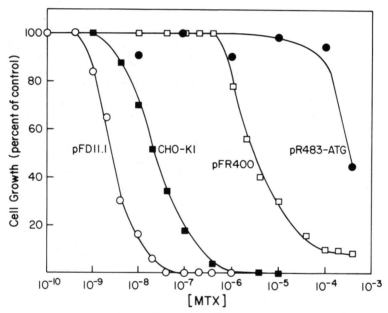

Figure 11 Inhibition of growth of GHT-independent cells by MTX. DHFR⁻ CHO
cells were transfected with murine DHFR expression plasmids pFD11 and
pFR400 and with pR483-atg. DHFR⁺ colonies that arose were isolated with a
glass cloning cylinder and propagated as described in the text. After subcloning, the
cells were passaged into larger tissue culture dishes. The various clonal cell lines
were then passaged into 60 mm dishes containing 10⁵ cells per dish in 5 ml of medium
and the indicated concentration of MTX. Cells transfected with the pR483-ATG were
supplemented with 6 μM dihydrofolate. After 3 days, the cells were trypsinized and
counted. The data are expressed as the percent of cells present in each dish relative
to cells grown in the absence of MTX. (GHT: glycine, hypoxanthine, and thymidine.)

required for cell growth. Attempts to amplify the gene in the presence of limiting amounts of dihydrofolate proved unsuccessful because the cells were not effectively selected with high MTX levels. It was noted that the cells expressing the R483 DHFR were selected under these conditions, although the selection was based entirely on the lack of dihydrofolate in the medium. A growth curve of these cells was obtained under conditions of varying dihydrofolate levels. At levels below 50 nM dihydrofolate, growth of the R483 DHFR⁺ cells was severely retarded (Fig. 12).

This result suggested that it would be possible to discriminate between the expression of the bacterial DHFR and the murine DHFR vectors based on the inability of cells expressing the bacterial type 1 DHFR to survive in the absence of dihydrofolate, whereas cells expressing the murine DHFR cDNA would be able to satisfy their requirement for tetrahydrofolate by reducing folate directly. We accordingly transfected one of the DHFR⁺ cell lines that arose after transfection of CHO DUXB-11 DHFR⁻ cells with pR483-ATG with pFD11, pFR400, and pR483-ATG and selected for cells able to grow in the absence of dihydrofolate. The results are shown in Table 6. It can be observed that colonies arose only from cells expressing the murine DHFR vectors. It should thus be possible to utilize the bacterial type 1 DHFR as a selectable marker for cells prior to using the murine DHFR vectors in those same cells. Although it is not possible to amplify the R483 DHFR, it is so MTX-resistant that it is a dominant marker, similar to the bacterial neo and Ecogpt plasmids

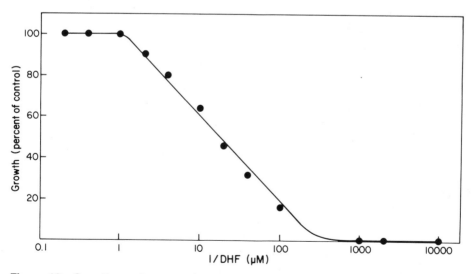

Figure 12 Growth requirement of cells expressing the type 1 DHFR for dihydrofolate. The CHO-R483-2 cell line used to generate the data shown in Fig. 11 was passaged under similar conditions into media containing decreasing levels of dihydrofolate. Three days later the cells were trypsinized and counted.

TABLE 6 Frequency of colony formation of DHFR⁻ cells transfected with Neo-DHFR expression plasmids

| | Transfection efficiency, colonies/μg DNA per 10^6 cells | | | |
	pFR400	pSVNeo	pSDS-ANH	pSDNH
g418[a]	0	550	495	85
MTX[b]	—	—	—	—
50 nM	—	32	155	980
100 nM	—	11	98	475
250 nM	—	0	23	123

[a]Cells were selected in 400 μg/ml G418.
[b]Cells arising from G418 treatment were split into the indicated level of MTX. The number of colonies that arose from 10^6 cells is shown.

(20, 66, 67, 97), with the added benefit that MTX is relatively inexpensive relative to selective agents such as G418.

A Dominant-Amplifiable Gene Based on DHFR-neo Coexpression Plasmids

Use of Cotransfected Plasmids to Introduce a Dominant Selectable Marker with an Amplifiable Marker

Initial attempts to use the wild-type murine DHFR vectors to directly select for MTX resistance proved unsuccessful due to the high rate of endogenous DHFR expression (as exemplified in Fig. 4). As the expression of the arginine, lysine, and glutamic acid position 22 variants and the type 1 bacterial DHFR were studied, it became obvious that the relatively low frequency of gene transfer using the calcium-phosphate coprecipitation technique was largely responsible for the inability of the wild-type murine DHFR vector to be used to directly transform cells to MTX resistance. Even though newly introduced sequences are apparently more easily amplified, it is difficult to use MTX to directly select DHFR⁺ cells after transfection with wild-type DHFR vectors because stable transfectants arise at frequencies ranging typically from less than 2×10^{-3} for the "efficient" cell lines and to less than 10^{-6} for the "difficult" cell lines. Even if cells expressing a plasmidborne DHFR of equal transcription and translation rates as the endogenous DHFR gene were 100-fold more likely to amplify the plasmid DHFR, most colonies arising after MTX selection of DHFR-transfected cells would contain amplified endogenous DHFR genes. A solution to this problem has been provided in the form of the MTX-resistant murine DHFR vectors (58, 90). Yet the MTX-resistant nature of the enzyme is not wholly satisfactory since the degree of amplification obtained is somewhat limited by the requirement for high levels of MTX.

It is likely that technical advances in gene transfer vectors may allow the use of wild-type DHFR vectors in DHFR⁺ cell lines. Retroviral vectors are often able to efficiently introduce a selectable marker into cells refractory to the calcium phosphate coprecipitation method, and it has been observed that genes within retroviral vectors are more efficiently expressed after introduction as packaged virus than as a calcium phosphate precipitate. In addition, improved gene-transfer techniques have been developed which allow the introduction of genes into virtually any cell type, and often very efficiently as well (e.g., see 64, 80). Yet it is a fact that the calcium phosphate coprecipitation method does not require specialized equipment, packaging cell lines, or vectors, and for many laboratories it will remain a standard method.

It was decided, therefore, to determine if it might be possible to utilize the propensity for cotransfected sequences to be taken up and expressed within a single cell to increase the percentage of cells taking up the plasmid-borne DHFR. A plasmid expressing the neo (aminoglycoside resistance) gene (20, 97) was coprecipitated with the wild-type and MTXʳ DHFR plasmids and introduced into several different cell lines. Cells were initially selected with G418 or MTX. G418-resistant colonies arose from cells transfected with the neo plasmid. MTX-resistant colonies arose only from cells transfected with the MTXʳ-DHFR plasmid pFR400, a result consistent with our previous findings (90). However, when the G418-resistant cells that arose after transfection with the neo expression plasmid were selected with 250-nM MTX, only those cells that had been cotransfected with a DHFR plasmid were able to form colonies.

Mammalian Cells Expressing a Polycistronic Dominant Vector

In order to increase the proportion of G418-resistant cells that could express the plasmidborne DHFR gene, we constructed a plasmid, pSDS-ANH, containing the DHFR transcription unit and a neo transcription unit on the same plasmid (Fig. 13a). A second plasmid, pSDNH (Fig. 13b), arose as an intermediate in the construction. It also contains the coding sequences for both neo and DHFR, but as a single transcription unit.

Table 6 shows the results when pADS-SNH and pSDNH were transfected into Ltk⁻ cells. Both plasmids were capable of creating G418-resistant colonies: pADS-SNH by virtue of the expression of the two independent cistrons contained within the same plasmid, and pSDNH by virtue of the expression of a single transcript containing the coding sequences of the two selectable markers. It should be noted that although mammalian mRNAs do not, as a rule, normally contain more than one translated sequence, polycistronic constructions can be expressed in mammalian cells (54). In such plasmids the second cistron is normally poorly expressed; since in pSDNH the neo gene is in the second position, it is more poorly expressed than DHFR.

When the G418-resistant cells arising from the transfection were pas-

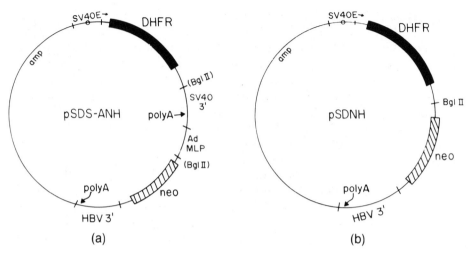

Figure 13 Schematic maps of neo-DHFR expression plasmids pSDS-ANH and pSDNH.

saged into MTX only those cells containing the DHFR cistron were able to form colonies (Table 6). It can also be observed that the resistance of the cells to MTX is greatest in cells expressing pSDNH, reflecting the expression from the relatively less efficient (for neo) polycistronic plasmid required to render the cells G418 resistant. On further propagation in the presence of higher levels of MTX, cells resistant to $5\mu M$ were obtained. Analysis of one of the amplified L-cell lines using a fluorescence-activated cell sorter showed an increase in cellular DHFR content (Fig. 14). These data indicated that it was indeed feasible to use the pSDNH vector as a highly amplifiable, dominant selectable marker for tissue culture cells.

Conclusions

The development of selectable markers for tissue culture cells has greatly facilitated the ability to introduce and manipulate DNA sequences in mammalian cell lines. In addition to enhancing the study of gene expression in mammalian cells, gene transfer methods promise to have a great impact in agriculture, disease treatment, and in the production of biologicals from recombinant cells lines. A series of experiments has been described in this chapter that has led to the development of dominant, amplifiable markers for mammalian cells based on various DHFR genes. Not only have these vectors proved useful as tools to produce polypeptides in mammalian cells, the use of these vectors has led to a finer appreciation of the subtle forces affecting a cell during drug selection.

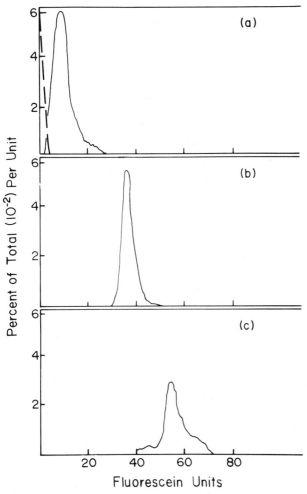

Figure 14 Generation of L cells containing amplified DHFR genes. (a) Fluorescence distribution of a clone of LTK cells containing pSDS-ANH resistant to 100 nM MTX. (b) The same cell line passaged to 500 nM MTX. (c) Distribution of cells resistant to 5 µM MTX. Cells were passaged for two generations before being incubated with fluoresceinated MTX(44).

Acknowledgments

Portions of this work were conducted while the author was at Genentech, Inc. The excellent technical assistance of J. Kennedy, F. Dunn (Invitron), and J. Dembroff and M. Walter (Genentech) are appreciated. The encouragement and advice offered by Drs. C. Benton, S. McIvor, and A. Levinson are also appreciated.

General References

SCHIMKE, R. T., KAUFMAN, R. J., ALT, F. W., and KELLEMS, R. F. 1978. Gene amplification and drug resistance in cultured mammalian cells, *Science* 202:1051–1055.

SIMONSEN, C. C., and LEVINSON, A. D. 1983. Isolation and expression of an altered mouse dihydrofolate reductase cDNA, *Proc. Nat. Acad. Sci. U.S.A.* 80:2495–2499.

WIGLER, M., PERUCHO, A., KURTZ, D., DANA, S., PELLICER, A., AXEL, R., and SILVERSTEIN, S. 1980. Transformation of mammalian cells with an amplifiable dominant-acting gene, *Proc. Nat. Acad. Sci. U.S.A.* 77:3567–3570.

References

1. Akusjarvi G., and Persson, H. 1980. Controls of RNA splicing and termination in the major late adenovirus transcription unit, *Nature (London)* 292:420–426.
2. Alt, F. W., Kellems, R. E., Bertino, J. R., and Schimke, R. T. 1978. Selective amplification of dihydrofolate reductase genes in methotrexate-resistant variants of cultured murine cells, *J. Biol. Chem.* 253:1357–1370.
3. Alt, F. W., Kellems, R. E., and Schimke, R. T. 1976. Synthesis and degradation of folate reductase in sensitive and methotrexate-resistant lines of S-180 cells, *J. Biol. Chem.* 251:3063–3074.
4. Andrulis, J. L., and Siminovitch, L. 1982. Amplification of the gene for asparagine synthetase, in *Gene Amplification* (R. T. Schimke, ed.), pp. 75–80, Cold Spring Harbor Laboratory, Cold Spring Harbor, N. Y.
5. Aymes, S. G. B., and Smith, J. T. 1974. R-factor trimethoprim resistance mechanism: An insusceptible target site, *Biochem. Biophys. Res. Commun.* 58:412–418.
6. Baccanari, D., Phillips, A., Smith, S., Sinsk, D., and Burchall, J. 1975. Purification and properties of *Escherichia coli* dihydrofolate reductase, *Biochemistry* 24:5267–5273.
7. Baccanari, D. P., Stone, D., and Kuyper, L. 1981. Effect of a single amino acid substitution on *E. coli* dihydrofolate reductase catalysis and ligand binding, *J. Biol. Chem.* 256:1738–1747.
8. Barth, P. T., Datta, N., Hedges, R. W., and Grinter, N. J. 1976. Transposition of a deoxyribonucleic acid sequence encoding trimethoprim and streptomycin resistances from R483 to other replicons, *J. Bacteriol.* 125:800–810.
9. Benoist, C., and Chambon, P. 1981. In vivo sequence requirements of the SV40 early promoter region, *Nature (London)* 290:304–310.
10. Berget, S. M., Moore, C., and Sharp, P. A. 1977. Spliced segments at the 5′ terminus of adenovirus 2 late mRNA, *Proc. Nat. Acad. Sci. U.S.A.* 74:3171–3175.
11. Berman, P., Dowbenko, D., Lasky, L., and Simonsen, C. 1983. Detection of antibodies to Herpes Simplex Virus with a continuous cell line expression cloned glycoprotein D, *Science* 222:524–527.
12. Biedler, J. L., and Spengler, B. A. 1976. Metaphase chromosome anomaly: Association with drug resistance and cell specific products, *Science* 191:185–187.
13. Birnboim, H. C., and Doly, J. 1979. A rapid alkaline extraction procedure for screening recombinant plasmid DNA, *Nucleic Acids Res.* 7:1513–1523.
14. Bitar, K. G., Blankenship, D. T., Walsh, K. A., Dunlop, R. B., Reddy, A. V., and

Freisham, J. H. 1977. Amino acid sequence of dihydrofolate reductase EC-1.5.1.3 from an amethopterin resistant strain of *Lactobacillus casei, FEBS Lett.* 80:119–122.

15. Blakley, R. L. 1969. *The Biochemistry of Folic Acid and Related Pteridines*, p. 139, Elsevier North-Holland Publishing Co., Amsterdam.

16. Bostock, C. J., and Tyler-Smith, C. 1981. Gene amplification in methotrexate resistant mouse cells. II. Rearrangement and amplification of non-dihydrofolate reductase gene sequences accompany chromosomal changes, *J. Mol. Biol.* 153:219–236.

17. Breathnach, R., and Chambon, P. 1981. Organization and expression of eukaryotic split genes coding for proteins, *Annu. Rev. Biochem.* 50:349–383.

18. Christman, J. K., Gerber, M., Price, P. M., Flordellis, C., Edelman, J., and Acs, G. 1982. Amplification of expression of hepatitis B surface antigen in 3T3 cells co-transfected with a dominant-acting gene and cloned viral DNA, *Proc. Nat. Acad. Sci. U.S.A.* 79:1815–1819.

19. Cline, M. J., Stang, H., Mercola, K., Morse, L., Ruprecht, R., Browne, J., and Salser, W. 1980. Gene transfer in intact animals, *Nature (London)* 284:422–425.

20. Colbere-Garapin, F., Horodniceanu, F., Kourilsky, P., and Garapin, A. C. 1981. A new dominant hybrid selective marker for higher eukaryotic cells, *J. Mol. Biol.* 150:1–14.

21. Collins, M. L., Wu, J. R., Santiago, C. L. Hendrickson, S. L. and Johnson, L. F. 1983. Delayed processing of dihydrofolate reductase heterogeneous nuclear RNA 6. *Mol. Cell. Biol.* 3:1792–1802.

22. Cowell, J. K., and Miller, O. J. 1983. Occurrence and evolution of homogeneously staining regions may be due to breakage-fusion-bridge cycles following telomere loss, *Chromosoma* 88:216–222.

23. Crouse, G. F., Simonsen, C. C., McEwan, R. N. and Schimke, R.T. 1982. Structure of amplified normal and variant dihydrofolate reductase genes in mouse sarcoma S180 cells, *J. Biol. Chem.* 257:7887–7897.

24. Davis, R. W., Botstein, D., and Roth, J. R. 1980. *Advanced Bacterial Genetics*, pp. 138–158, Cold Spring Harbor Laboratory, Cold Spring Harbor, N. Y.

25. DeBoer, H. A., Comstock, L. J., Yansura, D. G., and Yansura, H. L. 1981. Construction of a tandem trp-lac promoter and a hybrid trp-lac promoter for efficient and controlled expression of the human growth hormone in *Escherichia coli, Promoters, Structure, and Function.* in: M. J. Chamberlain and R. Rodriguez, eds., pp. 462–481, Praeger Publishers, New York.

26. Doolittle, R. F. 1981. Similar amino acid sequences: Chance or common ancestry? *Science* 214:149–159.

27. Fitzgerald, M., and Shenk, T. 1981. The sequence 5'-AAUAAA-3' forms part of the recognition site of polyadenylation of late SV40 mRNAs, *Cell* 24:251–260.

28. Fleming, M. P., Datta, N., and Gruneberg, R. 1972. Trimethoprim resistance determined by R factors, *Brit. Med. J.* 1:726–728.

29. Fling, M. E., and Elwell, L. P. 1980. Protein expression in *Escherichia coli* minicells containing recombinant plasmids specifying trimethoprim-resistant dihydrofolate reductases, *J. Bacteriol.* 141:779–785.

30. Fling, M. E., and Richards, C. 1983. The nucleotide sequence of the trimethoprim-resistant dihydrofolate reductase gene harbored by Tn7, *Nucleic Acids Res.* 11:5147–5148.

31. Flintoff, W. F., Davidson, S. V., and Siminovitch, L. 1976. Isolation and partial

characterization of three methotrexate-resistant phenotypes from Chinese hamster ovary cells, *Somatic Cell Genet.* 2:245–261.

32. Frost, E., and Williams, J. 1978. Mapping temperature-sensitive and host-range mutations of Adenovirus type 5 by marker rescue, *Virology* 91:39–50.

33. Gasser, C. S., Simonsen, C. C., Schilling, J. W., and Schimke, R. T. 1982. Expression of abbreviated mouse dihydrofolate reductase genes in cultured hamster cells, *Proc. Nat. Acad. Sci. U.S.A.* 79:6522–6526.

34. Gething, M. J., and Sambrook, J. 1981. Cell-surface expression of influenza haemagglutinin from a cloned DNA copy of the RNA gene, *Nature (London)* 293:620–625.

35. Graham, F., and Van der Eb, H. 1973. A new technique for the assay of infectivity of human adenovirus 5 DNA, *Virology* 52:456–467.

36. Graf, L. H., Urlaub, G., and Chasin, L. 1979. Transformation of the gene for hypoxathine phosphoribosyltransferse, *Somatic Cell Genet.* 5:1031–1044.

37. Gray, P. W., Leung, D. W., Pennica, D., Yelverton, E., Najarian, R., Simonsen, C. C., Derynck, R., Sherwood, P., Wallace, D. M., Berger, S. L., Levinson, A. D., and Goeddel, D. V. 1982. Expression of human immune interferon cDNA in *E. coli* and monkey cells, *Nature (London)* 295:503–508.

38. Haber, D. A., Beverley, S. M., Kiely, M., and Schimke, R. T. 1981. Properties of an altered dihydrofolate reductase encoded by amplified genes in cultured mouse fibroblasts, *J. Biol. Chem.* 256:9501–9510.

39. Haber, D. A., and Schimke, R. T. 1981. Unstable amplification of an altered dihydrofolate reductase gene associated with double-minute chromosomes, *Cell* 26:355–362.

40. Haynes, J., and Weissman, C. 1983. Constitutive long term production of human interferons by hamster cells containing multiple copies of a cloned interferon gene, *Nucleic Acids Res.* 11:687–706.

41. Hillcoat, B. L., and Blakley, R. L. 1966. Dihydrofolate reductase of *Streptococcus faecalis*. I. Purification and some properties of reductase from the wild strain and from stain A, *J. Biol. Chem.* 241:2995–3001.

42. Kaufman, R. J. 1985. Identification of the components necessary for adenovirus translational control and their utilization in cDNA expression vectors, *Proc. Nat. Acad. Sci. U.S.A.* 82:689–693.

43. Kaufman, R. J., Bertino, J. R., and Schimke, R. T. 1978. Quantitation of dihydrofolate reductase in individual parental and methotrexate-resistant murine cells, *J. Biol. Chem.* 253:5852–5860.

44. Kaufman, R. J., Brown, P. C., and Schimke, R. T. 1979. Amplified dihydrofolate reductase genes in unstably methotrexate-resistant cells are associated with double minute chromosomes, *Proc. Nat. Acad. Sci. U.S.A.* 76:5669–5673.

45. Kaufman, R. J., and Sharp, P. A. 1982. Amplification and expression of sequences cotransfected with a modular dihydrofolate reductase cDNA gene, *J. Mol. Biol.* 159:601–621.

46. Kaufman, R. J., and Sharp, P. A. 1982. Construction of a modular dihydrofolate reductase cDNA gene: Analysis of signals utilized for efficient expression, *Mol. Cell Biol.* 2:1304–1319.

47. Kaufman, R. J., and Sharp, P. A. 1983. Growth-dependent expression of dihydrofolate reductase mRNA from modular cDNA genes, *Mol. Cell. Biol.* 3:1598–1608.

48. Kaufman, R. J., Sharp, P. A., and Latt, S. A. 1983. Evolution of chromosomal regions containing transfected and amplified dihydrofolate reductase sequences, *Mol. Cell. Biol.* 3:699–711.

49. Kaufman, R., Wasley, L., Spilotes, L. Gossels, S., Latt, S., Larsen, G., and Kay, R. 1985. Coamplification and coexpression of human tissue-type plasminogen activator and murine dihydrofolate reductase sequences in Chinese hamster ovary cells, *Mol. Cell. Biol.* 5:1750–1759.

50. Kellems, R. E., Alt, F. W., and Schimke, R. T. 1976. Regulation of folate reductase synthesis in sensitive and methotrexate-resistant sarcoma 180 cells, *J. Biol. Chem.* 251:6987–6993.

51. Kellems, R. E., Leys, E. J., Harper, M. E., and Smith, L. E. 1982. Control of dihydrofolate reductase DNA replication and mRNA production, in: *Gene Amplification* (R. T. Schimke, ed.) pp. 81–87, Cold Spring Harbor Laboratory, Cold Spring Harbor, N. Y.

52. Kozak, M. 1978. How do eucaryotic robosomes select initiation regions in messenger RNA?, *Cell* 15:1109–1123.

53. Kozak, M. 1984. Possible role of flanking nucleotides in recognition of the AUG initiator codon by eukaryotic ribosomes, *Nucleic Acids Res.* 9:5233–5253.

54. Liu, C. C., Simonsen, C. C. and Levinson, A. D., 1984. Initiation of translation at internal AUG codons, *Nature (London)* 309:82–85.

55. Lusky, M., and Botchan, M. 1981. Inhibition of SV40 replication in simian cells by specific pBR322 DNA sequences, *Nature (London)* 293: 79–81.

56. Matthews, D. A., Alden, R. A., Bolin, J. T., Freer, S. T., Hamlin, R., Xuong, N., Kraut, J., Poe, M., Williams, M., and Hoogsteen, K. 1977. Dihydrofolate reductase: X-ray structure of the binary complex with methotrexate, *Science* 197:452–455.

57. McCormick, F., Trahey, M., Innis, M., Dieckmann, B., and Ringold, G. 1984. Inducible expression of amplified human beta interferon genes in CHO cells, *Mol. Cell. Biol.* 4:166–172.

58. McGrath, J. P., and Levinson, A. D. 1982. Bacterial expression of an enzymatically active protein encoded by RSV src gene, *Nature (London)* 295:423–425.

59. McGrogan, M., Simonsen, C. C., Smouse, D. T., Farnham, P. J., and Schimke, R. T. 1985. Heterogeneity at the 5′ termini of mouse dihydrofolate reductase mRNAs, *J. Biol. Chem.* 260:2307–2314.

60. Messing, J., Crea, R., and Seeburg, P. H. 1981. A method for shotgun DNA sequencing, *Nucleic Acids Res.* 9:304–321.

61. Milbrandt, J. D., Azizkhan, J. C., Greisen, K. S., and Hamlin, J. L. 1983. Organization of a Chinese hamster ovary dihydrofolate reductase gene identified by phenotypic rescue, *Mol. Cell. Biol.* 3:1266–1273.

62. Milbrandt, J. D., Azizkhan, J. C., and Hamlin, J. L., 1983. Amplification of a cloned Chinese hamster dihydrofolate reductase gene after transfer into a dihydrofolate reductase-deficient cell line, *Mol. Cell Biol.* 3:1274–1282.

63. Milbrandt, J. D., Heintz, N. H., White, W. C., Rothman, S. M., and Hamlin, J. L. 1981. Methotrexate-resistant Chinese hamster ovary cells have amplified a 135 kb sequence which includes the gene for dihydrofolate reductase, *Proc. Nat. Acad. Sci. U.S.A.* 78:6043–6047.

64. Miller, A. D., Eckner, R. J., Jolly, D. J., Friedman, T., and Verma, I. M. 1984. Expression of a retrovirus encoding human HPRT in mice, *Science* 225:630–632.

65. Mount, S. 1982. A catalogue of splice junction sequences, *Nucleic Acids Res.* 10:459–472.

66. Mulligan, R. C., and Berg, P. 1980. Expression of a bacterial gene in mammalian cells, *Science* 209:1422–1427.

67. Mulligan, R. C., and Berg, P. 1981. Selection for animal cells that express the Escherichia coli gene coding for xanthine-guanine phosphoribosyltransferase, *Proc. Nat. Acad. Sci. U.S.A.* 78:2072–2076.

68. Murray, M. J., Kaufman, R. J., Latt, S. A., and Weinberg, R. A. 1983. Construction and use of a dominant, selectable marker: A Harvey sarcoma virus-dihydrofolate reductase chimera, *Mol. Cell. Biol.* 3:32–43.

69. Novick, P., Stone, D., and Burchall, J. J. 1983. R plasmid dihydrofolate reductase with a dimeric subunit structure, *J. Biol. Chem.* 258:10956–10958.

70. Nunberg, J. H., Kaufman, R. J., Chang, A. C. Y., Cohen, S. N., and Schimke, R. T. 1980. Structure and genomic organization of the mouse dihydrofolate reductase gene, *Cell* 19:355–364.

71. Nunberg, J. H., Kaufman, R. J., Schimke, R. T., Urlaub, G., and Chasin, L. A. 1978. Amplified dihydrofolate reductase genes are localized to a homogenously staining region of a single chromosome in a methotrexate-resistant Chinese hamster ovary cell line, *Proc. Nat. Acad. Sci. U.S.A.* 75:5553–5556.

72. O'Hare, K., Benoist, C., and Breathnach, R. 1981. Transformation of mouse fibroblasts to methotrexate resistance by a recombinant plasmid expressing a prokaryotic dihydrofolate reductase. *Proc. Nat. Acad. Sci. U.S.A.* 78:1527–1531.

73. Pattishall, K. H., Acar, J., Burchall, J. J., Godstein, F. W., and Harvey, R. J. 1977. Two distinct types of trimethoprim-resistant sarcoma 180 cells, *J. Biol. Chem.* 252:2319–2323.

74. Patzer, E. J., Simonsen, C. C., Nakamura, G. R., Hershberg, R. D., Gregory, T. J., and Levinson, A. D. 1984. Characterization of recombinant-derived hepatitis B surface antigen secreted by a continuous cell line, in: *Viral Hepatitis and Liver Disease*, p. 477, Grune and Stratton, San Diego.

75. Pribnow, D. 1975. Bacteriophage T7 promoters: Nucleotide sequences of two RNA polymerase binding sites, *J. Mol. Biol.* 99:419–443.

76. Proudfoot, N., and Brownlee, G. G. 1976. 3′ Noncoding region sequences in eucaryotic messenger RNA, *Nature (London)* 263:211–214.

77. Ringold, G., Dieckmann, B. and Lee, F. 1981. Coexpression and amplification of dihydrofolate reductase cDNA and the Escherichia coli XGPRT gene in Chinese hamster ovary cells, *J. Mol. Appl. Genet.* 1:165–175.

78. Sanger, F., Nicklen, S., and Coulson, R. 1977. DNA sequencing with chain termination inhibitors, *Proc. Nat. Acad. Sci. U.S.A.* 74:5463–5467.

79. Scahill, S. J., Devos, R., Heyden, J. V., and Fiers, W. 1983. Expression and characterization of the product of a human interferon cDNA gene in Chinese hamster ovary cells, *Proc. Nat. Acad. Sci. U.S.A.* 80:4654–4658.

80. Schaffner, W. 1980. Direct transfer of cloned genes from bacteria to mammalian cells, *Proc. Nat. Acad. Sci. U.S.A.* 77:2163–2167.

81. Schimke, R. T. (ed.). 1982. *Gene Amplification*, Cold Spring Harbor Laboratory, Cold Spring Harbor, N. Y.

82. Schimke, R. T. 1984. Gene amplification in cultured animal cells, *Cell* 37:705–713.

83. Schimke, R. T. Kaufman, R. J., Alt, F. W., and Kellems, R. F. 1978. Gene

amplification and drug resistance in cultured mammalian cells, *Science* 202:1051–1055.

84. Setzer, D. R., McGrogan, M., Nunberg, J. H., and Schimke, R. T. 1980. Nucleotide sequence surrounding multiple polyadenylation sites in the mouse dihydrofolate reductase gene, *J. Biol. Chem.* 257:5143–5147.

85. Sheldon, R., and Brenner, S. 1976. Regulatory mutants of dihydrofolate reductase in *Escherichia coli* K-12. *Mol. Gen. Genet.* 147:91–97.

86. Shine, J., and Dalgarno, L. 1974. The 3-prime terminal sequence of *Escherichia coli* 16S ribosomal RNA complementarity to nonsense triplets and ribosome binding sites, *Proc. Nat. Acad. Sci. U.S.A.* 71:1342–1346.

87. Simonsen, C. C., Brown, P. C., Crouse, G. F., McGrogan, M., Setzer, D., Sweetser, D., Kaufman, R. J., and Schimke, R. T. 1981. Gene amplification as a mechanism for drug resistance in cultured animal cells, in: *Molecular Basis of Drug Action* (Singer T. P., and Ondarza, R. N., eds.)

88. Simonsen, C. C., Chen, E. Y., and Levinson, A. D. 1983. Identification of the type I trimethoprim-resistant dihydrofolate reductase specified by the *Escherichia coli* R-plasmid R483: Comparison with prokaryotic and eukaryotic dihydrofolate reductases, *J. Bacteriology* 155:1001–1008.

89. Simonsen, C. C., and Levinson, A. D. 1983. Analysis of processing and polyadenylation signals of the hepatitis B virus surface antigen gene using SV40-HBV chimeric plasmids, *Mol. Cell. Biol.* 3:2250–2253.

90. Simonsen, C. C., and Levinson, A. D. 1983. Isolation and expression of an altered mouse dihydrofolate reductase cDNA, *Proc. Nat. Acad. Sci. U.S.A.* 80:2495–2499.

91. Sirotnik, F. M., Kurita, S., and Hutchison, D. J. 1968. Relative frequency and kinetic properties of transport defective phenotypes among methotrexate-resistant L1210 clonal cell lines derived in vivo, *Cancer Res.* 28:75–80.

92. Skold, O., and Widh, A. 1974. A new dihydrofolate reductase with low trimethoprim sensitivity induced by an R factor mediating high resistance to trimethoprim, *J. Biol. Chem.* 249:4323–4325.

93. Smith, D. R., and Calvo, J. M. 1980. Nucleotide sequence of the E. coli gene coding for dihydrofolate reductase, *Nucleic Acids Res.* 8:2255–2274.

94. Smith, S. L., Stone, D., Novak, P., Baccanari, D., and Burchall, J. J. 1979. R plasmid dihydrofolate reductase with subunit structure, *J. Biol. Chem.* 254:6222–6225.

95. Smith, S. L., and Waterman, M. S. 1981. Identification of common molecular subsequences, *J. Mol. Biol.* 147:195–197.

96. Southern, E. 1975. Detection of specific sequences among DNA fragments separated by gel electrophoresis, *J. Mol. Biol.* 98:503–517.

97. Southern, P. J., and Berg, P. 1982. Transformation of mammalian cells to antibiotic resistance with a bacterial gene under control of the SV40 early promoter region, *J. Mol. Appl. Genet.* 1:327–341.

98. Stone, D., and Smith, S. L. 1979. The amino acid sequence of the trimethoprim resistant dihydrofolate reductase specified in Escherichia coli by R-plasmid R67, *J. Biol. Chem.* 254:10857–10861.

99. Stone, D., Paterson, S. J., Raper, J. H., and Phillips, A. W. 1979. The amino acid sequence of dihydrofolate reductase from the mouse lymphoma L1210, *J. Biol. Chem.* 254:480–488.

100. Stone, D., Phillips, A. W., and Burchall, J. J. 1977. The amino acid sequence of

the dihydrofolate reductase EC-1.5.1.3 of a trimethoprim resistant strain of *Escherichia coli, Eur. J. Biochem.* 72:613–624.

101. Swift, G., McCarthy, B. J., and Heffron, F. 1981. DNA sequence of a plasmid encoded dihydrofolate reductase, *Mol. Gen. Genet.* 181:441–447.

102. Toole, J. J., Knopf, J. L., Wozney, J. M., Sultzman, L. A., Bucker, T. L., Pittman, D. D., Kaufman, R. J., Brown, E., Shoemaker, C., Orr, E. C., Amphlett, G. W., Foster, W. B., Coe, M. L., Knutson, G. J., Fass, D. N., and Hewick, R. M. 1984. Molecular cloning of a cDNA encoding human antihaemophilic factor, *Nature (London)* 312:342–347.

103. Urlaub, G., and Chasin, L. A. 1980. Isolation of Chinese hamster ovary cell mutants deficient in dihydrofolate reductase activity, *Proc. Nat. Acad. Sci. U.S.A.* 77:4216–4220.

104. Volz, K. W., Matthews, D. A., Alden, R. A., Freer, S. T., Hansch, C., Kaufman, B. T., and Kraut, J., 1982. Crystal structure of avian dihydrofolate reductase containing phenyltriazine and NADPH. *J. Biol. Chem.* 257:2528–2536.

105. Wahl, G. M., Vincent, B. R. S., and DeRose, M. L. 1984. Effect of chromosomal position on amplification of transfected genes in animal cells. *Nature (London)* 307:516–520.

106. Wahl, G. M., Vitto, L., Padgett, R. A., and Stark, G. R. 1982. Single-copy and amplified CA genes in Syrian hamster chromosomes localized by a highly sensitive method for in situ hybridization. *Mol. Cell. Biol.* 2:308–319.

107. Wigler, M., Pellicer, A., Silverstein, S., Axel, R., Urlaub, G., and Chasin, L. 1979. DNA-mediated transfer of the adenine phosphoribosyltransferase locus into mammalian cells, *Proc. Nat. Acad. Sci. U.S.A.* 76:1373–1376.

108. Wigler, M., Perucho, A., Kurtz, D., Dana, S., Pellicer, A., Axel, R., and Silverstein, S.1980. Transformation of mammalian cells with an amplifiable dominant-acting gene, *Proc. Nat. Acad. Sci. U.S.A.* 77:3567–3570.

109. Wood, W. I., Capon, D. J., Simonsen, C. C., Eaton, D. L., Gitschier, J., Keyt, B., Seeburg, P. H., Smith, D. H., Hooingshead, P., Wion, K. L., Delwart, E., Tuddenham, E. G. D., Vehar, G. A., and Lawn, R. M. 1984. Expression of active human factor VIII from recombinant DNA clones, *Nature (London)* 312:330–337.

110. Zoller, M. J., and Smith, M. 1982. Oligonucleotide-directed mutagenesis using M13-derived vectors: An efficient and general procedure for the production of point mutations in any fragment of DNA, *Nucleic Acids Res.* 10:6487–6500.

Questions for Discussion with the Editors

1. *What experimental approach would you suggest be employed to determine whether other mammalian genes display DHFR-like amplification?*

Gene amplification in mammalian cells is a general phenomenon which likely encompasses the entire genome, but is observed only when an appropriate selection is placed upon the cells. Thus, cells treated with drugs such as MTX PALA (which inhibits aspartate transcarbamylase), and deoxycoformycin (inhibiting adenosine deaminase) all show the same characteristics: increased levels of the affected enzyme, increased mRNA and DNA copy numbers for the protein in question. Accompanying these molecular changes are alterations in the chromosomal locations whereupon these genes reside. Expanded homogeneously staining regions (HSRs) and acentric double minute (DM) chromosomes have been observed. The characteristic presence of DM chromosomes in neuroblastoma cell lines suggested to workers in the field that these cells might contain amplified genes. Subsequently, Bishop and his colleagues among others have shown the presence of amplified N-myc oncogenes in these tumor cells.

2. *It is obvious why one might want to express mammalian genes in bacteria. Please list, however, the potential advantages of being able to express bacterial genes in mammalian cells.*

Undoubtedly, the major benefit of expressing bacterial genes in mammalian cells has been the generation of dominant selectable markers for tissue culture cells. In particular, the aminoglycoside phosphotransferase (neo) gene has been a boon to molecular biologists seeking to introduce genes into a variety of cells. The chloramphenicol acetyltransferase (CAT) gene is another example of a non-mammalian gene which has been used in many labs as a marker for gene expression, since it has a unique enzyme property not normally found in mammalian cells, and the assay can be performed against a low background. In addition to these more practical benefits, it can be of interest to study the expression of bacterial genes in mammalian cells in order to study sequence effects (codon bias, RNA processing, nearest-neighbor effects) on genes whose structure has not been optimized for expression in animal cells by evolution.

CHAPTER 6

Amplification of the Dihydrofolate Reductase Gene

Nancy A. Federspiel and Robert T. Schimke

IN RECENT YEARS, it has become apparent that the genomes of living cells, from bacteria to mammals, are much less stable than had been previously believed. The discoveries of transposable elements in organisms ranging from *Escherichia coli* to *Drosophila* to maize, of retroviruses and their mechanism of infection and chromosomal integration, of mating type switches in yeast, of antigenic variation in trypanosomes, of rearrangements at the immunoglobulin loci, and of gene amplification in mammalian cells have all combined to uproot the notion of a constant invariant genome in somatic as well as germ line cells. The instability of the genome can, in many cases, be ascribed to changes that affect the expression of certain genes, either in a developmentally prescribed manner or in response to some environmental stimulus. Thus, immunoglobulin gene rearrangement occurs only in a small set of lymphoid cells in a precise, developmentally programmed manner. Sequence-specific rearrangements at the mating type locus in yeast control the expression of pheromones that are necessary for the sexual reproductive cycle and sporulation.

It is particularly interesting to consider the phenomenon of gene amplification, in its various manifestations, as another mechanism for regulation of gene expression. Some organisms' seem to have achieved a required level of expression of particular genes by their amplification in the course of evolutionary time. The chorion genes in the silkworm *Bombyx mori* exist as a multigene family whose expression is regulated developmentally (15). In *Drosophila melanogaster*, however, the chorion genes are amplified within the follicle cells at a particular stage of oocyte development, to allow for sufficient chorion gene expression during this period of rapid protein synthesis (36). In *Xenopus* oocytes, the presence of multiple copies of the ribosomal RNA genes in the germ line in conjunction with developmental amplification is utilized to provide the required quantities of ribosomal RNA (8). Thus, the mechanisms that allow for gene amplification may be a part of the normal repertoire of strategies a cell or organism can use to regulate expression of particular genes.

On the other hand, cellular transformation may result from aberrant utilization of these same mechanisms (whatever they may be) to amplify a gene whose product greatly upsets the balance of normal cellular metabolism. In accordance with this idea, tumor cells have often been found to contain chromosomal markers (homogeneously staining regions, or HSRs) or extrachromosomal elements (double minute chromosomes), which are characteristic of amplified DNA sequences (33). Indeed, it has recently been demonstrated that the cellular oncogene c-Kirsten ras is amplified in mouse adrenocortical tumor cells on both HSRs and double minute chromosomes (35). In human primary tumors or tumor cell lines, amplification of oncogene-related sequences has been reported for c-myc in acute promyelocytic leukemia (14) and in a human colon carcinoma (1), as well as for N-myc in neuroblastomas (24). The epidermal growth factor receptor gene has also been found to be amplified and overexpressed in primary human tumors of glial origin (25). Thus, while there must certainly be additional mechanisms that are important in the malignant process, it appears that, in some cases, gene amplification may play a significant role.

A final general class of gene amplification events which has been characterized in recent years is that obtained in response to the presence of a specific drug. This type of amplification, as well as those cited above relating to oncogenesis, may or may not involve the same mechanisms as those involved in developmental processes. Thus, in response to an external stress, such as exposure to a drug, a population of cells may arise that have amplified the gene coding for the protein whose activity alleviates that stress. One well-studied example is the amplification of the gene coding for dihydrofolate reductase (DHFR) in response to the drug methotrexate (MTX). DHFR converts dihydrofolate to tetrahydrofolate, an essential cofactor in the synthesis of thymidylate, purines, and glycine. In the presence of MTX, which is an analog of folic acid that binds the enzyme with high affinity, the cell is deprived of a continuous supply of tetrahydrofolate. Thus, precursors essential for DNA synthesis cannot be made and the cell dies. In response to stepwise increases in MTX concentration, however, cells may become resistant to the drug by everincreasing amplification of the DHFR gene. The cells thereby can synthesize enough DHFR protein from the extra gene copies to allow survival even in the presence of previously lethal doses of MTX (32).

Likewise, amplification of other genes for which a suitable selection scheme could be devised has been reported, and representative examples are given in Table 1. The proteins encoded by these genes are overexpressed in the amplified cell lines, and in many cases the levels of expression can be completely accounted for by the increase in gene copy number. It is becoming evident that tissue culture cells can readily develop resistance to many drugs by virtue of overproduction of the protein to which the drug is directed; this overproduction is often achieved by gene amplification. The implications

TABLE 1 Examples of gene amplification in response to stepwise drug selection

Amplified gene	Inducing drug	Cell line	Gene copy no.	Stability; localization	References
Dihydrofolate reductase (DHFR)	Methotrexate (MTX)	Murine S180	180	Unstable; DMs	(2, 23)
		Murine fibroblasts 3T6	50	Unstable; DMs	(9)
		Chinese hamster ovary (CHO)	150	Stable; HSR	(31)
			500	Stable; HSR	(30)
Aspartate transcarbamylase (CAD)	N-(phosphonacetyl)-L-aspartate (PALA)	Syrian hamster	6–100	Stable; HSR	(42, 43)
3-Hydroxy-3-methylglutaryl coenzyme A reductase	Compactin	CHO (UT-1)	15	Stable	(27)
Adenosine deaminase	Deoxycoformycin and selection in adenosine, alanosine, and uridine	Rat hepatoma	15–300	ND	(19)
		Derivatives of LMTK⁻	2600	ND	(44)
UMP-synthase	Pyrazofurin	Rat hepatoma	14	ND	(21)
Metallothionein	Cadmium		10	Unstable	(4)

Note: ND, not determined.

of this phenomenon extend far beyond the realm of *in vitro* cell culture. The possibility of drug resistance in native, intact organisms has recently been demonstrated in the protozoan parasite *Leishmania,* which was shown to develop resistance to MTX by the amplification of the gene for the endogenous bifunctional thymidylate synthetase-DHFR fused protein (13). This result in an organism that is a major public health problem in Third World countries points out the need for careful and appropriate use of antibiotics, herbicides, pesticides, and so on so that widespread drug resistance does not create a larger problem than that of the original pest or pathogen.

Similarly, many of the drugs now used for chemotherapy in human neoplastic disease are by nature potent antimetabolites, and resistance to these drugs can be developed in the very tumor cells that are the targets of the treatment. Several cases have now been reported that show clinical and/or biochemical evidence of MTX resistance accompanied by a 3- to 6-fold amplification of the DHFR gene in the patients' tumor cells compared to pre-treatment samples (12, 18, 38). In light of this new data, strategies for tumor management may have to be revised to minimize the risk of gene amplification. Thus, not only is it possible that amplification of certain key cellular genes can initiate or potentiate the process of neoplastic transformation, perhaps in response to an environmental stress, it is also apparent that other amplification events can occur in response to the drug intended to eradicate the tumor cells.

As yet, primary tumor cells with amplification of both an oncogene and the target gene of therapy have not been reported, but the possibility is not unlikely, since two different amplified regions have been found in *Leishmania* resistant to MTX. In fact, the question may be raised as to whether cells that already possess proven "amplification potential," such as tumor cells with amplified oncogenes, can more easily become resistant to a drug by virtue of a second amplification event.

Gene amplification appears not to be a rare phenomenon confined to a few obscure organisms or laboratory cell lines, but is apparently a process that occurs in many types of cells. Basic questions concerning the processes involved in gene amplification need to be answered to achieve an understanding of its role in development and in environmental adaptations. For instance, what is/are the mechanism(s) utilized by cells to amplify specific genes, both in the initial and subsequent events? Overreplication or rereplication of a portion of the genome is a possibility, as is sister-chromatid exchange. The regulation of DNA replication is a central issue, both in the control of normal developmental amplification and in the aberrant processes involved in oncogene amplification and drug resistance. Thus, what are the controlling factors for maintaining the normal complement of genetic material, and how are these parameters altered when amplification of genes occurs? What is the role of the drug when resistance due to specific gene amplification develops? The drug could act as an inducer or facilitator of aberrant processes that lead to gene amplification, or it could merely act as a selecting agent to allow the growth of only those cells which have spontaneously amplified a specific gene. Are there other factors, such as viral infection or forms of immunological.

stress, which also facilitate amplification and which may be important in on-cogenic processes?

In this chapter, we will use gene amplification in response to drug selection in cultured mammalian cells as a model system to investigate some of the questions raised above. In particular, data from the authors' laboratory and from other groups dealing with DHFR gene amplification in response to MTX will be presented and the implications of these findings with respect to the mechanisms involved and the stability of the genome will be discussed.

Role of the Selecting Drug in the Amplification Process

The classical method of obtaining cells with amplified DHFR genes is to subject the cells in culture to stepwise increases in MTX concentration. At low MTX concentrations, a few cells grow into colonies while most cells die. If the cells that are resistant to a low dose of MTX are subsequently plated in a higher concentration of drug, again most cells are killed while some maintain their growth. If these stepwise increases in MTX concentration are continued, cells may become resistant to drug concentrations up to 1000-fold higher than the original lethal dose. When these resistant cells are analyzed for the presence of increased numbers of DHFR gene copies, they characteristically contain amplified DHFR genes.

One view of how drug resistance due to gene amplification develops is depicted in Fig. 1, using DHFR as an example. It has been determined that the expression of DHFR is regulated over the cell cycle, with increased DHFR mRNA being produced in early S phase followed closely by increased levels of DHFR protein (28). It is also known that the DHFR gene is replicated during the first 2 hours of S phase (Fig. 1A) (27). When cells are subjected to metabolic stress, such as lack of precursors for DNA synthesis due to the presence of MTX, those cells that are in the S phase of the cell cycle will arrest in this stage (Fig. 1B). Cells that are in other parts of the cell cycle e.g., mitosis or G1 will continue progressing through the cycle until they are arrested at the beginning of S phase due to the metabolic block preventing DNA synthesis. Those cells that were arrested in mid-S phase, after the replication of the DHFR gene, may continue to transcribe and translate the extra copies of the gene, and after a period of time, accumulate enough DHFR enzyme to partially relieve the MTX block enabling the resumption of DNA synthesis. However, rather than continuing replication from the points at which the arrest occurred, the cells may return to the origins of replication used at the beginning of S phase, thus trying to go through a new complete S phase and rereplicating portions of the genome previously synthesized (Fig. 1C). The cells that can proceed through S phase and complete mitosis with this increased DNA load may have a selective advantage in the next cell cycle, because the extra copies of the DHFR gene could again allow increased amounts of DHFR synthesis alleviating the MTX block.

Therefore, according to this theory, the drug MTX plays a direct role in the amplification process by causing the cells to arrest during DNA synthesis

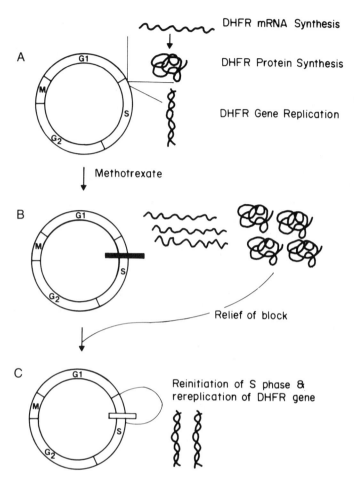

DHFR mRNA Synthesis

DHFR Protein Synthesis

DHFR Gene Replication

Methotrexate

Relief of block

Reinitiation of S phase &
rereplication of DHFR gene

Figure 1 Hypothetical model of the effect of methotrexate (MTX) on DHFR expression and replication during the cell cycle.

and subsequently to rereplicate portions of their genome, including the DHFR gene. Alternatively, rather than rereplicating portions of the genome during the same cell cycle, the increased levels of expression of DHFR protein in those cells that were blocked in S phase subsequent to DHFR induction may simply allow them to finish progressing through the cell cycle and impart temporary selective advantage for growth. In this case, the cells that can maintain this increased expression, including those that may have spontaneously amplified the DHFR gene, will be able to continue growth in MTX.

If resistance to MTX develops due to infrequent rereplication of stretches of DNA, other agents that block DNA synthesis might be predicted to increase the frequency of gene amplification. Indeed, it has been shown that sublethal doses of ultraviolet irradiation or carcinogen treatments prior to MTX selection enhance the frequency of DHFR amplification (37). Such treatments transiently inhibit DNA synthesis. In addition, treatment of cells with hydroxy-urea or aphidicolin, which arrest cells in S phase, also increases this amplification frequency (10). However, the frequency of MTX resistance due to mechanisms other than DHFR amplification (i.e., mutations, altered transport) is also comparably increased with these treatments, indicating that this general class of cytotoxic agents acts relatively nonspecifically to generate drug resistance.

It is difficult to obtain information about the initial events in gene amplification when a logarithmically growing culture of cells is used, because each cell is at a different point in its cell cycle and thus encounters the metabolic block at a different time. A variety of methods can be used to obtain a population of cells moving synchronously through the cell cycle, and one of the most accurate and least disruptive to the cells is that of mitotic shakes. Fibroblasts normally grow as a monolayer attached to a plastic dish, and during a brief period in mitosis, they round up and become unfastened from the plastic; cells can thus be collected that are within 30 minutes of each other in their progression through the cell cycle (Fig. 2A). When such a synchronized set of cells is treated with an agent such as hydroxyurea, increased frequencies of resistance to MTX are observed only following treatment with hydroxyurea at specific points in the cell cycle. Using a protocol of an 8-hour hydroxyurea treatment during G1 followed by a 12-hour recovery or release period before plating in MTX (Fig. 2B), no enhancement of MTX resistance was seen. On the other hand, hydroxyurea treatment in early S to mid-S phase followed by the same recovery period (Fig. 2C) increased the frequency of MTX resistance at least 25-fold. Treatment of cells in late S phase with this protocol drastically decreased their viability, but the frequency of MTX resistance also was increased in the surviving cells (27).

How is this increase in frequency of MTX resistance achieved? If cells are incubated in the presence of the density label bromodeoxyuridine (BrdU), newly synthesized DNA molecules will contain one heavy (BrdU-labeled) strand and one light (unlabeled) strand after a single round of replication (Fig. 3A). If more than one round of DNA replication has occurred, some of the DNA molecules will have incorporated BrdU into both strands; all three DNA species (LL, HL, and HH) can be separated on the basis of their density in CsCl gradients. When synchronized cells were blocked in early S phase (after normal DHFR replication) by hydroxyurea and then released, all in the presence of BrdU, a significant fraction of DNA was found labeled in both strands (i.e., HH DNA) when cells were harvested during the G2-M phase of the cell cycle. When all three densities of DNA were examined for DHFR sequences, DHFR genes were present in the HH DNA fraction (Fig. 3B III), and the total number of DHFR genes per cell was increased by 40 percent. This indicates that rereplication of the DHFR gene may have occurred in the same cell cycle following the release of hydroxyurea inhibition of DNA synthesis.

Thus, it may be important that cells be in the process of DNA replication when the metabolic block is applied in order for the subsequent release to be effective in promoting gene amplification. It also appears that the drug is not specific in causing the rereplication of a particular gene; rather, a large pro-

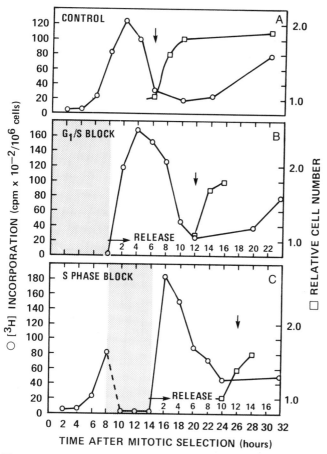

Figure 2 Rates of [³H]thymidine incorporation in CHO cells synchronized by mitotic selection and treated with hydroxyurea (28). Synchronized monolayers of CHO cells were labeled for 15 minutes at 37°C with [³H] thymidine at various times in the cell cycle. (A) Pattern of [³H]thymidine incorporation and increase in cell number of un-treated synchronized cells. (B) Cells were plated directly into 0.3 mM hydroxyurea at the time of synchronization (shaded area); after 8 hours, the medium was changed to normal medium. This protocol yielded no increase in the number of MTX-resistant colonies upon subsequent selection in MTX. (C) Synchronized cells were allowed to progress through the first 2 hours of S phase, at which time further DNA replication was blocked by the addition of 0.3 mM hydroxyurea. Six hours later, the medium was replaced with normal medium. On subsequent selection in MTX, cells subjected to this treatment had a 25-fold increased frequency of resistance to MTX. (Reprinted with permission from The American Society of Biological Chemists, Inc.)

portion of the DNA that was replicated prior to the block was rereplicated. This extra DNA was rapidly lost if the cells were not subjected to subsequent selection with MTX. It is this further drug selection that specifically requires

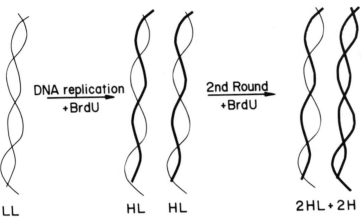

Figure 3A Illustration of the incorporation of the density label bromodeoxyuridine (BrdU) during DNA replication.

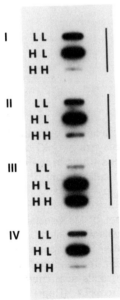

Figure 3B Distribution of DHFR genes between DNA density fractions from (I) cells allowed to progress through 2 hours of S phase; (II) cells subjected to a mid-S phase treatment with hydroxyurea according to the protocol described for Fig. 2C with no recovery period; and (III) cells treated as in (II) but allowed to recover in normal medium for 12 hours before harvesting; (IV) cells grown one full cell cycle and harvested at the time indicated by the arrow in Fig. 2A. In all protocols, the cells were labeled continuously with BrdU (28). (Reprinted with permission from The American Society of Biological Chemists, Inc.)

extra DHFR gene sequences for cell survival and thereby results in retention of extra copies of a particular gene.

Furthermore, small increments in drug dosage promote faster rates of gene amplification compared to high initial drug concentrations and large steps (31). This is compatible with the idea that rereplication of a particular DNA sequence occurs only once in a single cell cycle, so that the relatively small increases in gene copy number which are produced enable cell survival and growth in the presence of slightly elevated drug concentrations.

All of the above results are consistent with the idea that gene amplification can occur due to partial or complete blockage of DNA synthesis followed by a recovery in which rereplication of DNA sequences occurs. An alternative idea is that a specific drug merely selects a preexisting population of cells that have spontaneously duplicated or amplified a particular gene.

Indeed, the following experiments illustrate that spontaneous DHFR gene amplification can occur without any drug selection. The fluorescein derivative of MTX (F-MTX) can be used to quantitate the amount of DHFR enzyme per cell with the fluorescence-activated cell sorter (FACS) (21). Chinese hamster ovary (CHO) cells were grown in the presence of thymidine, glycine, and hypoxanthine to counteract any selection that might be imposed by exposure of the cells to F-MTX. If cells that had never been exposed to any MTX selection were stained with F-MTX and subsequently analyzed on the FACS, a distribution of fluorescence was seen corresponding to a range of DHFR enzyme concentrations among different cells (Fig. 4A). The cells displaying the extremes of fluorescence corresponding to the upper 2 to 5 percent of the population were collected and cultured in nonselective medium. This cell population was then reanalyzed on the FACS as above, and again the highest fluorescing cells were collected. After 10 cycles of this treatment, a population of cells (S-10) was obtained with a mean fluorescence 50-fold higher than the parental population (S-1) (Fig. 4B). When cells from each stage of the sorting process were analyzed for DHFR gene copy number, they were found to contain ever-increasing numbers of DHFR gene copies, up to 40 times the parental copy number in the most highly fluorescent cells (Fig. 4C). This correlated with a comparable increase in MTX resistance if these cells were plated in MTX. Thus, a spontaneous variation in DHFR gene copy number does occur in cells even in the absence of metabolic stress.

Another example of spontaneous gene amplification is that of a transfected gene coding for a cell surface protein in mouse, Leu2. Cells expressing the transfected gene were originally selected by using the FACS to sort cells stained with a fluorescent-tagged antibody directed against Leu2. Amplification of the transfected Leu2 gene was obtained by successive sorts on the FACS of the highest fluorescing cells (23). It is clear, therefore, that cells in culture do spontaneously vary the copy number of their genes. Thus, while drugs that impose a metabolic block on cells may indeed increase the likelihood of aberrant DNA replication leading to gene amplification, these processes must be considered to be an additional factor over and above a spontaneous rate of changes in gene copy number.

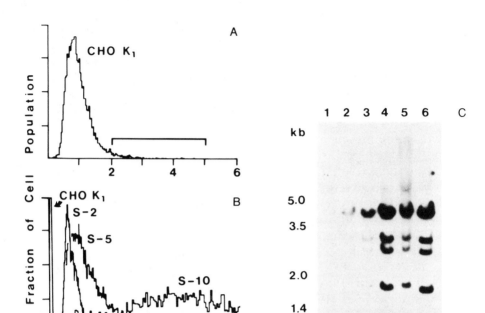

Figure 4 Fluorescence profile of CHO K1 cells stained with fluoresceinated methotrexate (F-MTX) and analyzed on the fluorescence-activated cell sorter (FACS). (A) The cells having fluorescence within the range indicated by the bracket (3 percent of the total population) were collected from the FACS and subsequently grown in nonselective medium. (B) This sorted population was grown and resorted 10 times in succession, yielding the new populations (S-2, S-5, and S-10). (C) Southern blot of DNA from the populations obtained in A and B, showing the increase in DHFR gene copy number. Lane 1, CHO K1 DNA; lane 2, Sort-2 DNA; lane 3, Sort-5 DNA; lane 4, Sort-10 DNA; lane 5, DNA from an amplified cell line derived from CHO K1 containing about 10 copies of the DHFR gene; lane 6, DNA from an amplified cell line derived from CHO K1 containing about 50 copies of the DHFR gene.

Structure of the Amplified Region

The above discussion provides evidence that in some cases disproportionate DNA replication may lead to gene amplification. A related question is how cells regulate the replication of their DNA so that each sequence is normally replicated only once per cell cycle. In studying the DNA structure of the products of the abnormal cell cycles that must have occurred during drug selection in the generation of amplified cells, it might be possible to derive conclusions concerning the normal cellular events.

When looking at the structure of the amplified region in various systems, one basic question is whether there actually is an amplified "unit" as such, where all the copies are identical, with precise end points. If so, this implies that there are specific sequences that facilitate amplification, i.e., recombinational hot spots, which are almost always used as end points. Related to this question of defined ends is the arrangement of the extra copies: Are they linear molecules, are they tandemly linked in direct or inverted arrays, are they circular? The possible products that might be found in an analysis of amplified DNA are diagrammed in Figs. 5 and 6, utilizing the DHFR gene as an example. Any excision or recombination event that occurs in the process of amplification will always produce new restriction fragments not present in the parental unamplified cell lines. If there are recombinational hot spots or precisely excised end points, all copies of the amplified sequences will be identical (Fig. 5B). Thus, if all the copies are monomeric linear molecules, restriction analysis of the amplified cellular DNA should show two new fragments (one at each end) compared to the parental DNA. The abundance of each of these new fragments should be equal to the copy number of all the

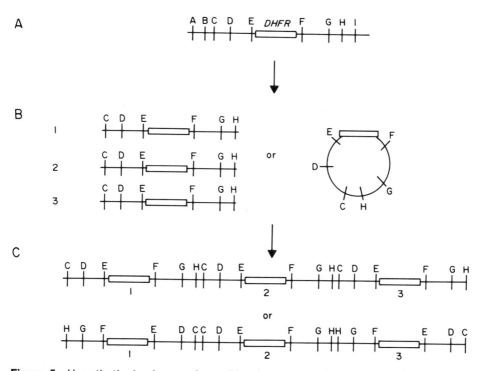

Figure 5 Hypothetical scheme of possible chromosomal structures associated with an amplified unit having defined ends.

Gene amplification with undefined end points

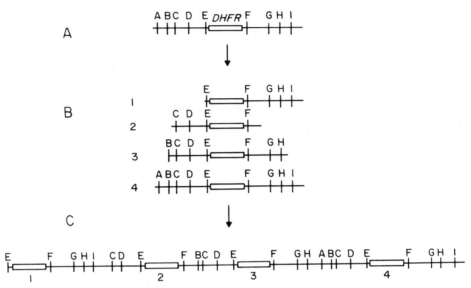

Figure 6 Hypothetical scheme of possible chromosomal structures associated with an amplified region with undefined ends.

other restriction fragments in the amplified unit. If the identical copies are linked in tandem direct repeats or circles, the new end fragments would be joined together in most of the copies to generate a single new type of junction fragment, C-H (Fig. 5B and C). By this type of analysis, monomeric circles, multimeric circles, and tandem linear arrays would all give the same type of restriction pattern and fragment abundance for the junction region, making it difficult to distinguish these different amplification products. This is what is observed in the amplified oocyte ribosomal genes of amphibians, due to a rolling circle mechanism of replication (8). If the amplified sequences are arranged in inverted repeats (Fig. 5C), two new amplified restriction fragments would be produced, C-C and H-H, and pallindromic-type restriction maps would be observed surrounding the junctions. In the "simple" case of amplification of the gene coding for the bifunctional thymidylate synthetase-DHFR protein in *Leishmania*, all three of these alternatives (direct repeats, indirect repeats, and circles) are characteristic of the amplified region (5).

On the other hand, if the amplified region does not have precise ends, then the term "amplified unit" is not appropriate because the multiple copies are heterogeneous. They might be expected to contain the same core sequences corresponding to the selected gene, but each copy may have different ends (Fig. 6B). In this case, many unique restriction fragments would be produced at the ends of unlinked copies. If the copies are linked, many unique junctions not present in the parental cells would still be observed. In this type of arrangement, the recombined fragments would be present as only a subset of the total number of copies of the amplified sequences, while the remainder of copies would still contain the parental-sized fragment (Fig. 6C). For example, in Fig. 6C, all five copies contain fragments E and F flanking the DHFR gene, so that the gene copy number and the copy number of fragments E and F are identical. However, between gene copies 1 and 2 a new fragment, I-C, has been produced which is not present in any of the other amplified copies. The model that predicts no unique ends appears to apply to the chorion gene amplification in *Drosophila*, where multiple rounds of initiation of DNA synthesis within a single cell cycle at an origin near the chorion genes leads to amplification of the gene, with representation of the flanking sequences decreasing farther from the gene (36). However, in *Drosophila*, these cells never divide and so require no recombinational events. Only when cells "must" divide and retain the extra gene copies to continue growth will recombination be required.

In studying the structure of the amplified regions in mammalian cells, the differences in the cytological locations of these sequences in different cell types and species must be kept in mind. CHO and Syrian hamster cells almost exclusively form HSRs, with very few or no extrachromosomal elements present after prolonged selection of cells. Thus, in these cells, some kind of tandem array of amplified sequences must exist. On the other hand, murine lines often contain double minute chromosomes as the mode of maintaining amplified sequences. The size of double minutes is not constant, so it is difficult to make a quantitative estimate of the number of copies of the amplified region present on each. However, depending on the cell line, a crude estimate based on the number of gene copies and the number of double minutes predicts that more than one up to several copies should be present per double minute. If the average size of a double minute is 1000 kb, then the size of the amplified region would correspondingly range from several hundred to 1000 kb. This is consistent with size estimates based on cytological studies of HSR-containing lines as well. Since the actual genes that so far have been shown to be amplified are less than 50 kb, it is apparent that vast stretches of flanking or otherwise unrelated DNA are coamplified with the gene in question in both double minutes and HSRs. Whether there are basic mechanistic differences in the initial amplification processes for the production of HSRs and double minutes or only differences in the later maintenance stages is unknown.

Several research groups have attempted to examine the structure of the amplified region in different systems. The two most highly studied genes are the CAD gene in Syrian hamster cells, and the DHFR gene in hamster and mouse cells. The structural studies dealing with these examples of gene amplification will be discussed in some detail, as their similarities and differences may exemplify the characteristics of mammalian gene amplification in general.

Amplification of the CAD Gene

The acronym CAD stands for the multifunctional enzyme involved in the first steps of UMP biosynthesis, and consists of carbamyl phosphate synthetase, aspartate transcarbamylase, and dihydroorotase activities. The aspartate transcarbamylase activity can be specifically inhibited by PALA [N-(phosphonacetyl)-L-aspartate]. Stepwise selection in this drug leads to amplification of the CAD gene in Syrian hamster cell lines, and the extra gene copies are exclusively found in a chromosomal location, usually a distinctive HSR (41). The amplified genes are relatively stable in these established cell lines in the absence of PALA. Cytogenetic evidence, together with determination of gene copy number, has indicated an average size of 500 kb for the amplified region in PALA-resistant cells, which is much larger than the 25 kb coding for the structural CAD gene.

The structural analysis of this amplified region was undertaken utilizing a method developed with the anticipation of isolating the entire set of amplified sequences. This method, known as differential screening (7), takes advantage of the fact that the cell line containing amplified CAD genes should contain approximately 200-fold more copies of these and all other amplified sequences than its parental cell line, while all other unamplified sequences should be represented equally in both cell lines. As diagrammed in Fig. 7A, two probes were prepared from genomic DNA, one from the amplified cell line and one from the parental cell line. After denaturation, each DNA probe was reannealed to itself until all sequences repeated 200-fold or more were rehybridized. The double-stranded DNAs were isolated on hydroxyapatite and labeled by nick translation; the highly repetitive fractions from both amplified and parental DNAs were then removed by hybridization to immobilized genomic DNA from the parental cells. Plaques that contained amplified sequences could be identified by comparing the hybridization of each of these two labeled probes to replicate filters of a cloned phage library from amplified cells. Those plaques corresponding to sequences in the amplified region should hybridize much more strongly to the probe from the amplified line, since more copies of that sequence are present in the probe (Fig. 7B).

Utilizing this method, 160 kb of DNA was isolated from one amplified cell line and 68 kb from another (3). In the first case, approximately 65 kb of the amplified sequences were linked to the CAD itself, while the organization of the remaining 99 kb relative to the CAD gene could not be determined. This cloned DNA was used to examine the structure of the amplified region in other independently derived CAD-amplified cell lines, and several conclusions could be drawn. First, some fragments that were amplified in one cell line were not amplified in others. Only a core set of fragments (44 kb) including and immediately surrounding the CAD gene were amplified in all cell lines examined. Second, within a single cell line, not all amplified sequences were present at the same copy number (e.g., one fragment was amplified 50-fold over parental levels, while another was amplified only 10-fold). Third, several amplified fragments were found that were not present at all in the

A. Preparation of probe

Genomic DNA

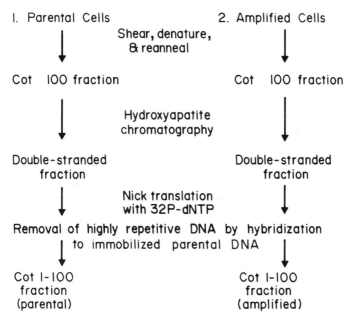

1. Parental Cells   2. Amplified Cells

Shear, denature,
& reanneal

Cot 100 fraction   Cot 100 fraction

Hydroxyapatite
chromatography

Double-stranded   Double-stranded
fraction   fraction

Nick translation
with 32P-dNTP

Removal of highly repetitive DNA by hybridization
to immobilized parental DNA

Cot 1-100   Cot 1-100
fraction   fraction
(parental)   (amplified)

B. Library screening

Replica filter lifts of phage plaques from amplified DNA library

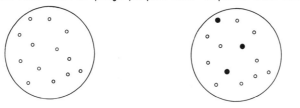

Hybridization to parental probe   Hybridization to amplified probe

Figure 7 Technique of differential screening to obtain amplified sequences (7).

parental cell lines. These novel junction fragments were unique to the cell line from which they were isolated; in addition, the wild-type parental fragments from which these unique fragments were derived were found at levels less than or equal to that of the novel fragment. This indicates that a simple model of a linear array of identical repeating units cannot apply to these amplified cell lines (see Figs. 5 and 6). Further implications of these results are discussed below.

Because the two groups of amplified sequences could not be linked together, it is unclear whether two amplification events might have occurred in these cells. The more likely explanation is tied to the fact that the frequency of obtaining positive clones with this technique of differential screening was much lower than expected. This was probably due to the ubiquitous presence of middle repetitive DNA dispersed throughout the amplified region which could not be removed by the selective hybridization steps and which gave strong signals with both parental and amplified probes. Thus, the amplified sequences that normally link the two amplified regions perhaps could not be distinguished from the background repetitive DNA hybridization.

Amplification of the DHFR Gene

Unlike the exclusively chromosomal location of amplified CAD genes in Syrian hamster cells, both homogeneously staining regions (HSRs) and double minute chromosomes can be found in the different systems that have been studied to define the structure of the amplified region encompassing the DHFR gene. In spite of these inherent differences in the cytological location of the amplified sequences, many properties are shared among the hamster and mouse cell lines having increased DHFR gene copy number. The characteristics of the structure of the amplified region in these cells together with the results cited above utilizing CAD gene amplification may lead to some interesting conclusions concerning the processes involved in gene amplification.

CHOC 400 CELLS

An MTX-resistant CHO cell line (CHOC400) was developed which contained approximately 1000 copies of the DHFR gene (500 times the parental level) in three chromosomal locations, two of which were recognizable as HSRs (15). This cell line was so highly amplified that ethidium bromide staining after agarose gel electrophoresis was sufficient to visualize a unique set of restriction fragments that could not be detected in the parental cell line. When the DNA was Southern blotted and hybridized to radioactive DHFR cDNA, a subset of those visible fragments corresponding to the DHFR gene hybridized to the probe. When the sizes of all ethidium bromide-staining bands (i.e., amplified restriction fragments) were totaled, the amplified region in these cells was estimated to be 135 kb. This figure is likely to underrepresent the true size of the amplified region, because the method assumes that all amplified restriction fragments can be detected and resolved on the gel, and that all of the fragments are amplified to the same extent. In this regard, it was estimated that ethidium bromide staining was sensitive enough only to detect a sequence present at 700 to 1400 times the level of a single copy sequence, so that regions amplified as much as 500-fold might not be detected by this method. Also, *in situ* hybridization studies of mitotic chromosomes from this CHO cell line predicted an amplified region 2.5- to 3-fold larger than that estimated by the ethidium bromide staining. Nevertheless, these studies suggest that a smaller than average amplified region might exist in these cells, making it more amenable to structural analysis.

Utilizing CHOC400 cells that were synchronized by a double block protocol utilizing isoleucine deprivation followed by either hydroxyurea or aphidicolin treatment, short periods of labeling with [^{14}C-]thymidine resulted in the incorporation of label into only a small subset of the amplified fragments, termed early labeling fragments (ELFs) (17). Other amplified fragments were labeled sequentially at later times of S phase. Analysis of a cosmid library from CHOC400 cells using two of these ELFs as probes allowed the identification of several overlapping clones which defined a contiguous region of the amplified genomic DNA involved in the initiation of DNA synthesis at the earliest detectable times in S phase. This region has been found to be approximately 30 kb 3′-ward of the DHFR gene. Thus, it appears that replication in the amplified region may originate at a specific site within each repeated unit. However, the time required to complete replication of all the detectable amplified fragments is much longer than that which would be predicted by the apparent size of the amplified region and the average rate of DNA replication in eukaryotic cells (3 kb/min). Therefore, it may be that certain chromatin conformations alter the rate of DNA replication, or that parts of the amplified region are replicated from a different origin than that used at the earliest times of S phase.

Mouse lymphoma lines

In a similar fashion, several highly resistant mouse lymphoma cell lines were developed that contained 600 to 750 copies of the DHFR gene (40). The structure of the amplified region in these cell lines was examined in several ways. Cytological studies showed that the different lines contained variable numbers of double minutes or HSRs, as well as larger acentric chromosome fragments and ring chromosomes. Occasionally HSRs and double minutes could be found together in the same cell (37). Restriction analysis and gel electrophoresis of the DNA from these cell lines showed that, as in CHOC400, ethidium bromide-stained, multiple, discrete bands not visible in the parental cells. These bands varied considerably in intensity within a given cell line as well as differing in pattern between cell lines (41). Although Southern blotting with the DHFR cDNA as probe showed that all cell lines had amplified the DHFR gene as well as other shared restriction fragments, some fragments were amplified in only a few cell lines and others were unique to only one. This implies that each cell line apparently amplified its own unique set of restriction fragments, even those cell lines obtained from the same parental population. These patterns remained qualitatively identical in each cell line when subjected to higher levels of MTX selection; this was also true when individual clones derived from single cells of MTX-resistant lines were examined. This result indicated that within a population of a given cell line, most of the cells contained the same amplified region.

Another observation resulting from the use of DHFR cDNA as probe of Southern blots of these amplified DNAs was that rearrangements had occurred near the ends of the DHFR gene in some of the copies; again, each such rearrangement was unique to a particular cell line. To further study the amplified region, the double minute chromosomes from one of these MTX-resistant lines were purified 7-fold by centrifugation and filtration, followed by cloning into λ-1059. A modification of the differential screening method, termed the "plus/minus competitor" method, was used to identify clones that contained sequences represented at different repetition frequencies in the parental cells (11). Briefly, the method utilized a probe consisting of nick-translated total nuclear DNA purified by denaturation/reassociation and hydroxyapatite adsorption to remove sequences repeated more than 10,000-fold. Filter hybridization to Southern blots or phage plaque blots was carried out in the presence or absence of 100-fold excess of competing unlabeled parental (unamplified) DNA; the presence of the competitor effectively diluted by 1000-fold the specific activity of nonamplified sequences, while amplified sequences were only reduced by 50 percent.

This type of analysis confirmed and extended the previous observations based on the ethidium bromide-staining patterns. For example, clones representing parental single-copy sequences were found which were amplified in all MTX-resistant lines; however, other clones were amplified in only a subset of the cell lines. Similarly, variant forms of amplified sequences were unique to particular cell lines.

Finally, based on these analyses, it was concluded that, while rearrangements must have occurred early in the amplification process to generate the observed variants and unique restriction fragments in each cell line, the amplified region, once formed, was maintained without change. Thus, it was postulated that at some point, rearrangements must have ceased to occur because the amplified unit became structurally resistant to further change. This point will be discussed again below in conjunction with structural analysis of the amplified region in other MTX-resistant mouse cell lines.

The techniques summarized above which were used to study the structure of the amplified region have the advantage of allowing the analysis of large amounts of amplified DNA at a time. Differences or changes in the ethidium bromide banding pattern of the DNA, for example, can be easily observed, and differential screening methods permit the isolation of many amplified clones. However, as pointed out above, the former method suffers from a lack of sensitivity, in that sequences amplified several 100-fold may not be detectable or resolvable from other bands on the gel. Both techniques have the drawback that the physical relationship of a particular piece of amplified DNA to the selected gene cannot easily be determined. Thus, using these methods, it would be difficult to define the possibility of two or more separate regions coamplified in a cell, as well as to analyze the relationship of any novel junction fragments generated by amplification to the amplified gene.

"Chromosome walking" is a technique that has previously been used to study the structure of chorion gene families in *Bombyx mori* and to study the mechanism of chorion gene amplification in *Drosophila*. Subcloned fragments of DNA flanking the gene of interest on both the 5′ and 3′ sides are subcloned and used as probes of a phage genomic library containing all the sequences in that particular amplified cell line (Fig. 8). These probes should identify a set of overlapping phage whose insert sequences overlap with the gene and extend further 5′- and 3′-ward. In a similar way, the most distal ends of these sequences can again be subcloned and used as genomic probes to identify sequences even further from the gene. By utilizing this rather tedious method of "walking" away from the gene, the structure of the amplified DNA flanking the gene can be determined and compared by restriction analysis to the analogous sequences in the parental cell line. Thus, any "ends" or new recombinant junctions due to the amplification process could be identified and localized with respect to the selected gene, and insights into the mechanism of amplification might be obtained if the arrangement of the amplified copies could be determined (see Figs. 5 and 6).

CHROMOSOME WALKING

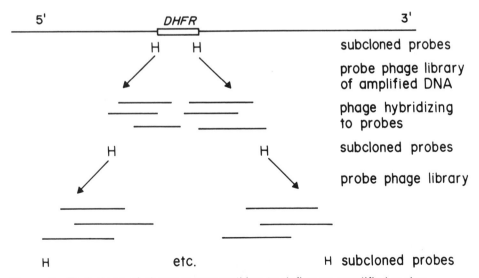

Figure 8 Technique of chromosome walking to define an amplified region.

M500 and CR200 Lines

Chromosome walking was used to analyze the structure of the amplified region in a murine cell line (M500) that contained several hundred copies of the DHFR gene, present for the most part on double minute chromosomes (16). Approximately 160 kb of DNA flanking and including the 32 kb of the DHFR gene itself was mapped by restriction analysis; this map, as far as could be determined, was colinear with the map in the parental, unamplified cell line (Fig. 9). However, an additional 80 kb of amplified DNA was defined which was found to be linked by several recombinant junctions to the normal sequences within and flanking the DHFR gene, generating divergent maps. Another unique junction region was found in a second DHFR-amplified cell line, CR200. The following discussion illustrates how these regions were discovered and characterized using the CR200 junction as an example.

As shown in Fig. 10A, when a probe (C+63) was utilized to probe the amplified genomic library for sequences 3' of the gene, a number of phages were identified that hybridized with the probe. When these phages were mapped by restriction enzymes, two were found whose maps corresponded fully to the previously defined map flanking the DHFR gene and extended it further 3'-ward (λCR-J112 and λCR-J109). However, two other phages were found whose maps corresponded to the same parental map in the region closer to the DHFR gene but, at some point, diverged into totally unrelated sequence (λCR-J111 and CR-J113). This indicated diversity within the CR200 genome. Figure 10B shows an enlarged map of the junction region represented by two of these phages. When the subclone from λCr-J109 which encompasses this junction was used to probe a Southern blot of BamHI-digested genomic DNA, the pattern shown in Fig. 10C was obtained. The probe hybridized to a common 4.1-kb Bam fragment, found in both phage, and in addition confirmed the presence in the CR200 genome of the two divergent frag-

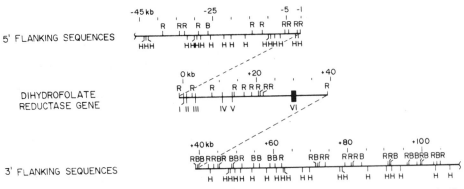

Figure 9 Restriction map of the DHFR gene and amplified flanking sequences in the M500 cell line obtained by chromosome walking. The Roman numerals below the vertical bars represent the six exons of the DHFR gene (16).

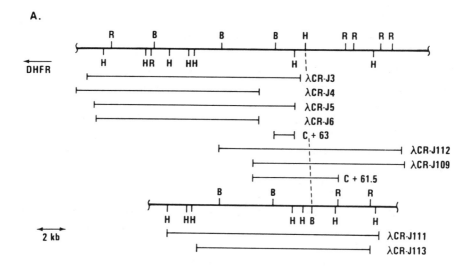

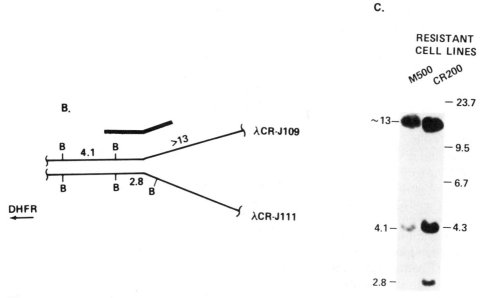

Figure 10 (A) Restriction maps illustrating an amplification-specific junction 3′-ward of the DHFR gene in the CR200 cell line. The subclone C+63 was used to probe the CR200 genomic library and the phages J112, J109, J111, and J113 were obtained; they are shown below their respective restriction maps. The divergence point is highlighted by the dashed line. (B) Enlarged map of the junction region of A showing the BamHI fragments expected to hybridize to the probe (C +61.5, depicted as a black bar). (C) Southern blot of DNA from the M500 and CR200 cell lines showing the amplification-specific junction present only in the CR200 line represented by the 2.8-kb fragment (16). (Reprinted with permission from The American Society of Biological Chemists, Inc.)

ments of 13 kb and 2.8 kb. In contrast, when the DNA from the M500 cell line was hybridized to this same probe, no fragment corresponding to the 2.8-kb junction was observed. The fact that this was a unique rearrangement generated by the amplification process was confirmed by hybridization of the probe to DNA from the parental, unamplified cell line; in this case again only the 13-kb fragment was observed (data not shown).

These results are significant in several ways. First, on a technical point, the confirmation of the presence of the rearranged fragment in the CR200 genome shows that cloning artifacts did not produce the rearrangement in the phage. Second, as found in other systems described above, the junction fragments, which have arisen during the amplification process, are unique to the cell line in which they were isolated. This implies that there is not a required end point to the amplified unit, because while the same parental sequences can exist among different cell lines, only one cell line will contain a given junction fragment.

In addition, within a given cell line, a particular junction fragment is present in only a fraction of the amplified copies, with the remaining copies containing the parental sized fragment. That is, both the 4.1- and 13-kb fragments were found together. If the 4.1-kb fragment contained a required end point, the 13-kb parental fragment should have been lost. This result could be interpreted in two ways. In one case, within a population of cells, a fraction of the cells could contain the junction fragment on all the copies of their amplified DNA, while the remaining cells contain only the parental type fragment on all their copies. On the other hand, within each cell of a population, a fraction of the copies might have rearranged to form the junction fragment while the remainder of the copies retain their parental organization. While the question is difficult to resolve conclusively, isolation of cellular clones from a population of cells containing a particular junction showed that all of the derived clonal lines contained both the junction and parental fragments in the same relative proportions as was found in the original population. This is consistent with results cited above examining the ethidium bromide banding pattern of amplified lymphoma cell lines which indicated that clonally derived lines again exhibited the same pattern of amplified fragments as did the parental amplified line. The conclusion that can be drawn from these results is that there is no discrete amplified unit as such in these cells; instead, a rather heterogeneous amplified region exists inclusive of and surrounding the DHFR gene, having no precise or required end points and containing multiple rearrangements within some of the amplified copies (see Fig. 6).

In further analysis of the regions surrounding the junctions described above, a rather interesting result was obtained. If a fragment distal to the junction fragment was used as a probe of the genomic library to walk further away from the rearrangement, the phages that were obtained invariably fell into two categories. One group corroborated and extended the restriction map of the region in a linear fashion; these sequences continued to diverge from the parental map distal to the DHFR gene. The other group of phages also contained some of the sequences found in the first group, but at a point several kilobases from the junction, their restriction maps again diverged from both the parental map and the map newly defined by the first group of phages. These types of diverging maps are illustrated in Fig. 11. In other words, a second junction was present in relatively close proximity to the first in all cases examined. Since the different phages isolated did not contain coincident junctions, it is unlikely that these results could be explained by a single recombination event. Rather, at least two rearrangements appear necessary to account for the branched maps that are predicted by these results. It is well known that repetitive elements are abundant in the mammalian genome, and the amplified region surrounding the DHFR gene is no exception. Thus, it is plausible that recombination events may occur between homologous repetitive sequences, and once one rearrangement has occurred, a structure or sequence might be formed that facilitates another event in the vicinity.

In most of the systems where the structure of the amplified region has been studied, highly amplified cell lines have been used to maximize the signal-to-noise ratio and to facilitate the isolation of clones from genomic libraries. When rearranged junction fragments have been identified in these cells, it is usually assumed that these rearrangements were formed during the selection process, and that the genome stabilizes once an optimal number of gene copies has been produced. However, using the same cell lines and probes generated during the chromosome walking experiments, evidence was obtained that leads to the conclusion that rearrangements continually occur in amplified murine cell lines, even under maintenance selection conditions.

Using several different probes, Southern analysis was performed on DNA isolated from one cell line after 2 years of growth in its normal selective medium and compared to DNA isolated at the beginning of that time span. Likewise, DNAs from four other cell lines were examined over a 6-month interval. In the first case, many probes that were utilized showed strong hybridization to new fragments after the 2 years in culture under maintenance selection conditions, both within and flanking the DHFR gene (Fig. 12A and B). In the

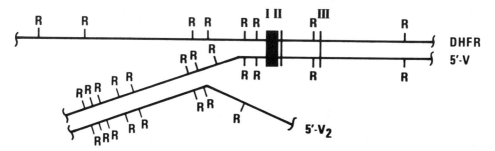

Figure 11 Restriction maps showing two amplified junctions 2 kb 5'-ward of the DHFR gene, illustrating the relatively close proximity of these rearrangements (16). (Reprinted with permission from The American Society of Biological Chemists, Inc.)

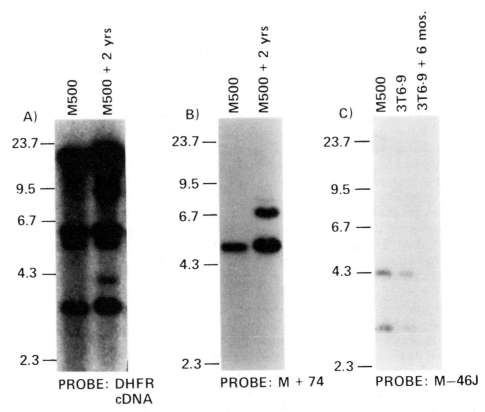

Figure 12 Temporal variability of amplified DHFR sequences. (A and B) DNA isolated from M500 cells or from M500 cells carried in culture an additional 2 years were digested with EcoRI, electrophoresed on 0.8 percent agarose gel, and transferred to diazotized paper. The filter was then hybridized to nick-translated DHFR cDNA (A). After regeneration, the filter was hybridized to nick-translated M − 74 probe which is derived from sequences in the 3′ flanking region of the DHFR gene. (C) DNA isolated from 3T6-9 cells either immediately after selection of the cells in 640 nM MTX or after the cells were carried in culture an additional 6 months, as well as M500 DNA as a control, were digested with BamHI, electrophoresed on 0.8 percent agarose gel, and transferred to diazotized paper. The filter was hybridized to nick-translated M–46J probe, which is derived from the 5′-flanking region of the DHFR gene (16). (Reprinted with permission from The American Society of Biological Chemists, Inc.)

second case, no newly hybridizing fragments were observed in the four cell lines after 6 months. However, in one of these lines, 3T6-9, a fragment that had been highly amplified was now found to be present at essentially single-copy levels (Fig. 12C). Thus, both structural changes in the amplified region (including within the DHFR gene itself) and copy number changes were found to occur in amplified cells even under conditions that should not have caused additional stress.

Conclusions

The structural analyses of amplified DNA discussed above lead to several conclusions that are common to the different systems. The first is that the structure of the actual gene that is amplified is identical with the parental gene. "Pseudogene" type structures lacking the normal introns have never been found as the amplified gene copies. This observation eliminates a mechanism involving reverse transcriptase copying of mRNA as a means for generating extra gene copies. However, all copies of the larger amplified region are not identical, because unique junction sites not present in the parental cells are observed in a fraction of the amplified copies; these rearrangements can occur in the flanking regions of the gene as well as within the gene itself. Also in different independently derived cell lines originating from the same parent, the rearrangements of the amplified regions occur in differing locations. Thus, unique sequences have not been found that function as required end points of an amplified unit, at least in the highly amplified cell lines that have been used for these structural studies. If such required sequences were utilized during the initial events of amplification, subsequent rearrangements at apparently random sites obscures their locations by the time amplification can be detected. From the wide distribution of junctions that have been observed, any such required sites, capable of being utilized in the process of gene amplification, may be widely distributed throughout the genome and would not be the limiting factor in this process.

If a rereplication model of gene amplification is considered, the origins of replication may be the important sites as the origins of amplification. Random termination points for the replication fork during additional rounds of DNA synthesis would provide that the extra gene copies would not have defined ends. The distribution of replication origins relative to a particular gene would then be important in the ease with which the gene may be amplified. On the other hand, the role of cell cycle regulation of expression and the relationship between transcription and replication are also intriguing questions. In the case of DHFR, initiation of transcription and DHFR protein production precede the replication of the gene. It is possible that this induction of DHFR expression at the start of S phase is intimately related to the ability of some cells to overcome an initial replication block and proceed through mitosis with elevated levels of DHFR enzyme; subsequent processes may then lead to selection of cells with extra gene copies.

Portions of the amplified region have been observed to vary over time even when the cells are maintained in their normal selective medium with no additional stress. Both changes in copy number of a particular sequence and new structural rearrangements have been observed over a period of time ranging from 6 months to 2 years (Fig. 12). This result indicates that the cellular genome has much greater plasticity than had previously been believed. Thus, any particular rearrangement in a highly amplified cell line could have arisen at various times during the amplification process. An early recombination between two copies of the nascent amplified region could subsequently become highly amplified during further selection steps; the actual copy number of this junction relative to the gene would depend on when this rearrangement took place. However, a recombination between two copies of the amplified region of a highly amplified line would be nearly impossible to detect within the background of several hundred unrearranged copies; the fact that new rearrangements are observed over time without further selection implies that some form of gene conversion might be occurring to increase the copy number of these new junctions to amplified levels.

Alternatively, a type of competition might exist among different sequences within a cell so that a smaller or more rapidly replicating functional unit would gradually constitute a larger proportion of the intracellular population of double minute chromosomes over multiple cell doublings. The loss of amplified sequences over time could be explained by a mechanism that functions to rid the cell of extraneous genomic material not required for survival in the selective medium. This process may not be very efficient considering the amount of DNA flanking the amplified gene that is coamplified; however, this "extra" DNA may play an as yet undefined structural role, such as in the maintenance of the amplified gene copies in a stable chromatin conformation. The amplified region thus appears to be a dynamic structure, and this continual generation of variability would allow the formation of structures that would be ever more efficient in allowing cell survival and growth during exposure to drug.

General References

SCHIMKE, R. T. 1980. Gene amplification and drug resistance, *Sci. Am.* 243:60–69.

SCHIMKE, R. T. 1984. Gene amplification in cultured animal cells, *Cell* 37:705–713.

References

1. Alitalo, K., Schwab, M., Lin, C. C., Varmus, H. E., and Bishop, J. M. 1983. Homogeneously staining chromosomal regions contain amplified copies of an abundantly expressed cellular oncogene (c-myc) in malignant neuroendocrine cells from a human colon carcinoma, *Proc. Nat. Acad. Sci. U.S.A.* 80:1707–1711.

2. Alt, F. W., Kellems, R. E., Bertino, J. R., and Schimke, R. T. 1978. Selective multiplication of dihydrofolate reductase genes in methotrexate-resistant variants of cultured murine cells, *J. Biol. Chem.* 253:1357–1370.

3. Ardeshir, F., Giulatto, E., Zieg, J., Brison, O., Liao, W. S., and Stark, G. R. 1983. Structure of amplified DNA in different Syrian hamster cell lines resistant to *N*-(phosphonacetyl)-L-aspartate, *Mol. Cell. Biol.* 3:2076–2088.

4. Beach, L. R., and Palmiter, R. D. 1981. Amplification of the metallothionein-I gene in cadmium-resistant mouse cells, *Proc. Nat. Acad. Sci. U.S.A.* 78:2110–2114.

5. Beverley, S. M., Coderre, J. A., Santi, D. V., and Schimke, R. T. 1984. Unstable

DNA amplifications in methotrexate-resistant *Leishmania* consist of extrachromosomal circles which relocalize during stabilization, *Cell* 38:431–439.

6. Bostock, C. J., and Tyler-Smith, C. 1981. Gene amplification in methotrexate-resistant mouse cells. II. Rearrangement and amplification of dihydrofolate reductase gene sequences accompany chromosomal changes, *J. Mol. Biol.* 153:219–236.

7. Brison, O., Ardeshir, F., and Stark, G. R. 1982. General method for cloning amplified DNA by differential screening with genomic probes, *Mol. Cell. Biol.* 2:578–587.

8. Brown, D. D., and Dawid, I. B. 1968. Specific gene amplification in oocytes, *Science* 160:272–280.

9. Brown, P. C., Beverley, S. M., and Schimke, R. T. 1981. Relationship of amplified dihydrofolate reductase genes to double minute chromosomes in unstably resistant mouse fibroblast cell lines, *Mol. Cell. Biol.* 1:1077–1083.

10. Brown, P. C., Tlsty, T. D., and Schimke, R. T. 1983. Enhancement of methotrexate resistance and dihydrofolate reductase gene amplification by treatment of mouse 3T6 cells with hydroxyurea, *Mol. Cell. Biol.* 3:1097–1107.

11. Caizzi, R., and Bostock, C. J. 1982. Gene amplification in methotrexate-resistant mouse cells. IV. Different DNA sequences are amplified in different resistant lines, *Nucleic Acids Res.* 10:6597–6618.

12. Cardman, M. D., Schornagel, J. H., Rivest, R. S., Srimathandada, S., Portlock, C. S., Duffy, T., and Bertino, J. R. 1984. Resistance to methotrexate due to gene amplification in a patient with acute leukemia, *J. Clin. Oncol.* 2:16–20.

13. Coderre, J. A., Beverley, S. M., Schimke, R. T., and Santi, D. V. 1983. Overproduction of a bifunctional thymidylate synthetase-dihydrofolate reductase and DNA amplification in methotrexate-resistant *Leishmania tropica*, *Proc. Nat. Acad. Sci. U.S.A.* 80:2132–2136.

14. Dalla-Favera, R., Wong-Stahl, F., and Gallo, R. C. 1982. *Onc* gene amplification in promyelocytic leukaemia cell line HL-60 and primary leukaemia cells of the same patient, *Nature (London)* 299:61–63.

15. Eickbush, T. H., and Kafatos, F. C. 1982. A walk in the chorion locus of *Bombyx mori*, *Cell* 29:633–643.

16. Federspiel, N. A., Beverley, S. M., Schilling, J. W., and Schimke, R. T. 1984. Novel DNA rearrangements are associated with dihydrofolate reductase gene amplification, *J. Biol. Chem.* 259:9127–9140.

17. Heintz, N. H., and Hamlin, J. L. 1982. An amplified chromosomal sequence that includes the gene for dihydrofolate reductase initiates replication within specific restriction fragments, *Proc. Nat. Acad. Sci. U.S.A.* 79:4083–4087.

18. Horns, R. C., Dower, W. J., and Schimke, R. T. 1984. Gene amplification in a leukemic patient treated with methotrexate, *J. Clin. Oncol.* 2:2–7.

19. Hunt, S. W., and Hoffee, P. A. 1983. Amplification of adenosine deaminase gene sequences in deoxycoformycin-resistant rat hepatoma cells, *J. Biol. Chem.* 258:13185–13192.

20. Johnston, R. N., Beverley, S. M., and Schimke, R. T. 1983. Rapid spontaneous dihydrofolate reductase gene amplification shown by fluorescence-activated cell sorting, *Proc. Nat. Acad. Sci. U.S.A.* 80:3711–3715.

21. Kanalas, J. J., and Suttle, D. P. 1984. Amplification of the UMP synthase gene and enzyme overproduction in pyrazofurin-resistant rat hepatoma cells, *J. Biol. Chem.* 259:1848–1853.

22. Kaufman, R. J., Bertino, J. R., and Schimke, R. T. 1978. Quantitation of dihydrofolate reductase in individual parental and methotrexate-resistant murine cells, *J. Biol. Chem.* 253:5852–5868.

23. Kaufman, R. J., Brown, P. C., and Schimke, R. T. 1981. Loss and stabilization of amplified dihydrofolate reductase genes in mouse sarcoma S-180 cell lines, *Mol. Cell. Biol.* 1:1084–1093.

24. Kavathos, P., Herzenberg, L. A., Sukhatme, U., and Parnes, J. A. 1984. Isolation of the gene encoding the human T-lymphocyte differentiation antigen Leu-2 (T8) by gene transfer and cDNA subtraction, *Proc. Nat. Acad. Sci. U.S.A.* 81:7688–7692.

25. Kohl, N. E., Kanda, N., Schuck, R. R., Bruns, G., Latt, S. A., Gilbert, F., and Alt, F. W. 1983. Transposition and amplification of oncogene-related sequences in human neuroblastomas, *Cell* 35:359–367.

26. Libermann, T. A., Nusbaum, H. R., Razon, N., Kris, R., Lax, I., Soreq, H., Whittle, N., Waterfield, M. D., Ullrich, A., and Schlessinger, J. 1985. Amplification, enhanced expression and possible rearrangement of EGF receptor gene in primary human brain tumors of glial origin, *Nature (London)* 313:144–147.

27. Luskey, K. L., Faust, J. R., Chin, D. J., Brown, M. S., and Goldstein, J. L. 1983. Amplification of the gene for 3-hydroxy-3-methylglutaryl coenzyme A reductase, but not for the 53-KDa Protein, in UT-1 Cells, *J. Biol. Chem.* 258:8462-8469.

28. Mariani, B. D., and Schimke, R. T. 1984. Gene amplification in a single cell cycle in Chinese hamster ovary cells, *J. Biol. Chem.* 259:1901-1910.

29. Mariani, B. D., Slate, D. L., and Schimke, R. T. 1981. S phase specific synthesis of dihydrofolate reductase in Chinese hamster ovary cells, *Proc. Nat. Acad. Sci. U.S.A.* 78:4985–4989.

30. Millbrandt, J. D., Heintz, N. H., White, W. C., Rothman, S. M., and Hamlin, J. L. 1981. Methotrexate-resistant Chinese hamster ovary cells have amplified a 135-kilobase-pair region that includes the dihydrofolate reductase gene, *Proc. Nat. Acad. Sci. U.S.A.* 78:6043–6047.

31. Nunberg, J. H., Kaufman, R. J., Schimke, R. T., Urlaub, G., and Chasin, L. A. 1978. Amplified dihydrofolate reductase genes are localized to a homogeneously staining region of a single chromosome in a methotrexate-resistant Chinese hamster ovary cell line, *Proc. Nat. Acad. Sci. U.S.A.* 75:5553–5556.

32. Rath, H., Tlsty, T., and Schimke, R. T. 1984. Rapid emergence of methotrexate resistance in cultured mouse cells, *Cancer Res.* 44:3303–3306.

33. Schimke, R. T. 1980. Gene amplification and drug resistance, *Sci. Am.* 243:60–69.

34. Schimke, R. T. (ed.) 1982. *Gene Amplification*, Cold Spring Harbor Laboratory, Cold Spring Harbor, N. Y.

35. Schimke, R. T. 1984. Gene amplification in cultured animal cells, *Cell* 37:705–713.

36. Schwab, M., Alitalo, K., Varmus, H. E., Bishop, J. M., and George, D. 1983. A cellular oncogene (c-Ki-ras) is amplified, overexpressed, and located within karyotypic abnormalities in mouse adrenocortical tumor cells, *Nature (London)* 303:497–501.

37. Spradling, A. C. 1981. The organization and amplification of two chromosomal domains containing *Drosophila* chorion genes, *Cell* 27:193–201.

38. Tlsty, T. D., Brown, P. C., and Schimke, R. T. 1984. UV radiation facilitates methotrexate resistance and amplification of the dihydrofolate reductase gene in cultured 3T6 mouse cells, *Mol. Cell. Biol.* 4:1050–1056.

39. Trent, J. M., Buick, R. M., Olson, S., Horns, R. C., and Schimke, R. T. 1984.

Cytologic support for gene amplification in methotrexate-resistant cells obtained from a patient with ovarian adenocarcinoma, *J. Clin. Oncol.* 2:8–15.

40. Tyler-Smith, C., and Alderson, T. 1981. Gene amplification in methotrexate-resistant mouse cells. I. DNA rearrangement accompanies dihydrofolate reductase gene amplification in a T-cell lymphoma, *J. Mol. Biol.* 153:203–218.

41. Tyler-Smith, C., and Bostock, C. J. 1981. Gene amplification in methotrexate-resistant mouse cells. III. Interrelationships between chromosome changes and DNA sequence amplification or loss, *J. Mol. Biol.* 153:237–256.

42. Wahl, G. M., Padgett, R. A., and Stark, G. R. 1979. Gene amplification causes overproduction of the first three enzymes of UMP synthesis in *N*-(phosphonace-tyl)-*L*-aspartate-resistant hamster cells, *J. Biol. Chem.* 254:8679–8689.

43. Wahl, G. M., Vitto, L., Padgett, R. A., and Stark, G. R. 1982. Single copy and amplified CAD genes in Syrian hamster chromosomes localized by a highly sensitive method for in situ hybridization, *Mol. Cell. Biol.* 2:308–319.

44. Yeung, C. -Y., Frayne, E. G., Al-Ubaidi, M. R., Hook, A. G., Ingolia, D. E., Wright, D. A., and Kellems, R. E. 1983. Amplification and molecular cloning of murine adenosine deaminase gene sequences, *J. Biol. Chem.* 258:15179-15185.

Questions for Discussion with the Editors

1. *Please speculate on the possibility that once the amplification machinery for DHFR has been synthesized by a cell, other (as yet undetected) gene amplifications may also take place? That is, the genome of MTX-treated cells may differ from that of control cells at several (e.g., remote) loci?*

It is very likely that other regions of the genome may also undergo amplification during the processes that lead to drug resistance. Treatments of cells with drugs or other environmental stresses may cause widespread chromosome fragmentation, increased sister chromatid exchange, and rereplication of large quantities of DNA. However, the maintenance of these aberrant forms of DNA in the absence of selection may constitute a great metabolic load on the cell; the majority of the extra DNA would thus be expected to be lost from the population of cells, either by an active mechanism of DNA degradation or extrusion from the cell, or by the loss of cells due to their competitive disadvantage. On the other hand, one mechanism for the maintenance of these extraneous DNA sequences might be their recombination into a segment of DNA which is now physically linked to the selected gene. Thus, as we found for DHFR by chromosome walking, some of the amplified sequences in a cell are found adjacent to sequences that were not originally juxtaposed in the parental genome. Similar low-frequency recombinational events could in fact cause the integration of extra DNA sequences into chromosomal DNA at sites physically unlinked from the DHFR gene and thereby maintain their stability in the absence of selection. Indeed, it has been found that cells that are resistant to a particular drug can become resistant to another drug at an increased frequency relative to the parental cell line. This could occur through the mechanism described above whereby some unrelated sequences are maintained at a higher copy number through their new physical association with the original selected gene or in a new chromosomal location, and are thereby more easily amplified to a higher level. Alternatively, the initial events of gene amplification may involve the induction of higher levels of certain enzymes involved in DNA replication or recombination which are then primed for response to future stress.

2. *Could you imagine that gene amplification not unlike that associated with DHFR might be involved in "pattern regulation?" e.g., amphibian appendage regeneration or shark dentition replacement?*

While an analogy might be drawn between the repetitive units involved in gene amplification and the repetitive pattern of a predictable number of units of body segments or shark teeth, it is unlikely that a similar mechanism might be the basis for these phenomena. The only known case of gene amplification in a developmental system is that in *Drosophila* oocyte development. In this case, the cells that undergo chorion gene amplification are helper cells, do not further divide, and in fact die as the oocytes mature. It would seem that in as complex a system as limb regeneration or tooth replacement, intercellular interactions due to positional effects should play a primary role in regulating the expression of specific genes at the transcriptional and translational levels.

CHAPTER **7**

Expression of a Hormone-Inducible Gene: The MMTV Promoter

Bernd Groner, Helmut Ponta,
and Nancy Hynes

INTEREST IN THE REGULATION of gene expression by steroid hormones has inspired a large volume of research. Many different approaches have been taken to describe the mechanism of hormone action (2). The basic premise of these experimental systems is the stimulus of a cell with a specific steroid molecule followed by the measurement of a change in activity of defined genes. Within the framework of cellular genes responding to unknown growth and differentiation signals and the intricate interactions of these gene products, the steroid hormone-regulated genes seem to offer a manageable reduction in complexity. Research in this area has taken different directions, for example, characterization of the intracellular route and fate of hormones (33) or the molecular description of the cellular response to hormones (11).

Gene Regulation by Steroid Hormones

The specificity of steroids in eliciting defined cellular responses has been ascribed to the presence of specific receptor molecules. The complex of receptor and steroid molecule was postulated to be an intermediate in the pathway from the entry of the steroid into the cell to the alteration in gene expression (8). Receptor function has been studied and receptor molecules have been isolated and characterized. The large number of nuclear receptor binding sites and the presence of binding sites with different affinities for the receptor complex has made further advances difficult in the description of steroid hormone action through experiments that focus only on the receptor (38).

A second approach taken to study steroid hormone action started with the identification and quantitation of the induced gene products. These studies have to date benefitted the most from the technical advances in molecular biology. Initially, the hormonal response of particular target genes was measured by immunoprecipitation or enzyme assays of the pertinent cellular proteins. The development of *in vitro* translation and mRNA isolation procedures has allowed measurement of the mRNA concentrations (25). In many instances the changes in mRNA concentration are proportional to the stimulation determined on the protein level. The molecular cloning of hormonally regulated DNA sequences has allowed the most recent developments in the field. Using these gene probes, the rate of transcription of specific genes has been accurately measured before and after hormone stimulation. DNA sequences that harbor regulatory potential have been found in the immediate neighborhood of induced genes, and minimal sequence requirements for the hormonal regulation have been defined (10). *In vitro* receptor/DNA interactions allow the correlation between biologically functional hormone response signals on DNA and purified receptor molecules. A description of the molecular interactions involved in steroid hormone regulation and an understanding of the mechanisms of action is likely to evolve from the basis laid by the two key ingredients: the signal-responsive DNA and the signal-transducing receptor.

The prevalence of genetic methods is reflected in this chapter. Gene cloning and gene transfer techniques, coupled with *in vitro* recombination and *in vitro* mutation procedures, have allowed the identification of a short DNA sequence of 150 nucleotides which subjects heterologous promoters in its vicinity to induction by glucocorticoid hormones. This DNA fragment was found in the long terminal repeat sequence of mouse mammary tumor proviral DNA and is located 50 to 200 nucleotides upstream from the site at which viral RNA synthesis is initiated.

Genetic Approach to Steroid Hormone Action Using MMTV Proviral DNA

The choice of a favorable gene for the study of hormone action is obviously the initial task. Several parameters of the biology of mouse mammary tumor virus (MMTV) make it a desirable candidate. Infection of mice with MMTV leads to a high incidence of mammary tumors (3). These tumors can be cultured *in vitro* and express MMTV. The infection with and production of MMTV does not have cytopathic effects on the cultured tumor cells and virus-producing lines can be grown indefinitely. The extent of virus production can be modulated by the addition of glucocorticoid hormones to the culture medium (Fig. 1). A 10-fold stimulation in the number of virus particles produced and released into the culture medium can be observed. This stimulation in viral

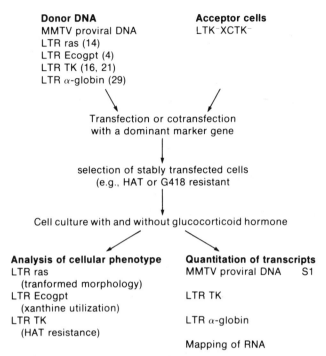

Donor DNA
MMTV proviral DNA
LTR ras (14)
LTR Ecogpt (4)
LTR TK (16, 21)
LTR α-globin (29)

Acceptor cells
LTK⁻XCTK⁻

Transfection or cotransfection
with a dominant marker gene

selection of stably transfected cells
(e.g., HAT or G418 resistant

Cell culture with and without glucocorticoid hormone

Analysis of cellular phenotype
LTR ras
  (tranformed morphology)
LTR Ecogpt
  (xanthine utilization)
LTR TK
  (HAT resistance)

Quantitation of transcripts
MMTV proviral DNA   S1

LTR TK

LTR α-globin

Mapping of RNA

Figure 1 Analysis of the regulatory potential in the MMTV LTR. Proviral DNA or MMTV LTR chimeric genes were used to analyze the regulatory potential. DNA was introduced into cultured cells by transfection or cotransfection. The properties of indicator genes allowed in the case of, for example, LTR-ras, LTR-Ecogpt or LTR-TK for probing the hormonal effects directly at the level of cellular phenotype. Transfection of, for example, proviral DNA or LTR-α-globin DNA do not confer a selectable, hormone-dependent phenotype. Quantitation of LTR or α-globin transcripts has to be used to describe hormonal responsiveness.

RNA synthesis occurs very rapidly after hormone addition and cannot be blocked by inhibition of protein synthesis (20, 31). Although the evidence is circumstantial, these early observations suggested that there might be a direct mechanism linking glucocorticoid hormone addition and MMTV production.

In addition to the availability of a steroid-responsive tissue culture system, another advantage is provided by the defined structure of the MMTV proviral gene. Like all retroviruses, MMTV undergoes a life cycle that includes extracellular and intracellular stages. The transcription of proviral DNA yields viral mRNA and genomic viral RNA. The translation of viral mRNA produces structural proteins that are packaged together with genomic viral RNA into viral particles that are shed from the cell and infect susceptible host cells. Reverse transcription of viral RNA leads to the production of a double-stranded DNA molecule that is circularized and integrated into the genomic DNA of the host cell.

The integrated proviral DNA is distinguished from the viral RNA, the template for reverse transcription, by the presence of the long terminal repeats (LTRs). These repeats at the ends of the proviral DNA are composed of sequence information derived from the 5' and 3' ends of the viral RNA. The function of the LTRs is 2-fold. They contain signals important for the integration of viral DNA into the genomic DNA of the host cell and for the control of viral transcription (36). LTR sequences preceding the proviral promoter are candidates for use in the functional investigation of sequences regulating transcription. The border between the flanking host DNA and the viral DNA proper is well defined. If there are transcriptional regulator sequences present in the proviral genome and if these signals are located 5' to the RNA initiation site, the U3 region of the LTR must contain them.

The proviral and the LTR structures are schematically shown in Fig. 2.

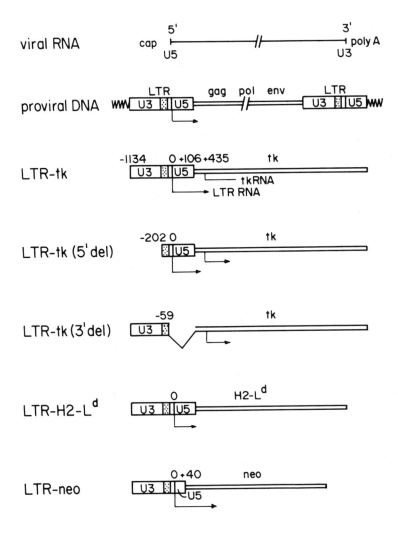

Sequence comparison between the proviral LTR DNA and cellular genes suggest that regulatory signals are present. Two types of upstream control elements found in a number of eukaryotic protein-coding genes can be observed in the U3 region upstream of the start site of transcription. A TATAAA element is located at position −26 to −31 in the LTR. This sequence is thought to be involved in the fixation of a defined start point about 30 nucleotides downstream from its own position. A CCTAT element at position −83 to −87 which might be important in determining the level of transcription in various RNA polymerase II transcribed genes can also be found (18). The hormone response region of MMTV was suspected to reside in the U3 region of the LTR.

Identification of control sequences not common to many different genes and not evident by mere sequence comparison, however, is a more difficult task. The availability of a defined gene sequence does not permit conclusions to be drawn about the presence of regulatory potential. Multicomponent interactions, which certainly occur in hormone-regulated gene transcription, cannot be recognized on the primary DNA level. They are also difficult to reconstruct *in vitro*. The phenotype of a hormone response signal is restricted to the amount of transcript from a particular gene. This phenotype can only be accurately measured if all other components potentially affecting the efficiency of transcription are held constant. A situation in which only one variable is changed (i.e., the nature of the DNA template being transcribed), and in which the entire cellular environment is constant, can be achieved by transfecting DNA into cells. Stably transfected cells that have acquired the exogenously supplied genes can then be subjected to a hormonal stimulus and

Figure 2 Proviral structure and chimeric gene constructs used to define the hormone response element (HRE) in the MMTV LTR. Genomic viral RNA is reverse transcribed in MMTV infected cells into proviral DNA. The U5 and U3 region of the RNA is copied twice by the reverse transcriptase, thus giving rise to the LTRs. Proviral DNA is hormone inducible in infected cells (31). LTR DNA is recombined with a TK gene *in vitro* (LTR-TK). Transcription in LTR-TK transfected cells can be quantitated in RNA derived from cells grown in the absence and presence of glucocorticoid hormones. RNA initiated in the LTR and in the TK gene is subjected to hormonal induction (16). Stepwise deletion of U3 sequences from the LTR [LTR-TK (5′ del)] defines the 5′ border of the hormone response element (16, 21). It is possible to delete U3 sequences up to position −202 without affecting the induction of LTR or TK transcripts.

Stepwise deletion of U5 sequences from the LTR [LTR-TK (3′del)] defines the 3′ border of the HRE (29). The LTR RNA initiation site up to position −59 can be deleted without affecting the hormonal responsiveness of the TK mRNA transcription. The constructs LTR-H2-L^d and LTR-neo are examples for chimeric constructs that allow the hormonal effect to be visualized at the level of the cellular phenotype. Addition of dexamethasone to LTR-H2-L^d transfected cells causes the enhanced expression of H2-L^d at the cell surface. LTR-neo transfected cells can be selected for in G418 resistance at a higher frequency when dexamethasone is present (Ponta et al., unpublished results). The arrows indicate RNA initiation sites.

the gene transcripts from the introduced gene can be quantitated. The RNA concentration of a transcript originating at a promoter site in the vicinity of the regulatory sequence that is being studied thus provides a measure of regulatory potential.

Introduction of MMTV Proviral DNA and MMTV LTR Chimeric Constructs into Cultured Cells Results in Hormone-Responsive Transcription

The strategy that is based on cloned MMTV proviral DNA, *in vitro* manipulation and recombination of MMTV fragments with other genes, transfer into cultured cells, and analysis of phenotypic changes as a consequence of the introduced DNA has been extensively applied (11). Figure 1 summarizes the strategies of several groups that have made use of the MMTV DNA for the study of glucocorticoid hormone action.

The first indication that a hormonally responsive DNA signal is associated with and part of the MMTV proviral DNA came from experiments in which the intact, cloned proviral DNA was introduced into recipient cells. The Herpes Simplex virus thymidine kinase (HSV TK) gene was cotransfected with MMTV proviral DNA into Ltk$^-$ cells, which were subsequently selected for the TK$^+$ phenotype. Since cotransfected DNA is ligated intracellularly and integrated into nonspecific chromosomal loci, this procedure provided potential advantages. The isolation of TK$^+$ cells in HAT medium (hypoxanthine, aminopterin, thymidine) selects for cells that have stably acquired the TK gene and are expressing it sufficiently to overcome the metabolic inhibition of aminopterin (37). The presence of the proviral MMTV gene in the vicinity of the transcriptionally active TK gene favors the probability that the MMTV gene is accessible for components of the transcription machinery. The nonspecific chromosomal location of DNA acquired by transfection is merely restricted by "active loci," i.e., all integration events that subsequently allow the transfected DNA to be transcribed can be selected in HAT medium. This is an advantage because it allows elimination of possible effects that are conferred by the integration region onto the acquired gene when several clones of transfectants are compared with one another.

A number of the TK$^+$ cell clones that contained the proviral MMTV gene were grown in the absence and presence of the synthetic glucocorticoid hormone dexamethasone. RNA extracted from these cells was then analyzed by gel electrophoresis and Northern blotting for MMTV-specific mRNAs. The intensity of the characteristic MMTV mRNA bands and quantitation by solution hybridization showed an increase of intracellular MMTV RNA in the presence of hormone (15).

The viral genome of MMTV comprises about 10 kb of genetic information coding for the viral structural proteins and reverse transcriptase (Fig. 2). *In vitro* recombination combined with gene transfer and phenotypic analysis allowed a delimitation of the proviral DNA which is actually responsible for the hormonal effect. This delimitation was carried out in three steps and utilized chimeras in which (a) the promoter of the indicator gene was removed and replaced by the MMTV LTR promoter (21), (b) the MMTV LTR sequence was added to an indicator gene that retained its own promoter (16), and (c) a segment obtained from the MMTV LTR with regulatory potential but without a promoter function was recombined with an indicator gene (29).

The replacement of a promoter with the MMTV LTR was used to demonstrate the conferral of hormonal induction on several indicator genes. The genes were chosen to exhibit a phenotype at the cellular level, i.e., induction of the gene product could be demonstrated by a property of the intact cell. Introduction of an LTR Ha-ras into NIH3T3 cell chimera resulted in transformed NIH3T3 cells when they were incubated in the presence of dexamethasone. The Ha-ras gene product, p21, was induced and the transformed foci of NIH3T3 cells could be scored (13). Combination of the MMTV LTR promoter with a prokaryotic gene, Ecogpt, resulted in hormone-inducible expression of the xanthine-guanine phosphoribosyltransferase gene and allowed cells to be grown in limiting concentrations of xanthine when dexamethasone was simultaneously present (4). Finally, the MMTV LTR was combined with a gene coding for a cell surface protein (H-2L^d). The expression of this gene can be quantitated with fluorescent antibodies directed against H-2L^d and flow cytometry and was found to be inducible in transfected cells (Ponta et al., unpublished results). All three constructs (LTR Ha-ras, LTR Ecogpt, and LTR H-2L^d) clearly showed that no viral gene product is required for the hormone effect. The regulatory potential is encoded in a DNA sequence near the proviral promoter and various heterologous gene products can be brought under its control. Two constructs used in these studies (LTR H-2L^d and LTR neo) are schematically shown in Fig. 2.

The addition of the MMTV LTR to a gene with its own promoter (LTR TK, Fig. 2) provided further insights. Although this chimeric construct provides phenotypic selection to transfected LTK⁻ cells, the phenotype does not allow conclusions about hormonal inducibility (16). The TK mRNA is constitutively expressed, independent of the presence of the MMTV LTR sequences. In order to assess hormonal inducibility, quantitative mapping of mRNA molecules initiated at the LTR cap site has been done. The S1 mapping procedure showed that correctly initiated LTR transcripts are induced by dexamethasone. Since the introduction of the LTR TK plasmid and the selection of the transfected cells does not depend on LTR transcription and induction, this construct proved to be a suitable basis for an analysis of the minimal DNA requirements for hormonal regulation. The MMTV LTR is 1328 nucleotides long, and 1194 nucleotides precede the RNA initiation site. These sequences comprise a long open reading frame of 960 nucleotides (7, 18). Removal of U3 sequences in a controlled fashion yielded deletion mutants with decreasing LTR sequences preceding the RNA start site [LTR TK (5′ del), Fig. 2]. Hormonal inducibility of LTR transcripts was retained in all deletion mutants with more than 200 nucleotides of LTR sequence preceding the RNA start site (position −200). Further deletion to position −137 severely reduced the inducibility and finally abolished it altogether (−37). The analysis of the 5′ deletion series indicates that a short DNA sequence of about 200 nucleotides 5′ of the RNA start site contains sufficient regulatory information to cause hormone-inducible transcription (16).

The chimeric construction used as a basis for the 5′ deletion series contains an LTR sequence linked to a TK gene that retained its promoter region and its structural information. TK transcripts are initiated at the authentic TK start site also in the absence of hormone. S1 mapping of TK transcripts made in cells grown in the presence of hormone showed that the TK transcripts can also be hormonally induced. The mere presence of the regulatory LTR sequences in the vicinity of a normally unresponsive gene leads to inducibility. This unexpected result allowed a further delimitation of the essential regulatory DNA. Since the inducibility could be scored by quantitation of the TK transcripts, deletions in the region of the LTR promoter and the LTR DNA initiation site still could be tested biologically. A 3′ deletion series was constructed in which LTR sequences around the LTR RNA initiation site were removed [LTR TK (3′ del), Fig. 2]. Hormonal inducibility of TK transcripts could still be measured when the LTR promoter region up to position −59 was deleted from the 3′ side. This result defines the hormonal regulation sequence as a purely regulatory sequence which functions independently of the promoter and initiation region of the MMTV LTR (29).

The Hormone Response Element (HRE) in the MMTV LTR Modulates the Rate of Transcription and Has Enhancerlike Properties

The experiments that were carried out to delimit the minimal sequence required to confer hormonal inducibility to the TK promoter allowed the definition of a 143-nucleotide HRE. The presence of the DNA sequence located at position −59 to −202 5′ of the RNA site led to an increased concentration of LTR and TK transcript when transfected cells were cultured in dexamethasone for about 18 hours. The mechanism by which this increase in the steady-state mRNA concentration occurred was investigated by measuring the LTR transcription as a function of time after hormone addition. This can be done in a sensitive fashion since no accurately initiated LTR transcripts were found in cells transfected with LTR indicator gene constructs grown in the absence of hormone. This is an important advantage in kinetic studies because the "preexisting RNA" (i.e., RNA molecules that are present before hormone addition) can be neglected. The presence of preexisting RNA molecules usually makes it necessary to distinguish between them and RNA newly synthesized on stimulation. S1 nuclease mapping was used to follow the increase in concentration of accurately initiated LTR RNA molecules on hormone addition. The earliest time when LTR RNA could be demonstrated was 7.5 minutes after hormone addition. The rate of transcription increased steadily and after about 15 minutes a maximum was reached. Fifteen minutes are also required to achieve a saturation in nuclear binding of hormone. The nuclear accumulation of steroid receptor complex is thus proportional to the rate of LTR transcription, suggesting a direct mechanism of action (10).

The idea that the glucocorticoid receptor complex is directly responsible for the increased rate of LTR transcription was supported by two additional observations. First, when the accumulation of LTR transcripts was monitored after the simultaneous addition of dexamethasone and cycloheximide to the growth medium, the observed accumulation kinetics were indistinguishable from the ones observed in the experiment from which the protein synthesis inhibitor had been omitted. The accumulation of LTR RNA is independent of ongoing protein synthesis at the time of hormone addition. This rules out the possibility that the primary induction event is dependent on the synthesis of a regulatory protein. The second observation is based on the effect of RU486. RU486 is a glucocorticoid antagonist that has a high affinity for the glucocorticoid receptor and prevents the formation of an active dexamethasone glucocorticoid receptor complex. The simultaneous addition of RU486 and dexamethasone to transfected cells inhibits the accumulation of LTR transcripts. The sequential addition—dexamethasone followed by RU486 administration—leads to a rapid decay of LTR RNA. These observations suggest that an active hormone receptor complex is directly involved in the transcriptional regulation process. The continued presence of the receptor complex is required and upon its removal the process of transcriptional activation is readily reversible (10).

The biological observations that link the hormone receptor complex to the transcription machinery are supported by experiments that measure the *in vitro* interaction of the MMTV LTR DNA with the receptor protein (27). Several methods have been devised that can be used to recognize protein DNA interactions. The association can be described on the individual nucleotide level by the protection of DNA against DNase I attack or methylation by dimethylsulfate by the presence of binding proteins. Investigation of the DNA region that was found to be biologically significant (−202 to −59) has shown that this region contains four receptor binding sites. Two strong and two weak binding sites for activated glucocorticoid receptor have been found upstream of the RNA initiation site. These four sites are delimited by position −72 to −192. A hexanucleotide sequence 5′-TGTTCT-3′ is found in all four binding regions. The G residues of these hexanucleotides are protected by the receptor protein from methylation by dimethylsulfate (32). In addition, the presence of the 5′-TGTTCT-3′ sequence in the region essential for glucocorticoid receptor binding and regulation of the chicken lysozyme gene (30), and a similar sequence (5′-TGTTCC-3′) in the hormone receptor binding region of the human metallothionein IIA gene (17), make it likely that this motiff is truly important for the recepfor-DNA interaction.

The confluence of *in vivo* (gene transfer) and *in vitro* (DNA-protein binding) experiments strengthens the assumption that the interaction between the receptor and the MMTV LTR DNA is an essential step in the hormonal activation of the MMTV promoter. Experiments based on the reaction of intact cells to exogenously added hormone are necessarily limited in their extent of interpretation. Too many cellular components are involved in the signal transduction that culminates in the accumulation of specific mRNA, and certainly not all of them are known. The conclusions drawn from the above-described experiments, however likely and convincing, are indirect. The biological effect of *in vitro* manipulated genes on gene transfer and transcription in transfected cells is combined with *in vitro* observations utilizing purified DNA and partially purified glucocorticoid receptor. Proteins that are required for the transcription of genes and that bind to regions just upstream of the regulated promoter have been described for a heat shock gene (hsp 70) of *Drosophila* (26) and the SV40 early promoter (6). In analogy to these proteins the glucocorticoid receptor can be viewed as a gene-specific transcription factor that acts in concert with RNA polymerase II and possibly other general transcription factors. The ultimate proof of the proposed idea (i.e., the activity of activated glucocorticoid receptor as a specialized transcription factor that increases the rate of MMTV LTR RNA initiation), has not yet been shown *in vitro*. *In vitro* transcription systems are not as yet complex, efficient, and accurate enough to allow a reconstitution experiment in which the increase in the initiation of transcription of the MMTV LTR RNA can be measured in the presence of hormone receptor complex.

The description of the mechanism of hormone action, that is, the increase in the rate of promoter utilization, mediated by a DNA signal that is distinct from the proximal transcription regulatory sequence (TATA box) and the RNA initiation site, is reminiscent of the effect conferred by control sequences identified in a number of genes. These control elements can be classified into those whose action requires a position at the immediate 5′ end of the gene (22) and those that act relatively independently of distance and orientation with respect to the regulated transcript. The second class of regulatory signals has been termed enhancers (19). The autonomy of the HRE with respect to position orientation and distance from the inducible promoter has been tested. *In vitro* recombination of the mouse α-globin gene with the HRE yielded four constructs (Fig. 3). In two of the constructs the HRE is located at the 5′ side of the α-globin gene and assumes both possible orientations (syn and anti), and in the other two constructs the HRE is located at the 3′ side of the α-globin gene, again in both possible orientations. The 5′ and the 3. constructs maintain about the same distance of 1.1 kb between the HRE and α-globin promoter. All four constructs give rise to hormonally inducible α-globin transcripts after introduction into cultured cells. No appreciable difference in the constitutive level of α-globin transcription (in the absence of hormone) or in the extent of hormonal induction was observed (Fig. 3). In addition, the constitutive level of globin transcription was very similar to the transcription observed in a construct that did not contain HRE sequences (29).

These results allow the description of the HRE as a "conditional enhancer." In the absence of hormone no effect is exerted. The presence of hormone makes the HRE behave like an enhancer sequence stimulating transcription of heterologous promoters in a position- and orientation-independent fashion. Cellular, enhancer-interacting molecules have been

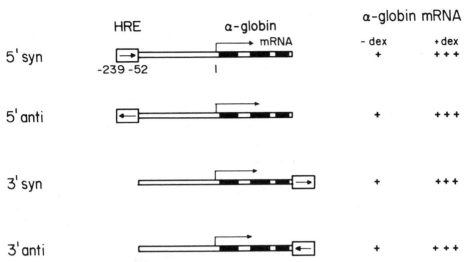

Figure 3 Hormone response element HRE-mediated gene induction is position and orientation independent *in vitro* recombination of LTR sequence -52 to -239 (HRE) with the α-globin gene was carried out. Two possible relative orientations were obtained by recombination at the 5' end (5' syn and 5' anti) and at the 3' end (3' syn and 3' anti). All four constructs were transfected into L cells and globin transcripts were quantitated. Independence of position and orientation hormonal inducibility was found (29). These features allow the definition of the HRE as a "conditional enhancer." The arrow within the HRE defines the orientation with respect to the α-globin gene. The transcriptional start of the α-globin gene is designated as 1.

postulated and visualized in their action on the SV40 enhancer by an *in vivo* competition assay (34). The nature of this cellular molecule is not known, but different enhancers can be expected to depend on different host factors. In the case of the HRE the conditional enhancement is mediated by at least one known cellular component—the activated glucocorticoid receptor.

A Model for Steroid Hormone Enhancement of MMTV Transcription

The action of glucocorticoid hormones on MMTV transcription can be described as an enhancement process of the rate of RNA initiation mediated by a nearby regulatory sequence. The functional similarity to the enhancer sequences makes it likely that mechanistic considerations applicable to enhancers are also important for the HRE. Unfortunately, no experimentally verified model for enhancer action is available, but a number of interacting components have been suggested. Specific DNA sequences probably bind cellular proteins, and specificity of action with respect to cell type could be conferred by the presence and absence of these interacting proteins (12). The role of the specific cellular protein can be assumed by the hormone receptor. This receptor, however, needs the hormone ligand to be active as an enhancement factor. The observation of relative independence of orientation and (in a limited range) of distance of enhancers, suggests a mode of action that affects a domain of DNA rather than a narrow promoter sequence. Distance and position independence of enhancers have been explained by a facilitated entry of RNA polymerase in conjunction with a scanning mechanism. In this model the enhancer sequence would provide an entry site for the polymerase searching for RNA initiation signals in nearby DNA sequences. The bidirectionality of the enhancement process would make it necessary for the RNA polymerase to change directions after encountering a promoter sequence. All of these considerations are based on a model that does not take into account nuclear organization and focuses only on the DNA template and the RNA polymerase.

Nuclear transcription, however, does not seem to take place in a "soluble transcription system," but the nuclear matrix has been recognized as a subnuclear structure with which heterogeneous nuclear RNA as well as specific precursor RNAs are associated (5). The nuclear matrix is composed of elements of the nuclear envelope and nuclear lamina, residual nucleoli, and an extensive granular and fibrous interchromatinic matrix structure which extends throughout the interior of the nuclear sphere. Not only are RNA and DNA synthesis complexed with this nuclear protein structure, but also steroid hormones are preferentially associated with the nuclear matrix. More than 60 percent of the steroid can reside in a matrix structure harboring only 1 to 2 percent of the total DNA (1, 28).

The model presented in Fig. 4 takes three considerations into account: (a) the transcription of nascent RNA is associated with the nuclear matrix, (b) the steroid hormones are localized preferentially in the nuclear matrix, and (c) the steroid hormone complex can bind to DNA sequences in the vicinity of a regulated promoter. The following assumptions are made in the model: (a) it is not only the steroid hormone but the steroid hormone receptor complex that binds to the nuclear matrix; (b) this binding is independent of the receptor complex association with DNA, and (c) the binding of the steroid receptor complex takes place in the vicinity of a "transcription site," that is a site on the nuclear matrix that contains (or allows easy access to) RNA polymerase and transcription factors. The nuclear acceptor site could be imagined to be made up of three components, the steroid hormone receptor complex, the DNA binding site and specific chromatin proteins allowing the anchorage of a DNA sequence via the receptor complex at a specific site of the nuclear matrix (35).

The mode of steroid hormone action could then be pictured as the attachment of the hormone receptor to two sites: (a) the DNA sequence (HRE) in the vicinity of a hormonally regulated gene, and (b) the anchoring of the DNA at a nuclear matrix site particularly suited for initiation of transcription. If the DNA is held in place by two attachment sites of the hormone receptor, sequences in the vicinity of the HRE might serve as initiation signals for RNA

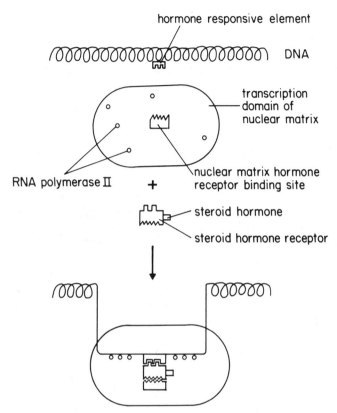

Figure 4 A model for steroid hormone responsiveness of MMTV transcription. This model tries to incorporate the position and orientation independence of HRE action described in Fig. 3. It is proposed that the HRE subjects a DNA domain to its 5' and 3' side to the access of RNA polymerase and transcription factors. This could be made possible by a simultaneous attachment of hormone-receptor complex to a binding site on the DNA (the HRE) and to a site on the nuclear matrix. The anchoring of DNA could than result in the efficient transcription of regions located in the vicinity of the HRE. The presence of hormone receptor complex as well as the presence of nuclear attachment sites could constitute determinants for cell-type specificity of steroid hormone action.

synthesis. Sequences upstream and downstream would be exposed and transcribed with similar efficiency.

This model proposes that the steroid hormone receptor complex acts as an anchor with two binding sites: the receptor complex binds simultaneously to a regulatory DNA sequence and a "transcription site" on the nuclear matrix. A number of features particular to steroid hormone and enhancer action could be accommodated. Specificity of hormone and enhancer action could be explained by the absence or presence of the interacting proteins. Furthermore, the number of affinity of nuclear matrix acceptor sites could be regulated. Different states of cellular differentiation could be distinguished by the number of acceptor sites, thus accounting for differential sensitivity of cells toward hormonal stimuli (23). The process of anchoring is probably readily reversible, and hormonal withdrawal or inactivation of the steroid hormone receptor complex results in a rapid shut-off of transcription (10). This would require that one or both binding interactions (on the DNA and/or the nuclear matrix) depend on a receptor-hormone complex rather than its individual components. The "predisposition" to induction, which could be pictured as the number of available nuclear matrix anchor sites, is probably a stable feature of differentiation and established during "primary" stimulation. Secondary stimulation of "primed" cells can result in a rapid resumption of specific mRNA transcription (23). In the case of glucocorticoid action most cells would be predisposed constitutively, in contrast, for example, estrogen or progesterone responsive oviduct cells which might have to undergo anchorage site establishment before specific egg white genes can be induced.

Although speculative, this model outlines the directions future experiments will take. One level, the regulatory potential in the DNA, has been clearly defined by experiments that take advantage of the advanced technology in gene manipulation and of whole cells as "transcription systems." *In vitro* recombination of MMTV LTR DNA with various indicator genes allowed the definition of the HRE. The defined DNA sequence of 150 nucleotides is necessary and sufficient for MMTV RNA induction and can be conferred to heterologous promoters. These experiments are nicely complemented by *in vitro* binding studies of the glucocorticoid receptor. The functional potential could only be demonstrated through introduction into the nuclear context. *In vitro* transcription experiments using soluble extracts do not take into account the structural complexity of nuclear organization. More sophisticated assay systems will probably be necessary to reconstitute hormonal effects *in vitro* and to explain more subtle aspects of steroid hormone action.

It will be particularly interesting to address questions that deal with the following features:

1. The potential of a gene to be induced by a specific steroid hormone is a differentiation parameter of a cell type. The mere presence of hormone-receptor complex is not a sufficient requirement for the gene induction. For example, the transferrin-conalbumin gene in chicken liver is not induced by estrogen although the liver is a hormonally responsive tissue. The same gene can be induced in another responsive tissue, the oviduct, by estrogen. The stable changes in cellular differentiation which accompany primary induction, and which can be described as the potential for inducibility, have not been defined in molecular terms (9, 24).

2. The extent of induction is gene-specific. Several responsive genes in a given cell may be coordinately controlled by steroid hormone but the extent of the inducibility might widely vary (14). The factors determining the quantitative aspects of induction are mainly unknown.

3. The components participating in the hormonal induction that have been defined so far are limited to the regulated DNA, the RNA polymerase II and the hormone-receptor complex. Additional components that could be functioning as general and gene-specific transcription factors, as determinants of the architecture and specificity of the nuclear matrix attachment sites for DNA and hormone receptor, and proteins that influence the gene conformation and chromatin structure are waiting to be identified.

General References

Groner B., Salmons B., Günzburg W. H., Hynes N. E. and Ponta H. 1984. Expression of proviral DNA of mouse mammary tumor virus and its transcriptional control sequences, in: *Oxford Surveys on Eukaryotic Genes*, vol. 1 (N. Maclean, ed.), pp. 82–110, Oxford Univ. Press, Oxford, U. K.

Kingston R. E., Baldwin A. S., and Sharp P. A. 1985. Transcription control by oncogenes, *Cell* 41:3–5.

References

1. Barrack, E. R., and Coffey, D. S. 1982. Biological properties of the nuclear matrix: Steroid hormone binding, *Recent Progr. Horm. Res.* 38:133–195.
2. Baxter, J. D., and Rousseau, G. G. 1979. Glucocorticoid hormone action, in: *Monographs on Endocrinology*, vol. 12, Springer Verlag, Berlin.
3. Bentvelzen, P., and Hilger, J. 1980. The murine mammary tumor virus, in: *Viral Oncology* (G. Klein, ed.), pp. 311–355, Raven Press, New York.
4. Chapman, A. B., Costello, M. A., Lee, F., and Ringold, G. M. 1983. Amplification and hormone-regulated expression of a mouse mammary tumor virus Eco gpt fusion plasmid in mouse 3T6 cells, *Mol. Cell. Biol.* 3:1421–1429.
5. Ciejek, E. M., Nordstrom, J. L., Tsai, M. J., and O'Malley, B. 1982. Ribonucleic acid precursors are associated with the chick oviduct nuclear matrix, *Biochemistry* 21:4945–4953.
6. Dynan, W. S., and Tijan, R. 1983. The promoter specific transcription factor Sp1 binds to upstream sequences in the SV40 early promoter, *Cell* 35:79–87.
7. Fasel, N., Pearson, K., Buetti, E. and Diggelman, H. 1982. The region of mouse mammary tumor virus DNA containing the long terminal repeat includes a long coding sequence and signals for hormonally regulated transcription, *EMBO J.* 1:3–7.
8. Gorski, J., Welshons, W., and Sakai, D. 1984. Remodeling the estrogen receptor model, *Mol. Cell. Endocrinol.* 36:11–15.
9. Grody, W. W., Schrader, W. T., and O'Malley, B. W. 1982. Activation, transformation and subunit structure of steroid hormone receptors, *Endocrinol. Rev.* 3:141–163.

10. Groner, B., Hynes, N. E., Rahmsdorf, U., and Ponta, H. 1983. Transcription initiation of transfected mouse mammary tumor virus LTR DNA is regulated by glucocorticoid hormones, *Nucleic Acids Res.* 11:4713–4725.

11. Groner, B., Ponta, H., Beato, M., and Hynes, N. E. 1983. The proviral DNA of mouse mammary tumor virus: its use in the study of the molecular details of steroid hormone action, *Mol. Cell. Endocrinol.* 32:101–116.

12. Gruss, P. 1984. Magic enhancers, *DNA* 3:1–5.

13. Huang, A. L., Ostrowski, M. C., Berard, D., and Hager, H. C. 1981. Glucocorticoid regulation of the Ha Mu SV p21 gene conferred by sequences from mouse mammary tumor virus, *Cell* 27:245–255.

14. Hynes, N. E., Groner, B., Sippel, A. E., Ngyen Han, M. C., Schütz, G. 1977. A comparison of total mRNA complexity and ovalbumin, ovomucoid, and lysozyme mRNA content in the oviduct of laying hens and estradiol-stimulated and withdrawn chicks, *Cell* 11:923–932.

15. Hynes, N. E., Kennedy, N., Rahmsdorf, U., and Groner, B. 1981. Hormone responsive expression of an endogenous proviral gene of mouse mammary tumor virus after molecular cloning and gene transfer into cultured cells, *Proc. Nat. Acad. Sci. U.S.A.* 78:2038–2042.

16. Hynes, N. E., van Ooyen, A., Kennedy, N., Herrlich, P., Ponta, H., and Groner, B. 1983. Subfragments of the large terminal repeat cause glucocorticoid responsive expression of mouse mammary tumor virus and of an adjacent gene, *Proc. Nat. Acad. Sci. U.S.A.* 80:3637–3641.

17. Karin, M., Haslinger, A., Holtgreve, H., Richards, R. T., Krauter, P., Westphal, H. M., and Beato, M. 1984. Characterisation of DNA sequences through which cadminum and glucocorticoid hormones induce human metellothionein-IIA gene, *Nature (London)* 308:513–519.

18. Kennedy, N., Knedlitschek, G., Groner, B., Hynes, N. E., Herrlich, P., Michalides, R., and van Ooyen, A. 1982. The long terminal repeats of an endogenous mouse mammary tumor virus are identical and contain a long apon reading frame extending into adjacent sequences, *Nature (London)* 295:622–624.

19. Khoury, G., and Gruss, P. 1983. Enhancer elements, *Cell* 33:313–314.

20. Lee, F., Mulligan, R., Berg, P., and Ringold, G. 1981. Glucocorticoids regulate expression of dihydrofolate reductase cDNA in mouse mammary tumor virus chimeric plasmids, *Nature (London)* 294:228–232.

21. Majors, J., and Varmus, H. E. 1983. A small region of the mouse mammary tumor virus long terminal repeat confers glucocorticoid hormone regulation on a linked heterologous gene, *Proc. Nat. Acad. Sci. U.S.A.* 80:5866–5870.

22. McKnight, S. L., Kingsbury, R. C., Spence, A., and Smith, M. 1984. The distal transcription signals of the Herpes virus tk gene share a common hexanucleotide control sequence, *Cell* 37:253–262.

23. Moen, R. C., and Palmiter, R. D., 1980. Changes in hormone responsiveness of chick oviduct during primary stimulation with estrogen, *Develop. Biol.* 78:450–463.

24. O'Malley, B. 1984. Steroid hormone action in eukaryotic cells, *J. Clin. Invest.* 74:307–312.

25. Palmiter, R. D., Moore, P. B., and Mulvihill, E. R. 1976. A significant lag in the induction of ovalbumin messenger RNA by steroid hormones: A receptor translocation hypothesis, *Cell* 8:557–572.

26. Parker, C. S., and Topol, J. 1984. A Drosophila RNA polymerase II transcription factor binds to the regulatory site of an hsp 70 gene, *Cell* 37:273–283.

27. Pfahl, M., McGinnis, D., Hendricks, M., Groner, B., and Hynes, N. E. 1983. Correlation of glucocorticoid receptor binding sites on MMTV proviral DNA with hormone inducible transcription. *Science* 222:1341–1343.

28. Pietras, R. J., and Szego, C. M. 1984. Specific internalisation of estrogen and binding to nuclear matrix in isolated uterine cells, *Biochem. Biphys. Res. Commun.* 123:84–91.

29. Ponta, H., Kennedy, N., Skroch, P., Hynes, N. E., and Groner, B. 1985. Hormonal response region in the mouse mammary tumor virus long terminal repeat can be dissociated from the proviral promoter and has enhancer properties, *Proc. Nat. Acad. Sci. U.S.A.* 82:1020–1024.

30. Renkawitz, R., Schütz, G., von der Ahe, D., and Beato, M. 1984. Sequences in the promoter region of the chicken lysozyme gene required for steroid regulation and receptor binding, *Cell* 37:503–510.

31. Ringold, G. M., Yamamoto, K. R., Tomkins, G. M., Bishop, J. M., and Varmus, H. E. 1975. Dexamethasone mediated induction of mouse mammary tumor virus RNA: A system for studying glucocorticoid action, *Cell* 6:299–305.

32. Scheidereit, C., and Beato, M. 1984. Contacts between hormone receptor and DNA double helix within a glucocorticoid regulatory element of mouse mammary tumor virus, *Proc. Nat. Acad. Sci. U.S.A.* 81:3029–3033.

33. Schmidt, T. J., and Litwack, G. 1982. Activation of the glucocorticoid receptor complex, *Physiol. Rev.* 62:1131–1192.

34. Schöler, H. R., and Gruss, P. 1984. Specific interactions between enhancer containing molecules and cellular components, *Cell* 36:403–411.

35. Spelsberg, T. C., Gosse, B. J., Littlefield, B. A., Toyoda, H., and Seelke, R. 1984. Reconstitution of native-like nuclear acceptor sites of the avian oviduct progesterone receptor: Evidence for involvement of specific chromatin proteins and specific DNA sequences, *Biochemistry* 23:5103–5113.

36. Varmus, H. E. 1983. Retroviruses, in: *Mobile Genetic Elements* (J. A. Shapiro, ed.), pp. 411–503, Academic Press, New York.

37. Wigler, M., Silverstein, S., Lee, U. S., Pellicer, A., Chang, Y. C., and Axel, R. 1977. Transfer of purified Herpes virus thymidine kinase gene to cultured mouse cells, *Cell* 11:223–232.

38. Yamamoto, K. R., and Alberts, B. M. 1976. Steroid receptors: elements for modulation of eukaryotic transcription, *Annu. Rev. Biochem.* 45:721–746.

Questions for Discussion with the Editors

1. *Please speculate further on the role the nuclear matrix might play in steroid regulated gene expression. Do you favor an active (e.g., gene-specific) or passive (e.g., structural) role?*

The anchorage of genes to the nuclear matrix by, for instance, the steroid receptor complex or in a more general fashion by enhancer interacting molecules is being proposed to accommodate the observation that domains of DNA are affected by enhancers rather than individual gene promoters. An important "*in vivo*" example is the activation of the int-1 and int-2 loci by MMTV in murine mammary tumors.[1] In these examples the transcription of cellular genes in the vicinity of the proviral integration site is being activated over a distance of up to 20 kb. This activation in concert with the proposed model would strongly argue for a passive role of the nuclear matrix association, i.e., the physical proximity to the "transcription domain" is sufficient to activate gene transcription and no gene specificity is observed. A definitive answer to this question, however, will require investigations to determine how general the association of nascent transcripts with the nuclear matrix is and how this association can be correlated with the presence of enhancers in the vicinity of these genes.

2. *How metabolically stable do you expect are the components that constitute the "predisposed" state? Do you imagine that the primary stimulation of hormone responsiveness in an individual cell must constantly be renewed, or do you visualize that stimulation as being a stably differentiated characteristic?*

Further speculation on this subject would make us distinguish between the "predisposed" state and the "differentiation-specific" gene expression. Our prejudice would say that the "predisposed state" is a stable characteristic of a cell that has responded to a variety of transient differentiation signals (i.e., the continued presence of these signals is not required to maintain this state). Evidence for this notion would come from the observation that many differentiation-specific characteristics or potentials are maintained in tissue culture. Especially certain established, transformed cells seem to have a "memory" for their tissue and cell type of origin. *In vivo*, the "primary stimulation" of, for example, chicken oviduct cells can be distinguished clearly from "secondary stimulation" with respect to the time required for a response and the steroid hormone specificity. Potentials to express genes can therefore be stably maintained in the absence of the inducer for specific transcription. Experiments with hepatoma cells into which albumin gene chimeras have been introduced argue that the potential for cell type-specific gene expression is controlled by a single gene.[2] This single gene product could constitute a nuclear matrix binding site or an enhancer interacting protein in our model.

It is, however, well known that differentiation-specific gene expression can be lost rapidly upon culturing of primary cells. For example, mammary gland cells rapidly lose prolactin receptor upon *in vitro* cultivation and egg white proteins in chicken oviduct cells are only inducible by steroid hormones for a short period after cultivation.[3] These observations could be explained by the necessity of interaction of several discrete DNA elements, which are simultaneously responsible for differentiation-specific gene expression. Only the interaction of the modular regulatory sequences and their binding proteins would allow gene expression. This "coding of regulation" might be composed of stable and unstable elements, i.e., if only

[1]Nusse R., Van Ooyen A., Cox D., Fung Y. K. T., and Varmus H. E. 1984. Mode of proviral activation of a putative mammary oncogene (int-1) on mouse chromosome 15, *Nature (London)* 307: 131–136.
Peters G., Lee A. E., and Dickson C. 1984. Activation of a cellular gene by mouse mammary tumor virus may occur early in mammary tumor development, *Nature (London)* 309:273–275.
[2]Ott, M. O., Sperling L., Herbomel P., Yaniv M., and Weiss M. C. 1984. Tissue-specific expression is conferred by a sequence from the 5′ end of the rat albumin gene, *EMBO J.* 3:2505–2510.
[3]Renkawitz R., Schütz G., von der Ahe, D., and Beato, M. 1984. Sequences in the promoter region of the chicken lysozyme gene required for steroid regulation and receptor binding, *Cell* 37:503–510.

one of the required components is missing gene expression would stop. It can, however, be resumed rapidly by provision of this element since the other elements (which define the "predisposed" state) are stable. What is required for a definitive answer? A definition of the proposed "discrete, interacting" elements, their number and mode of interaction are important questions for the future study of control of gene expression.

Expression of Heterologous Genes

CHAPTER **8**

Egg White Genes: Hormonal Regulation in Homologous and Heterologous Cells

Michel M. Sanders and G. Stanley McKnight

UNDERSTANDING THE REGULATION of eukaryotic gene expression remains one of the major challenges of modern biology. This includes defining how alterations in hormonal, dietary, and environmental status control the selective activation or repression of genes during development and cellular differentiation. As an integral part of this, the objective of the research described in this chapter is to identify the molecular mechanisms by which steroid hormones regulate the transcriptional activity of target genes.

The current hypothesis for the mechanism of action of steroid hormones proposes that the steroid first binds to its receptor and allosterically increases the affinity of the receptor for nuclear binding sites. The steroid-receptor complex is then thought to interact with specific DNA or chromatin sites to elicit the steroid-specific response. The net effect of steroid action is an alteration in the biosynthesis of specific proteins, triggered primarily by changes in the rate of transcription of their genes. An implicit assumption of this model is that the receptor-DNA (chromatin) interaction is the primary event in the induction of transcription of a particular gene. The binding of the steroid-receptor complex in the vicinity of a gene is then thought to facilitate attachment of the transcriptional machinery and the initiation of RNA synthesis. Experimental examination of this hypothesis has generated new insights into the molecular events involved in the induction of specific genes by steroids; some of these will be discussed in detail in this chapter.

The chick oviduct has been exploited extensively to examine the hormonal regulation of eukaryotic gene expression. As a result, it probably represents the best characterized nonviral model system for studying the mode of action of steroid hormones. Considerable information exists about the transcriptional regulation of the egg white genes by four classes of steroids and about the organization and structure of these genes. However, the molecular pathways connecting the binding of steroid to its receptor and gene activation remain elusive. In this chapter we have attempted to review (a) information acquired within the last 5 years on the biological regulation and organization of the endogenous egg white genes, and (b) experiments designed to define specific DNA sequences that play a role in the regulation by steroid hormones.

Biological Regulation

The biosynthesis of the major egg white proteins (ovalbumin, transferrin, ovomucoid, and lysozyme) constitutes about 80 percent of the total protein synthesis of the tubular gland cells in the oviduct of the laying hen. For ovalbumin, this represents about 3×10^{19} molecules synthesized per day. As will be discussed in more detail below, transcriptional and posttranscriptional events apparently regulate the extent and specificity of this response to complex hormonal stimuli.

Regulation by Steroid and Peptide Hormones

The initial or primary induction of egg white mRNAs and their resultant proteins in immature chicks requires estrogen and is accompanied by the proliferation of tubular gland cells in the magnum portion of the oviduct [(46) and references therein]. Upon withdrawal of estrogen, the tubular gland cells regress and the synthesis of egg white proteins ceases (46). The secondary administration of estrogens, progestins, androgens, or glucocorticoids [(30) and references therein] reinitiates ovalbumin and transferrin (also referred to as conalbumin or ovotransferrin) production, although estrogen is still required for proliferation of the tubular gland cells.

Until recently, most of the studies on the hormonal regulation of the egg white genes were done in whole animals. These indicated that estrogen is sufficient for the primary and secondary induction of these genes. However, with the development of a method for culturing tissue explants from hormone-withdrawn oviduct (30), it became apparent that the permissive effect of a nonsteroidal component of serum is required to obtain a significant induction of the ovalbumin gene (16). This component was subsequently identified as insulin (15); estrogen can increase the rate of transcription of the ovalbumin gene only when insulin is present (Fig. 1A and B). Other insulinlike peptides such as proinsulin and multiplication-stimulating activity are 10-fold less potent than insulin. The induction of the ovomucoid gene also requires insulin (Fig. 1E); however, at least in these relatively short incubations, the transferrin gene does not (Fig. 1C and D). This indicates that insulin is not exerting its effects by altering the functioning of the estrogen receptor. Thus, insulin ap-

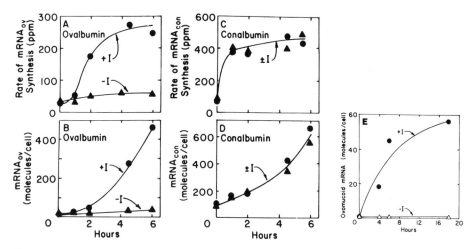

Figure 1 Effects of insulin on transcription rates and accumulation of egg white mRNAs. Withdrawn oviduct tissue fragments were incubated for the indicated times with 10^{-7} M estradiol in the presence or absence of insulin (T) ($5\mu g/ml$) (16).

parently generates a signal to specifically regulate the induction of the ovalbumin and ovomucoid genes.

Surprisingly, 8-bromo-cAMP (8Br-cAMP) and agents that elevate endogenous cAMP mimic the permissive effects of insulin on the induction of the ovalbumin gene (Table 1) (15). These data suggest that both cAMP and insulin affect some process that is absolutely required for the estrogen-induced increase in the transcription of the ovalbumin gene. By investigating the effects of cycloheximide on the induction of the ovalbumin and transferrin genes, we demonstrated that the convergent effects of insulin and cAMP are not due to a generalized increase in protein synthesis (15). However, a role for the synthesis of a specific protein can not be ruled out. Interestingly, 8Br-cAMP and forskolin, a reversible activator of adenylate cyclase, initiate a slower increase in the rate of transcription of the ovalbumin gene in response to estrogen than does insulin. These differences in the kinetics of the response imply that insulin and cAMP are acting through alternate pathways to achieve the same result. For example, the interaction of insulin with its receptor, which is a tyrosine-specific protein kinase, may trigger a phosphorylation event(s) that is required for the subsequent induction of the ovalbumin gene by steroid hormones. A cAMP-dependent protein kinase may either play a direct role in these events or impinge on the pathway at some point to mimic the effects of insulin (15).

That the oviduct truly represents a multihormonal regulatory system with all the associated complexities this entails became evident during the development of a cell culture system from hormone-withdrawn oviduct (45). With both estrogen and insulin in the culture medium, the amount of ovalbumin mRNA increases only 2-fold to 8-fold during the first 2 days in culture. As

TABLE 1 Effects of cholera toxin and forskolin on the accumulation of ovalbumin mRNA

	Ovalbumin mRNA (molecules per cell)	
Incubation condition	Control	+100 mM Estradiol
No addition	10	53
Insulin (5 μg/ml)	40	772
8Br-cAMP (500 μM)	43	625
Cholera toxin (2 μg/ml)	36	406
Forskolin (33 μM)	22	1049

Note: Withdrawn oviduct tissue fragments were incubated with various compounds as indicated, and ovalbumin mRNA was measured by cDNA hybridization.

shown in Fig. 2, a second steroid (corticosterone, progesterone, or testosterone) is required to induce ovalbumin and mRNA to levels achieved *in vivo* (100-fold or more increase). This is also true of the transferrin gene (data not shown). Although this is in contrast to our data with tissue explants (15, 16, 30), it seems probable that the tissue chunks retain enough residual steroid to enhance the response to estrogen during the 8-hour incubations. In support of this interpretation, tissue explants from estrogen-primed chicks require corticosterone in addition to estrogen and insulin to maintain ovalbumin mRNA levels for more than 20 hours (G. Parrish and G. S. McKnight, unpublished results). Thus, the tissue explants and the cultured oviduct cells are probably very similar in their hormonal requirements.

Dose-response analyses (Fig. 2) revealed that suboptimal concentrations of corticosterone, progesterone, and androgen synergize with estrogen to induce ovalbumin mRNA in the oviduct cell cultures. Normal circulating concentrations of corticosterone (the major glucocorticoid in chickens) range from 10^{-8} to 4×10^{-8} M and since 10^{-8} M corticosterone substantially enhances the estrogen-induced increase in ovalbumin mRNA, we believe that endogenous corticosterone facilitates the induction of ovalbumin mRNA when estrogen is administered *in vivo*. Thus, the induction of the ovalbumin gene is considerably more complicated than originally thought from the studies in which the chickens received only estrogen. The transcriptional activity of the ovalbumin gene is regulated by a second steroid (45), probably corticosterone *in vivo*, and by insulin in addition to estrogen (15).

Role of Protein Synthesis and Acetylation

Although the molecular events involved in the increase in the rate of transcription of the egg white genes are poorly understood, it is known that protein synthesis is required (30, 32). When inhibitors of protein synthesis are

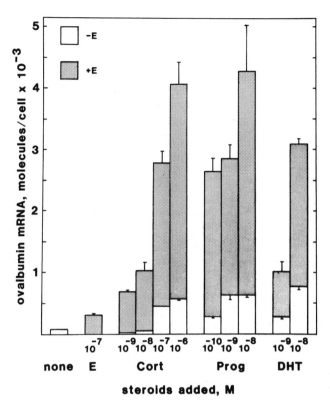

Figure 2 Requirement for a second steroid in addition to estradiol (E) for the induction of ovalbumin mRNA. Oviduct cells were cultured for 44 hours with insulin (50 ng/ml) and steroid hormones as indicated (45).

added to tissue explants, the induction of both ovalbumin and transferrin mRNAs is prevented without affecting the accumulation of nuclear estradiol receptors. If the inhibitor is removed, then induction of both genes occurs after the characteristic lag period. These results suggest that one or more intermediary proteins with short half-lives couple the binding of the steroid-receptor complex to activation of transcription.

Butyrate and other inhibitors of histone deacetylation also block the steroid-mediated induction of the egg white genes and terminate ongoing transcription (32). This inhibition is probably not due to effects on the accumulation of nuclear estrogen receptors, on total protein synthesis, or on total RNA synthesis. The inhibition by butyrate occurs within 30 minutes of administration and is reversed when the drug is removed. One interpretation of these results is that the intermediary protein(s) affected by inhibitors of protein synthesis is inactivated by acetylation. Alternatively, other accessory proteins such as steroid receptors or chromosomal proteins are plausible sites for covalent modification and for regulation via acetylation.

Differential Regulation

The transcriptional regulation of the ovalbumin and transferrin genes has been examined extensively, and some intriguing differences exist. As discussed above, induction of the ovalbumin gene requires the permissive effects of insulin, but induction of the transferrin gene does not, at least in short-term incubations. In addition, transcription of the transferrin gene occurs with virtually no lag after administration of estrogen, and the amount of transferrin mRNA synthesized is directly proportional to the number of nuclear estrogen receptors [(40) and references therein]. In contrast, transcription of the ovalbumin gene follows a 2-hour lag and is exponentially related to the number of nuclear estrogen receptors (40). In explant cultures, estrogen and progesterone act synergistically to increase transcription of the ovalbumin gene, yet progesterone has a transient inhibitory effect on the estrogen-stimulated transferrin gene (30, 40). This differential regulation of the two major egg white genes by the same steroids implies that distinct mechanisms are involved that allow each gene to respond individually to activated steroid receptors.

By analyzing differences in the response of the egg white genes to steroid hormones, information may be obtained about the mechanisms involved. Palmiter et al. (40) correlated the number of nuclear estrogen receptors with the rates of transcription of the ovalbumin and transferrin genes following administration of estrogen to chickens (Fig. 3). Transcription of the ovalbumin gene is only about 12 percent of maximum when nuclear estrogen receptors are half-maximal and receptors must reach 90 percent of maximum to get half-maximal transcription. A polynomial expression was derived to describe the relationship between the relative rate of transcription and the number of nuclear estrogen receptors. From this formulation, it appears that five estrogen receptor binding sites must be filled on the ovalbumin gene for complete induction (Fig. 3). Based on this model, filling one receptor binding site stimulates a low level of induction (about 15 percent of maximum) and no additional increase is observed until all five sites are filled, after which transcription becomes maximal. In contrast, a single site must be activated for induction of the transferrin gene.

This interpretation may explain differences in the lag times after stimulation with estrogen and in the disparate effects of progesterone and estrogen on the two genes. Because nuclear estrogen receptors are limiting (40), if the induction of the transferrin gene only requires that one receptor binding site be filled, then its induction may proceed faster than that for the ovalbumin gene where five sites must be filled. In addition, if both estrogen and progesterone compete for the single receptor binding site in the transferrin gene, then these steroids might have different effects on transcription due to different affinities and capacities to activate transcription. Progesterone may have an inhibitory effect on the transferrin gene because it competes with estrogen for the binding site yet cannot activate the gene as efficiently as estrogen. In contrast, progesterone may enhance the ability of estrogen to induce the ovalbumin

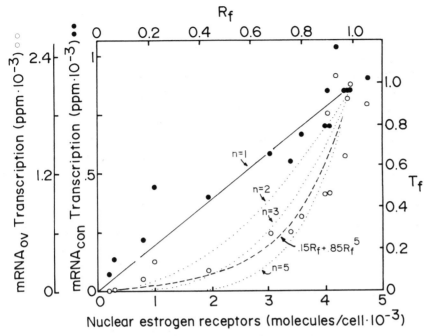

Figure 3 Correlation of nuclear estrogen receptor levels with rates of ovalbumin and conalbumin gene transcription. The *solid* and *dotted* lines are theoretical curves based on the equation $T_d = R_f^n$ where T_f is the fractional rate of transcription, R_f is the fractional nuclear receptor concentration, and n is the number of binding sites (40).

gene because it might help fill adjacent sites. Testing this hypothesis requires identification of the steroid-receptor binding sites for both estrogen and progesterone and analysis to test their functional relevance. Aspects of ongoing research that address these problems will be discussed later in this chapter.

Structure and Organization

Understanding the molecular mechanisms involved in differential gene activation in eukaryotic cells requires an intimate knowledge of the correlation between the structural organization of the DNA in chromatin and its resultant functional properties. Elucidation of the sequences of specific genes and their putative control regions is an integral part of this. The genes for the four major egg white proteins have been cloned and for the most part sequenced [(2, 3, 19, 24, 52) and references therein].

The Genes

The ovalbumin gene and its flanking sequences have been cloned and the structure has been extensively analyzed. As depicted in Fig. 4, the 7.6 kilobase (kb) gene is interrupted by seven introns and codes for a mature mRNA of 1.9 kb (52). Thus, intervening sequences comprise about 76 percent of the gene. The first exon, also called the leader, encodes most of the 5'-untranslated region of the mRNA. A region of 637 nucleotides separates the termination codon for translation from the poly(A) tail (35). As is the case with other eukaryotic genes transcribed by RNA polymerase II, the TATA box defines site-specific initiation of transcription of the ovalbumin gene (54). The processing of the transcript involves the generation of a series of intermediates that can occur through multiple pathways rather than by splicing in a specific temporal order (4).

Although ovalbumin is the major secretory protein from chicken oviduct, it lacks the transient NH_2-terminal signal sequence found on all other secretory proteins studied thus far. Kinetic experiments in a cell-free translation

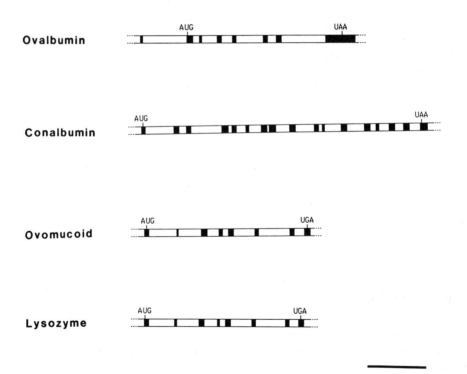

Figure 4 Structure of the four major egg white genes. Dark bands indicate the positions of the exons. The start (AUG) and the stop sites for translation (UAA or UGA) are indicated. [Compiled from (3, 19, 24, 52) and references therein.]

system revealed that translation of 50 to 60 amino acid residues is sufficient to bind ovalbumin-synthesizing polysomes to pancreas membranes (36). Because the amino acid residues between 26 and 45 are all either hydrophobic or uncharged, Meek et al. (36) have postulated that they are involved in the interaction of the protein with membranes prior to secretion. This putative signal sequence would thus be located in the second exon.

The ovalbumin gene appears to be part of a larger gene family that includes angiotensinogen, antithrombin III, and α_1-antitrypsin [(14) and references therein]. Sequence analysis, the secondary structure as predicted by computer programs, the lengths of the proteins, and the positions of "invariant" residues indicate that these genes developed from a common ancestral gene. It has been estimated that the genes diverged before or during early vertebrate evolution, more than 500 million years ago. Interestingly, of the four proteins only ovalbumin has no known function; angiotensinogen is a precursor for a very potent hormone (angiotensin I) and the other two are protease inhibitors.

Transferrin is the second major secretory protein of the oviduct. The transferrin gene is divided into 17 exons, is 10.3 kb in length, and codes for a mature mRNA of 2.4 kb (8). The first exon of the transferrin gene encodes the majority of the signal sequence (8). There is a 76-nucleotide 5′-untranslated sequence (8) and a 182-nucleotide 3′-untranslated region (23). Amino acid and mRNA sequence studies suggest that the transferrin gene evolved by duplication of an ancestral gene of seven or eight exons [(23) and references therein]. Two blocks of sequence, from 134 to 934 base pairs (bp) and from 1134 to 1934 bp, are homologous. Highly repetitive sequences are present in the 5′-flanking region and in the second and third introns of the transferrin gene (8).

The transferrin gene in the oviduct appears to be identical to that in the liver (27). In the chicken, the proteins also seem to be the same, differing only in their attached glycan chain. However, the modulation of expression of the gene differs in the two tissues. As discussed above, the transferrin gene in the oviduct is extremely sensitive to estrogen, yet estrogen induces only about a 2-fold increase in the rate of transcription of the liver gene (26). In addition, the liver transferrin gene is regulated by nutritional iron deficiency (33). Thus, the transferrin gene is an exciting model for examining the factors involved in designating cell-specific expression and in defining the extent of response to inducers such as estrogen.

Ovomucoids comprise about 10 percent of the egg white proteins and are responsible for most of the antitrypsin activity of chicken egg white. As shown in Fig. 4, the ovomucoid gene is 5.6 kb and is divided into eight exons [(3) and references therein]. The major mRNA is 0.86 kb or about 15 percent of the gene in length. Several minor functional species exist as the result of two promoters and three polyadenylation sites (18). Transcription initiates at two sites separated by 85 nucleotides, and each is controlled by a TATA box. About 0.5 percent of the ovomucoid mRNAs initiate at −85 bp. Two additional polyadenylation sites are functional at about 170 and 800 bp downstream of the dominant site (18). About 5 percent of the ovomucoid mRNAs from laying hen oviduct are represented by these longer species. The ovomucoid gene was thus the second gene (after the mouse α-amylase 1A gene) discovered where heterogeneity exists in the mRNAs because of multiple promoter and polyadenylation sites. Because the multiple forms of ovomucoid mRNAs are present in the same proportion after hormonal stimulation, the multiplicity of forms is probably not correlated to the hormonal status. Gerlinger et al. (18) speculate that the additional promoter and polyadenylation sites may represent remnants of the evolutionary history of the gene with no particular relevance to present function.

Ovomucoid can be divided into three functional domains, each capable of binding one molecule of trypsin or serine protease [(48) and references therein]. Amino acid and DNA sequence studies indicate that this gene is the result of triplication of an ancestral gene containing only one intron. From analysis of the DNA sequence, Stein et al. (48) propose that the ancestral gene consisted only of exons 7 and 8 intron G.

The lysozyme gene is 3.9 kb and has four exons (Fig. 4). The gene codes for a mature mRNA is about 0.59 kb, which includes a 3' noncoding region of 113 nucleotides (24). As is the case with the ovomucoid gene, heterogeneity exists in the length of the 5' end of the mRNA, with the ends mapping 29, 31, and 53 nucleotides upstream of the initiation codon (19). At least five families of repeated sequences are present in or near the lysozyme gene (1). Most notable is an inverted repeat located about 1 kb upstream of the cap site and in the second intron. Baldacci et al. (1) postulate that the inverted repeats were involved in creating the gene from a piece inserted between the repeat and the other exons, either by transposition or by viruses.

The genes for ovalbumin, ovomucoid, and transferrin are located on different chromosomes (22). This implies that separate regulatory sites must be present for each gene. Because these genes are all regulated by steroid hormones, it was thought that comparison of the DNA sequences flanking the genes would be invaluable for identifying which regions interact with steroid hormone receptors. However, sophisticated analysis using computer programs has revealed few homologies among the genes. So far, only one region of partially conserved sequence has been identified, at about −140 bp in the ovalbumin, lysozyme, and transferrin genes (19).

Two other chicken genes, designated X and Y, of unknown function but containing sequences homologous to the ovalbumin gene, were discovered while attempting to clone the 5'-flanking region for the ovalbumin gene [(9, 20, 28) and references therein]. The three genes are oriented in the same direction in the order 5'-X-Y-ovalbumin-3' with about 8.7 kb between the Y and ovalbumin genes and 5.4 kb between the X and Y genes (Fig. 5). Because the X, Y, and ovalbumin genes are each composed of eight exons, the genes are closely linked, and the sequence homology of the exons is extensive, these genes probably arose by gene duplication (20). The three genes are similar in length with the X gene being 7.88 kb and the Y gene 6.44 kb. The first seven exons of the X and Y genes are the same lengths as those for ovalbumin, but the last exon in the X gene is about 500 bp longer than the other two, reflecting a difference in length of the 3' untranslated region. The introns, with the

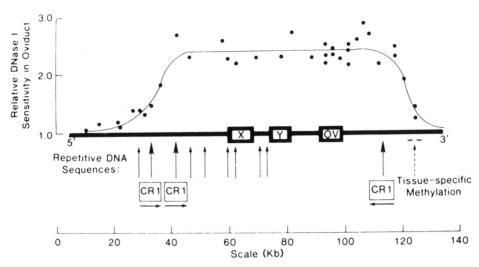

Figure 5 Summary of the ovalbumin gene domain. The solid circles indicate the relative DNase I sensitivity in oviduct nuclei of DNA sequences at various positions in the ovalbumin gene domain. The solid vertical arrows denote the locations of repetitive DNA sequences within the domain. The bolder vertical arrows point out the specific locations of CR1 family repetitive DNA sequences. The orientations of these CR1 sequences are denoted by the horizontal arrows. The dashed arrow indicates a region where distinctive methylation patterns are noted for different chick tissues and where oviduct is specifically undermethylated [Reprinted with permission from Stumph et al. *Biochemistry* 22:306–315. Copyright 1983. American Chemical Society (50).]

exception of intron A, differ greatly in length and sequence among the three genes indicating a more rapid evolutionary divergence than for the exons. However, by comparing sequences in the regions of exons 5, 6, and 7, and by S_1 nuclease mapping of the 5' ends of these exons, it was possible to show exact correspondence in the location of the exon-intron junctions between the ovalbumin and X genes (20).

The X and Y genes are transcriptionally active in stimulated chick oviduct, although the concentrations of the X and Y transcripts are only about 1, and 6 to 10 percent, respectively, of that for ovalbumin (9). Measurements of the relative transcription rates for the three genes suggest that the differences in amounts of their transcripts reflect mostly differential rates of transcription rather than turnover of the mRNAs. Progesterone is as potent as estradiol in inducing the ovalbumin and Y mRNAs but has only a weak effect on expression of the X gene. In addition, dexamethasone barely affects the accumulation of the mRNAs for both the X and Y genes when compared to that for ovalbumin. While no functional role in developing or mature oviduct tissue has been ascribed to the X and Y genes, the observations that the members of this gene family exist in a similar environment, are extensively homologous, and are regulated at the level of transcription is of considerable interest because the rates of transcription in response to steroids are so different. However, sequence homology among the three genes persists for only about 30 nucleotides upstream of the TATA sequence (21).

The X and Y mRNAs are polyadenylated and their presence in laying-hen oviduct polysomes indicates that they are expressed as proteins (28). Translation of X and Y mRNAs in an *in vitro* translation system produces bands larger than ovalbumin, but the proteins are immunoprecipitated by antibodies to ovalbumin. By comparing the sequences of exons 5, 6, and 7 for the X and ovalbumin genes, a 232 amino acid sequence for the C-terminal end of the protein was derived with a calculated homology of 67 percent (20). By using standard assumptions for the rate of evolution of proteins that share a common ancestor, the estimated time of divergence of the X and ovalbumin genes is 55 million years.

Chromatin Structure

Detailed molecular knowledge of the structural conformation of chromatin is thought to be essential for an understanding of gene commitment during differentiation and development and during gene activation by inducers such as steroid hormones. Evidence from a number of sources suggests that genes having transcriptional potential exist in an altered conformation that renders them more susceptible to digestion by endonucleases than genes that are never expressed. This preferential sensitivity presumably reflects a modification in the packaging or the configuration of the nucleosomes in those regions of the chromatin that contain active genes.

Several types of experiments indicate that it is not transcription per se that is responsible for the increased sensitivity to endonuclease. Studies employing pancreatic deoxyribonuclease I (DNase I) revealed that the conformation of chromatin that is recognized selectively by this enzyme is retained even after transcriptional activity ceases, and it includes flanking and intervening sequences as well as coding sequences. The sensitivity of specific genes appears to be correlated more with the differentiated state of the cell rather than with transcriptional activity since the acquisition of sensitivity can precede expression in some developmental programs (47). In addition, genes that are transcribed at different rates are equally sensitive to selective degradation by DNase I. These data are consistent with the concept that sensitivity to endonucleases may be a necessary but not a sufficient condition for transcription.

The ovalbumin gene in laying-hen or in estrogen-stimulated chick oviduct chromatin is in a conformation that is selectively cleaved by micrococcal nuclease (MNase) and DNase I[(25) and references therein]. The ovalbumin genes in liver, erythrocytes, and other tissues are not preferentially recognized by these enzymes, indicating that recognition is related to the transcriptional activity of the gene. As depicted in Fig. 5, this generalized sensitivity to nuclease extends to the entire 100-kb ovalbumin domain, including the structurally related X and Y genes, the spacer DNA, and 20 kb of flanking DNA at both the 5'- and 3'-termini [(49, 50) and references therein]. At each end of the domain, a gradual transition from nuclease sensitivity to nuclease insensitivity occurs over 10 kb of DNA.

In order to determine what signals might be involved in defining or establishing the sensitivity of the ovalbumin domain to nucleases, the DNA in the transition region was examined by Stumph et al. (49). They identified a family of dispersed repetitive DNA sequences, termed CR1 sequences, located near or within the transition regions of nuclease sensitivity (Fig. 5). In addition, the CR1 sequences found at opposite ends of the domain exist in inverse orientation with respect to each other. Stumph et al. (49) suggest that the CR1 sequences may play a role in defining the ends of the nuclease-sensitive ovalbumin gene domain. To determine if the CR1 sequences have any functional significance, it will be necessary to ascertain if inverted repeats analogous to the CR1 sequences are associated with transition regions in other nuclease-sensitive domains.

The data on nuclease sensitivity discussed above are derived from experiments in which the DNA is exhaustively degraded. In addition to this generalized sensitivity, smaller regions of about 10-fold greater sensitivity, called DNase I hypersensitive sites, have been observed at specific locations in chromatin. These hypersensitive sites are frequently located within the 5'-flanking region of active genes and apparently represent highly localized alterations in chromatin structure. These sites probably represent local interruptions in the regular nucleosomal array and may identify regions where regulatory proteins gain access to signal sequences in the DNA and in this way play a role in the control of gene expression. Like the more generalized nuclease sensitivity described above, hypersensitive sites can reflect the potential of a gene to become active rather than actual transcriptional activity. Genes subject to complex regulatory control during development or to regulation by hormones generally have a more complicated pattern of DNase I hypersensitivity, with many sites upstream, downstream, and within the gene.

DNase I hypersensitive sites have been identified in the 5'-flanking region of the lysozyme gene [(17) and references therein]. These hypersensitive sites are centered at 2.4, 1.9, and 0.1 kb upstream of the promoter in the gene from estrogen-stimulated chicken oviduct. Functional correlates have been ascribed to some of these sites. For example, the site at −1.9 kb disappears upon withdrawal of estrogen and reappears upon administration of estrogen, implying that this site is associated with the activation of transcription of this gene by steroid hormones. The site at −0.1 kb is only present in cells that either express the lysozyme gene or that have the potential to do so, suggesting that this region specifies the transcriptional state of the gene. In chicken macrophages, which constitutively express the lysozyme gene at a low rate, neither site identified in the oviduct at −2.4 or −1.9 kb is present. Instead, two alternate sites at −2.7 and −0.7 kb were detected. Thus, two distinct modes of transcriptional regulation, constitutive and inducible, correlate with alterations in chromosome structure. These data support the contention that the structure of chromatin determines the activity of the gene.

Similarly, Kaye et al. (25) identified four regions of DNase I hypersensitivity upstream of the cap site in the ovalbumin gene in oviduct but not in erythrocyte nuclei. These are centered at -0.15, -0.8, -3.2, and -6 kb. MNase also cleaves oviduct DNA in these same regions, corroborating the hypothesis that the structure of chromatin is altered in these sites. However, functional correlates are not yet available for these hypersensitive sites.

Kay et al. (25) speculate that one or more of the hypersensitive regions located upstream in the ovalbumin gene are associated with an enhancer element that may be controlled by the binding of steroid receptors. This seems probable by analogy to the mouse mammary tumor virus (MMTV) system. Zaret and Yamamoto (53) demonstrated that a specific DNase I hypersensitive site is induced in MMTV DNA by dexamethasone, a synthetic glucocorticoid, with a time course that parallels the rate of increase of transcription of the gene. Moreover, the induced hypersensitive region coincides exactly with the region that binds glucocorticoid receptor *in vitro* and that acts as a steroid-dependent enhancer element *in vivo*. These investigators speculate that the interaction of steroid-receptor complexes with DNA initiates alterations in the structure of the DNA or chromatin near the binding sites to generate an active transcriptional enhancer. To ascertain if the steroid-induced hypersensitive site at -1.9 kb in the lysozyme gene, for example, is a steroid-dependent enhancer, experiments must be done to determine if the site overlaps with a receptor binding site, appears rapidly in response to steroids, and confers hormone-responsive expression when linked to another promoter.

Association with the Nuclear Matrix

Another level of organization of DNA involves interactions with the nuclear matrix. This matrix is a fibrillar network consisting primarily of protein and large supercoiled loops of DNA, and it is prepared by extraction of nuclei with high salt and/or detergent. The DNA of transcriptionally active genes appears to be arranged nonrandomly in supercoiled loops anchored to the nuclear matrix (41). These observations imply that the nuclear matrix may be involved in the regulation or in the physical mechanics of transcription.

Using restriction endonucleases to cleave DNA loops anchored to the nuclear matrix, Robinson et al. (44) demonstrated that the ovalbumin and transferrin genes are bound to the nuclear matrix from oviduct cells but not to that from brain. After cleaving the attached DNA into fragments ranging in size from 2.5 to 5 kb with a restriction endonuclease, several fragments from the ovalbumin gene are independently bound to the nuclear matrix. One of these fragments extends from approximately -2.5 to ± 1 kb while others are in the central and 3'-flanking regions. Following withdrawal of estrogen, significantly fewer fragments from the ovalbumin gene are associated with the nuclear matrix, suggesting that attachment may be reversible. Robinson et al. (44) speculate that steroid receptors mediate the association of the egg white genes with the nuclear matrix. Ciejek et al. (7) confirmed these results and also demonstrated that only the actively transcribed portions of the 100-kb ovalbumin gene domain, including the X and Y genes, are selectively associated with the nuclear matrix. These results suggest that several sites in the ovalbumin gene may have properties that allow the gene to preferentially attach to the nuclear matrix and thus be actively transcribed.

More recently, by using a low-salt procedure to extract the nuclear matrix Mirkovitch et al. (38) have identified specific, relatively small fragments from the histone gene clusters and the heat-shock genes that bind to the matrix. However, none of these specific fragments are from actively transcribed regions of these genes; all are in A + T−rich tracts in spacer DNA or upstream of the genes. By comparing the fragments attached after extraction with either high or low salt, they conclude that transcription-dependent association of active genes with the nuclear matrix may be the result of the high-salt treatment, which may cause sliding of the DNA or precipitation of transcription complexes onto the matrix. Thus, by extracting the nuclear matrix with low salt, it may be possible to define specific sequences of DNA within the ovalbumin and transferrin genes that serve as attachment sites. These sites may therefore designate functional domains in the eukaryotic genome rather than specify which genes are being actively transcribed.

Binding Sites for Steroid Receptors

Steroid hormones are thought to exert their effects through the binding of steroid-receptor complexes to specific DNA sequences in the hormone-responsive gene. Until fairly recently, specific binding of steroid-receptor complexes could not be detected over a high background of nonspecific binding. However, by using relatively small fragments of genes, a number of groups have demonstrated that steroid receptor complexes can bind to specific regions of target genes.

Mulvihill et al. (39) observed specific binding of crude or partially purified oviduct progesterone receptor to DNA fragments isolated from the ovalbumin, transferrin, ovomucoid, X, and Y genes in a competitive filter binding assay (Fig. 6). Many of these competing DNA fragments are from the 5′-flanking region. To define more specifically the sequences involved in binding the receptor, deletion mutants of the ovalbumin gene were constructed. Analysis of several fragments excised from these mutants identified a specific region 250 to 300 bp upstream from the mRNA start site that preferentially binds progesterone receptor in this assay. A degenerate consensus sequence of 19 bp (ATC$_{tt}^{cc}$ATT$_t^c$TCTG$_t^c$TTGTA) is present at least once in all the fragments that bind receptor from the five steroid responsive genes. In the 5′-flanking region of the ovalbumin gene, seven binding sites were found within 1 kb of the start site of transcription. Only one site was found in 250 bp of 5′-flanking DNA for the transferrin gene. While incomplete, this information correlates with the prediction obtained from the data in Fig. 3 (i.e., that more receptor binding sites exist in the 5′-flanking region of the ovalbumin gene than in the transferrin gene).

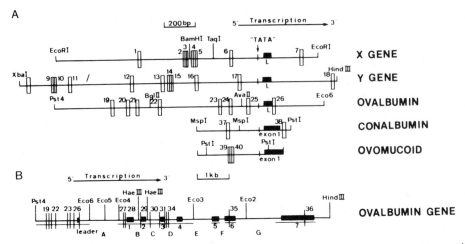

Figure 6 Location of the putative binding sites for progesterone-receptor complex in five hormonally regulated egg white genes. Maps of the sequenced 5'-end flanking regions included in the computer search. The five sequences have been aligned with respect to the TATA box. Open boxes, sequences of homology. L, positions of the leader exon of the X, Y, and ovalbumin genes. The locations of sequences of homology (vertical bars) found within and surrounding the ovalbumin gene are indicated relative to the eight exons (leader, 1 to 7) and seven introns (A to G). Horizontal lines, regions where sequence information is available. [Reprinted with permission from Mulvihill et al. (39). Copyright 1982 MIT.]

Comparable results were observed when a highly purified chicken oviduct progesterone receptor preparation was used in conjunction with a nitrocellulose filter binding assay. Compton et al. (10) identified several fragments in the 5'-flanking region of the ovalbumin, X, and Y genes that preferentially bind receptor. Characterization of a number of overlapping fragments in the 5'-flanking region of the ovalbumin gene revealed that they are all preferentially retained by the receptor when compared to plasmid sequences. However, in contrast to what Mulvihill et al. (39) observed, the strongest receptor binding site seems to be between −247 and −135 bp. A highly A + T−rich (90 percent) 18-bp sequence is present in this region, which is consistent with the documented DNA-binding properties of the receptor A subunit [(10) and references therein]. The A subunit prefers A + T-rich, single-stranded sequences and appears to be a helix-destabilizing protein. These investigators suggest that the receptor might bind with high affinity to a sequence between −247 and −135 bp and destabilize the DNA helix at a proximal A + T−rich region.

While both studies outlined above indicate that the progesterone receptor can bind to specific regions in the 5'-flanking region of the ovalbumin gene, the results are inconclusive as to which are the physiologically relevant sites. Both groups have identified a number of regions that preferentially bind receptor, but the strongest regions described by each do not overlap. Because the two assay systems and receptor preparations differ, comparison of their results is difficult. However, as discussed below, the region of the ovalbumin gene between −197 and −95 bp is required for the induction of an ovalbumin β-globin fusion gene by estrogen and progesterone in cultured oviduct cells (12, 13).

Experiments with the chicken lysozyme gene have provided additional information about sites required for regulation by steroids. Renkawitz et al. (43) assayed both for receptor binding sites on the lysozyme gene *in vitro* and for functional sites *in vivo*. (The data on which sites are functionally relevant will be discussed in a later section.) They prepared a gene constructed by fusing the 5′-flanking region of the lysozyme gene to the structural gene for the Simian virus 40 (SV40) T antigen. The nitrocellulose filter binding assays defined two regions that preferentially bind partially purified glucocorticoid receptor isolated from rat liver. The stronger binding site is between −74 and −39 bp and the weaker between −208 and −164 bp.

Within the stronger binding site the hexanucleotide 5′-AGAACA-3′ is present in the sense strand. This hexanucleotide is found in many glucocorticoid receptor binding sites, including the four sites in the well-characterized mouse mammary tumor virus long terminal repeat (MMTV LTR). It has been suggested that the hexanucleotide plays an essential role in the interaction of the glucocorticoid receptor with DNA; therefore its absence in the weaker binding site in the lysozyme gene is perplexing, especially as the weaker site appears to be the functionally relevant site as discussed below. Perhaps of more significance in the regulation of the lysozyme gene by glucocorticoids is a 10-fold nucleotide sequence, 5′-ATTCCTCTGT-3′, present in both binding sites. Also of interest is the observation that the decanucleotide sequence occurs in opposite orientation and on opposite strands in the two binding sites.

More recently, von der Ahe et al. (51) utilized exonuclease III "footprint" analysis to compare the binding sites in the lysozyme and MMTV promoter regions for partially purified progesterone receptor from rabbit uterus with those for the liver glucocorticoid receptor. With one exception, the sites covered by the two receptors are identical within both promoters. Again, two sites were identified in the lysozyme promoter, one between −80 and −54 bp which is identical for both receptors and the other with a 5′ terminus at −200 bp. The 3′ limit of this other site is at −174 bp for the glucocorticoid receptor and at −160 bp for the progesterone receptor. While the sequences recognized by the two steroid receptors overlap, their relative affinities differ (51). The site between −80 and −54 bp in the lysozyme gene is bound strongly by the glucocorticoid receptor but only weakly by the progesterone receptor. In contrast, the site between −200 and −160 bp shows a higher affinity for the progesterone receptor. Similar differences in the affinity of the two receptors for various binding sites were obtained with the MMTV promoter. Thus, the specificity of a given steroid might be determined by different affinities for binding sites or by the sequences within a particular site. Two steroid receptors might compete with each other for a site or enhance a given response depending on the number of binding sites or the particular sequence within a given region.

The consensus sequence determined by Mulvihill et al. (39) shows only a weak homology to the lysozyme gene, at position −290 bp, but deletion of this region has no effect on the regulation of this gene by progesterone or dexamethasone [(43) and below]. Renkawitz et al. (43) have noted a region of limited sequence homology at about −140 bp in the lysozyme, ovalbumin, transferrin, ovomucoid, and very low density apolipoprotein II genes. As all of these genes are regulated by estrogen, these authors speculate that this sequence ($\text{AAA}_C^T\text{ATGG}_C^A\text{C}$) may be involved in the control of gene expression by estrogen. Therefore, while the egg white genes are all regulated by four classes of steroids, only limited sequence homologies exist among the genes and do not readily explain how the steroid-receptor complexes recognize specific regions of DNA. One of the problems with analyzing steroid regulated genes for consensus sequences is that we do not know what the receptor proteins actually recognize. Footprinting has only a limited capacity to resolve specific aspects of the interaction between protein and DNA. With this approach it is therefore not possible to identify the nucleotide residues that make direct contact with the protein. More relevant may be the patterns of hydrogen-bond acceptors and donors in the grooves of the helix or other factors as yet undetermined.

Expression in Homologous Cells

Although steroid-receptor complexes can bind *in vitro* to specific regions in the egg white genes, the physiological role of such binding sites remains obscure. Recently, assessment of the functional significance of particular sequences of DNA has become possible by introducing mutated genes into cultured cells. Relatively few reports to date have documented the transfer of genes back into differentiated cells that normally express the endogeneous genes; cultured chicken oviduct cells represent one such system. Evidence from several groups has helped define specific regions in the chicken egg white genes that may play a crucial role in their regulation by steroids.

To ascertain whether sequences in the putative regulatory region of the chicken lysozyme gene are sufficient for regulation of transcription by steroid hormones, Renkawitz et al. (42) microinjected primary oviduct cell cultures with a gene constructed by fusing the 5′-flanking region of the lysozyme gene to the SV40 T-antigen structural gene. The number of cells expressing the lysozyme fusion gene was determined by indirect immunofluorescence staining for the T-antigen protein. When the calls are treated with dexamethasone or progesterone, a 5- to 10-fold increase in the number of fluorescent cells is observed without need for viral enhancer sequences to stimulate transcription. Control genes, such as the intact SV40 T-antigen gene, do not show a steroid-dependent increase in the number of positive cells. These experiments suggest that sequences in the 5′-flanking region of the lysozyme gene are sufficient to define this gene as a target for steroid hormones.

Deletion studies with the lysozyme fusion gene revealed that sequences between −208 and −164 bp are required for expression to be regulated by dexamethasone or progesterone (Fig. 7)(43). These sequences are not essential for constitutive levels of transcription; when the SV40 enhancer is inserted into the plasmid, HeLa cells express the lysozyme fusion gene deleted to −64 bp, indicating that this deletion mutant retains functional promoter elements. As discussed above, the region between −200 and −160 bp in the lysozyme gene also binds the glucocorticoid and progesterone receptors. Thus, the upstream receptor binding site overlaps with the nucleotide sequence required for regulated expression of the lysozyme promoter.

The downstream binding site, between −79 and −34 bp, however, is insufficient to obtain a response to steroids (Fig. 7). This is somewhat surprising as this site has a much stronger affinity for the glucocorticoid receptor than does the upstream site. It is possible that more than one receptor binding site is essential for induction of the lysozyme promoter by steroids. Most of the steroid-responsive genes examined thus far have more than one receptor binding site, and many of these sites are within the transcribed portion of the gene. While the transcribed portion of the lysozyme gene has not been assayed by footprint analysis for binding sites, it seems likely that they exist. Because the lysozyme fusion gene lacks these putative internal sites, the promoter may be dependent on both upstream sites to retain a response to steroids.

Using the calcium phosphate coprecipitation method to transfect cultured oviduct cells, Dean et al. (12) examined the steroid regulatory sequences in the ovalbumin gene. They constructed a gene (termed ovalglobin) with the 5′-flanking region of the ovalbumin gene fused to the chick β-globin structural gene. When the ovalglobin gene is transfected into oviduct cells cultured with progesterone, an 8-fold increase in ovalglobin transcripts is observed compared to cells cultured without progesterone. Deletion of sequences upstream of −95 bp eliminates the regulation by progesterone, but a 5-fold increase in response to progesterone is still observed when at least 222 nucleotides of the 5′-flanking region remain. More recently, Dean et al. (13) showed that deletion of 5′-flanking sequences to −197 bp has little effect on the induction of this fusion gene by either progesterone or estradiol. Thus, the region between −197 and −95 bp apparently has functional significance for regulation of the ovalbumin gene by steroids, which confirms the *in vitro* binding studies of Compton et al. (10) discussed earlier.

The results described thus far suggest that receptor binding sites within 200 bp upstream of the transcription start site can confer regulation by steroids on both the lysozyme and ovalbumin genes. In contrast to these results, Chambon et al. (5, 6) have found a negative regulatory effect of sequences between −420 and −132 bp on the ovalbumin and transferrin genes. Primary oviduct cell cultures were microinjected with recombinant genes prepared by fusing the 5′-flanking regions from the ovalbumin or transferrin genes to the structural gene for the SV40 T antigen, analogous to the method-

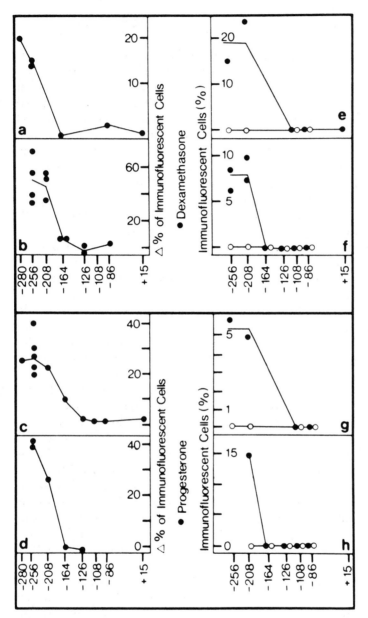

Figure 7 Number of steroid-induced immunofluorescent cells depends on lysozyme sequences between −208 and −164 bp. Plasmids deleted as indicated were microinjected into primary oviduct cells in amounts of 3000 copies (a, c) or 600 copies (b, d) per nucleus or isolated DNA fragments were injected with about 25 copies per nucleus (e through h). Cells were kept incubated with or without 10^{-7} M dexamethasone (upper panel) or 10^{-8} M progesterone (lower panel). As the overall expression varied from one oviduct culture to the next, the data are expressed as the increase in the percentage of immunofluorescent cells of cultures containing steroids versus cultures lacking steroids (a through d). Injections with isolated fragments (e through h) were always negative for immunofluorescence when the cultures were

ology of Renkavitz et al. (42, 43). When ovalbumin fusion genes deleted to −295 or −132 bp are microinjected, T-antigen protein is actively synthesized in the presence or absence of steroids (Table 2)(6). However, when the upstream sequence is extended to −420 bp, expression of the fusion gene decreases in the absence of steroids. Neither an estrogen (tamoxifen) nor a glucocorticoid (RU486) antagonist affect the expression of the −295 or −132 deletion mutants, yet that of the −420 mutant is dramatically reduced.

The above results have been interpreted as indicating that the ovalbumin promoter is subject to negative control by sequences located between −420 and −295 bp (6). This repression is relieved when steroids are present and the gene becomes actively transcribed. The negative control element is apparently overridden when viral enhancer sequences are included in the fusion gene, which explains why Dean et al. (12, 13) have not observed this effect. Without the sequences between −420 and −295 bp, the ovalbumin gene is constitutively expressed, even in the absence of steroid hormones.

The somewhat conflicting results just described makes it difficult to designate any particular DNA sequences as essential for regulation of the egg white genes by steroid hormones. Considering the complex molecular events that must be evoked when two steroids and insulin interact to regulate these genes (45), discrepancies are not too surprising. Variation in the oviduct cell cultures systems may contribute significantly to the differences (45). Another factor may relate to the relatively modest 2- to 10-fold increase in the expression of the exogenous genes in response to steroids as the steroid-induced increase in egg white mRNAs in the chicken is considerably greater. This may mean that these cells are basically refractory to stimulation when cultured under the conditions used in these experiments, that some unknown components are missing from the system, or that the sequences in the 5'-flanking region are insufficient to direct the dramatic increase in egg white mRNA in response to steroids observed with the intact gene. These questions could be addressed in part if the responsiveness of the endogenous genes was assessed in these model systems.

Expression in Heterologous Cells

In conjunction with most of the studies outlined above in which the egg white genes were transferred back into cultured oviduct cells, attempts were made to obtain expression of these genes in heterologous cells with the goals of defining sequences involved in directing cell-specific expression and regulation by steroid hormones.

Although the lysozyme gene is expressed constitutively in macrophages at a low level, it is inducible only in avian oviduct. To determine if the cell

kept without steroids (○). [Reprinted with permission from Renkawitz et al. (43) Copyright 1984 MIT.]

TABLE 2 Effect of estradiol and progesterone on the activity of the ovalbumin promoter in primary culture of chicken oviduct cells maintained in steroid hormone-stripped medium (SM) DNA: 100 copies per nucleus; 24 hours

Recombinant	Activity as percent of control[a]					
	SM	+ Estradiol (10⁻⁸ M)	+ Progesterone (10⁻⁹ M)	+ Tamoxifen (10⁻⁶ M)	+ RU486 (2 × 10⁻⁶ M)	+ Tamoxifen (10⁻⁶M) + RU486 (2 × 10⁻⁶ M)
Control (SV40)	100[b] (15)	100[c] (9)	100[d] (9)	100[e] (3)	100[f] (4)	100[g] (8)
pTOT-132	60[42−100](7)	74[58−84](4)	75[60−85](5)	75[58−91](2)	ND	60[52−71](4)
pTOT-420	26[10−42](13)	60[34−91](8)	55[34−75](7)	12[8−17](4)	13[5−20](4)	4[0−8](7)
pTOT-295	60[50−72](3)	64[44−83](2)	65[49−81](2)	ND	ND	53[42−65](2)

Note: The plasmids described were microinjected into primary cultures of oviduct cells that had been in culture for several days; 24 hours later the function of the ovalbumin promoter was monitored by indirect immunofluorescence against T-antigen. Numbers in brackets show range of percent activity. Numbers in parentheses are number of experiments. ND, not determined.

[a]Percent given as mean. Numbers in brackets show range of percent activity. Numbers in parentheses are number of experiments. ND, not determined.

[b]50% of the injected cells became T-antigen positive (extreme values: 43−60).

[c]49% of the injected cells became T-antigen positive (extreme values: 36−74).

[d]53% of the injected cells became T-antigen positive (extreme values: 34−68).

[e]50% of the injected cells became T-antigen positive (extreme values: 45−55).

[f]55% of the injected cells became T-antigen positive (extreme values: 45−67).

[g]48% of the injected cells became T-anigen positive (extreme values: 36−68).

Source: Endocrinology: Proc. of the 7th Intl. Cong. of Endocrinology, 1984. Reprinted with permission of Prof. Pierre Chambon, Institut de Chimie Biologique (6).

specificity of expression is conferred by sequences within the 5′-flanking region, the lysozyme fusion gene described above was microinjected into chicken macrophages, chicken fibroblasts, or rat fibroblasts (42). None of these cells express the fusion gene, yet they do express the intact SV40 gene. In a more definitive study, Matthias et al. (29) demonstrated that no expression of the lysozyme fusion gene occurs in HeLa cells, which contain glucocorticoid receptors, or in MCF-7 cells, which contain glucocorticoid and estrogen receptors, unless the SV40 enhancer sequences are in the plasmid. Furthermore, the expression observed with the enhancer present is not regulated by steroids in these cells. They concluded that the SV40 enhancer increases the level of expression of the lysozyme fusion gene several 100-fold and that without the enhancer it is not transcribed at a detectable level in HeLa or MCF-7 cells. Thus, the inducible expression of the chicken lysozyme gene appears to be restricted to oviduct cells. Because the chicken macrophages and fibroblasts are unable to express the lysozyme fusion gene, its inactivity in nonoviduct cells is not due to a problem of species specificity. These findings suggest that cell-specific factors are present in oviduct cells that recognize sequences in the 5′-flanking region of the lysozyme gene.

Chambon et al. (5) have reported similar results, though with some striking differences. Neither the ovalbumin nor the transferrin fusion gene discussed above is expressed in primary cultures of chicken fibroblasts, mouse fibroblasts (L cells), monkey kidney cells (CV1 cells), HeLa cells, or MCF-7 cells. This again suggests that the lack of expression is not due to problems of species specificity. However, both recombinants are expressed to various extents in primary cultures of chicken hepatocytes (Table 3)(5). These data imply that chicken hepatocytes contain factors that specifically recognize and express the ovalbumin and transferrin genes. This expression appears to be species-specific because the recombinants are not active in primary liver cultures from mouse. Because the ovalbumin gene is expressed when introduced into differentiated chicken hepatocytes, one interpretation is that some event during development that normally inactivates this gene is bypassed by introducing it at a later stage. One could postulate from studies done with this and other systems that this event might include methylation of the gene during development or alteration in the packaging of the DNA. It is also intriguing that the transferrin fusion gene deleted to −44 bp is expressed in a cell- and species-specific manner. This indicates that sequences in or near the TATA box are recognized by cell-specific factors (5).

Chicken hepatocytes, therefore, have cell-specific factors that allow constitutive expression of the microinjected ovalbumin and conalbumin promoters. This expression is independent of progesterone and estrogen. However, endogenous transferrin mRNA is only increased about 2-fold by estrogen (26). It is probable that a similar increase occurs with the exogenous genes without a detectable increase in the number of fluorescent cells or in the intensity of fluorescence. Because expression of the −420 bp deletion mutant of the ovalbumin fusion gene is not diminished in the absence of steroids, the factor responsible for negative control of the ovalbumin gene in oviduct is not present in hepatocytes (5, 6).

TABLE 3 Expression of conalbumin and ovalbumin promoter chimeric recombinants in primary culture of chicken hepatocytes (synthetic medium without steroid hormone)[a]

	Immunofluorescent cells (percentage of pSV1)[b]	
Chimeric recombinants	24-hour microinjection DNA (50 μg/ml)	6-hour microinjection DNA (100 μg/ml)
Conalbumin promoter:		
pTCT-44	5.5 (3)	7.9 (2)
pTCT-102	28.1 (12)	49.6 (6)
pTCT-182	29.7 (4)	29.5 (4)
pTCT-265	45.3 (7)	30.2 (6)
pTCT-400	48.4 (5)	50.3 (3)
pTCT-1050	62.5 (12)	58.3 (7)
pTCT-1930	48.4 (4)	38.8 (3)
pTCT-3600	18.7 (3)	7.2 (3)
Ovalbumim promoter:		
pTOT-132	28.1 (6)	28.8 (3)
pTOT-295	26.5 (4)	32.4 (3)
pTOT-430	37.5 (3)	27.3 (3)
pTOT-1348	40.6 (3)	41.7 (3)

[a] The primary cultured chicken hepatocytes were maintained in a synthetic medium in the absence of steroid hormones. For methods and representation of the results, see text and legend to Table 1. Approximately 70 percent of the cells microinjected with the reference recombinant pSV1 were T-antigen positive.

[b] Number of experiments given in parentheses.

Source: Chambon, P., et al. 1984. *Recent Progr. Horm. Res.* 40:1–42.

Expression of the Transferrin Gene in Transgenic Mice

The transfer of a foreign gene into a fertilized egg by microinjection allows one to assess the functional behavior of this gene during development and differentiation. In addition, a large number of offspring can be generated to study the effects of various inducers on the exogenous gene, the cell-specificity of its expression, and the role of particular DNA sequences in these processes. This approach has the advantage that the response of the transferred gene can be monitored in normal, differentiated tissues that are often difficult to manipulate in cell culture systems.

We introduced the entire chick transferrin gene with approximately 2.3 kb of 5'-flanking DNA into fertilized mouse eggs (34). Of the 44 mice that were generated initially, 11 acquired at least part of the chick transferrin gene, and six of the mice carried multiple copies. These six mice expressed the chick transferrin gene, making both chick transferrin mRNA and protein. The chick transferrin mRNA corresponded in size to that of authentic chick transferrin mRNA as shown by Northern blot analysis. Likewise, the chick transferrin synthesized by the mice migrated on SDS-polyacrylamide gels like the native protein. Thus, the transgenic mice are accurately expressing the chick gene and are actively synthesizing and secreting chick transferrin.

One of the most interesting observations with these transgenic mice relates to the cell-specificity of expression of the chick gene. In both the original transgenic mice and their offspring, the liver synthesizes significantly more chick transferrin mRNA than other tissues tested (Fig. 8). These results suggest that sequences within the transferrin gene or its flanking regions are sufficient to direct the cell-specific expression of this gene.

To determine if the transplanted chick gene could still respond to estrogen, transgenic mice were injected with estradiol benzoate. The rate of transcription of the chick transferrin gene was induced 4-fold over control values within 4 hours (Fig. 9). The rate of transcription remained elevated for 10 days of estrogen treatment. To determine if this increase was a specific effect of estrogen on the chick transferrin gene, the rate of transcription of the en-

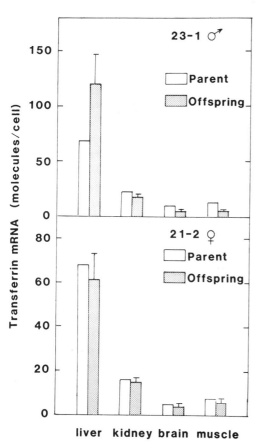

Figure 8 Concentration of chick transferrin mRNA in various tissues from the parents and offspring of two lines of transgenic mice.

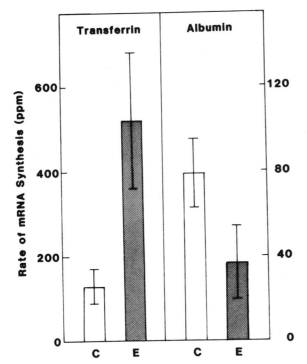

Figure 9 Rate of transcription of the chick transferrin gene in transgenic mice after treatment with estrogen. Mice were injected with estrogen (E) or vehicle (C) and liver nuclei were isolated 4 hours later. The transcription rates of the chick transferrin gene and the endogenous mouse albumin gene were determined and expressed as parts per million relative to total mRNA transcription. Estrogen had no effect on total transcription rates.

dogenous mouse albumin gene was examined and found to be diminished. These results demonstrate that estrogen is causing a specific induction of the chick gene in transgenic mice.

From the data described above, it is apparent that the chick transferrin gene can be expressed in mice in a cell-specific manner. The transferrin mRNA is then translated accurately, and the transferrin is secreted into the serum. The chick gene responds to estrogen in a direct and specific manner by increasing the rate of transcription of the gene followed by increases in chick transferrin mRNA levels. Thus, these transgenic mice are a useful model system for examining which sequences in the gene are required to signal cell-specific expression and the regulation of expression by steroid hormones.

Conclusions

Within the past 5 years our understanding of how steroid hormones control gene expression has increased enormously. Advances have come not only from elucidating the molecular events involved in activating the less complex viral systems but also from examining the extremely complicated multihormonal regulatory systems such as the chicken oviduct. Attention has focused on determining which regions of these genes interact with steroid-receptor complexes and on defining how this interaction modulates the rate of transcription.

The development of an oviduct explant model (30) and more recently of a cell culture system (45) has allowed careful analysis of the hormonal requirements for the induction of the egg white genes. Surprisingly, estrogen alone is insufficient to induce the ovalbumin gene. In addition to estrogen, insulin (15) and a second steroid (45), which is probably corticosterone *in vivo*, are required to achieve a significant increase in ovalbumin mRNA. The transferrin gene has a similar requirement for a second steroid but apparently does not need the permissive effects of insulin. Preliminary data suggest that insulin may regulate the induction of the ovalbumin gene through a protein phosphorylation event (15). Cyclic AMP and activators of adenylate cyclase can mimic insulin although the kinetics of induction are slower, suggesting an alternate pathway. These experiments have revealed that the hormonal regulation of the egg white genes is much more complex than previously thought.

The cloning and sequencing of the genes for the four major egg white proteins also represent a substantial contribution to this field. For the first time, the putative regulatory regions upstream of the cap site could be examined for consensus sequences. It was expected that homologies would exist to designate receptor binding sites for the four classes of steroid hormones that regulate these genes. While weak homologies are present, they do not seem to correlate with regions that bind receptors *in vitro* or that are required to obtain a response to steroids *in vivo*. Part of the difficulty in identifying receptor sites within a sequence may stem from our lack of understanding of what the receptor protein actually recognizes. The nucleotide sequence per se may be less relevant than the position of certain nucleotides or of hydrogen bond donors and acceptors within the grooves of the helix.

A more productive approach has been to compare the DNA sequences that bind steroid receptors *in vitro* to those required for a response to steroids in cultured oviduct cells. Dean et al. (12, 13) and Compton et al. (10) have identified a region in the ovalbumin gene between −197 and −95 bp that meets both of these requirements. Similarly, Renkawitz et al. (42, 43) found two such sites within the 5′-flanking region of the lysozyme gene, between −200 and −160 bp and between −74 and −39 bp. Footprint analysis revealed that partially purified progesterone and glucocorticoid receptors bind to each of these sites in the lysozyme gene, although with different affinities for the two receptors. These data suggest that positive regulatory elements reside within 200 bp upstream of the cap site in both the ovalbumin and lysozyme genes.

Data from Pierre Chambon's laboratory suggest an alternate hypothesis to explain steroid hormone action (5, 6). They have found negative control elements within the 5′-flanking region of the ovalbumin gene. A region between −420 and −295 bp in the ovalbumin gene represses transcription when steroids are absent, and this repression is relieved by steroids. They also find that when the ovalbumin promoter is deleted past −295 bp, it is constitutively expressed at a high level, with or without steroids present (6). These data are in contrast to those of Dean et al. (12, 13) who find that steroids induce an ovalbumin gene deleted to −200 bp and that little activity of the gene occurs without steroids. Because the two assay systems and oviduct cell culture models differ, these results are not directly comparable and may not actually be irreconcilable. One way to ascertain if the region of the ovalbumin gene between −200 and −95 bp is a positive steroid control element would be fuse it to a gene that is not normally regulated by steroids and to determine if this gene becomes inducible. A similar approach could be used to see if the region between −420 and −295 bp in the ovalbumin gene defines a negative control element.

Expression of the chick transferrin gene in mice has provided another means of assessing how this gene is regulated (34). The chick gene is accurately transcribed in mice, and the resultant protein is secreted into the serum. Interestingly, the chick gene is apparently regulated in the mouse in the same way as it is in the chick; the chick gene is preferentially expressed in the liver and is induced by estrogen. These data imply that at least some aspects of cell-specific expression and regulation of the gene by steroids are encoded by the DNA itself and can be recognized between species. By exploring the expression of various deletion mutants, it may be possible to define the sequences that specify cell-specific expression and inducibility by steroids.

An understanding of how genes are coordinately regulated by several hormones remains elusive. Clues to the molecular events involved may be forthcoming from the MMTV system. Yamamoto and colleagues (11) have evidence that combining pairs of heterologous enhancers can cause synergistic stimulation of a given promoter. Moreover, the specific promoter involved affects both the quantitative and qualitative response to enhancers and the interactions between enhancers. They speculate that interactions of several enhancers might be involved in multihormonal regulatory systems such as the ovalbumin gene. Thus, it might be necessary not only to determine which sites in the ovalbumin gene bind steroid hormones but also whether two such sites must be activated in concert and whether insulin directly or indirectly activates a comparable region of the gene. It may be that activation of multiple enhancer-like sequences is responsible for the magnitude and the kinetics of the response to a particular hormonal status.

Evidence from the chicken oviduct and MMTV systems suggests that steroid-receptor complexes stimulate transcription by specifically altering the local chromatin structure to create transcriptional enhancers. While, if true, this would represent a significant theoretical advancement, it still adds little to our understanding of the actual molecular events. As yet, the mechanism of action of enhancers in unknown but may include an alteration in the chromatin structure to allow greater access of the transcriptional machinery to the DNA and/or recognition of a particular sequence by some component of this machinery (5). Information about these mechanistic details may await more extensive characterization of the regulatory proteins involved in these interactions. To this end, primary sequence information for the chick progesterone receptor B antigen (55) and the rat liver glucocorticoid receptor (37) should be available soon.

The technology is now available, however, to answer many exciting questions about how steroid hormones regulate gene expression. For example, are multiple steroid-receptor binding sites associated with all responsive genes? Are several sites required for a maximal response? Are there sites in the transcribed portions of the egg white genes that are functionally relevant? Do steroid hormones increase the rate of transcription by relieving the action of a repressor in addition to or instead of acting at positive control elements? Are there different binding sites for each steroid receptor or do all receptors have some affinity for all sites? How does this relate to the additive and synergistic effects observed upon treatment with multiple steroids? How does progesterone have both stimulatory and inhibitory effects on the chick oviduct at different times and developmental stages? Many of these questions can now be meaningfully addressed by concomitant *in vitro* and *in vivo* experiments.

General References

DeFranco, D., Wrange, O., Merryweather, J., and Yamamoto, K. R. 1985. Biological activity of a glucocorticoid regulated enhancer: DNA sequence requirements and interactions with other transcriptional enhancers, *UCLA Symp. Mol. Cell. Biol. New Ser.* 20:305–321.

Palmiter, R. D., Mulvihill, E. R., Shepherd, J. H., and McKnight, G. S. 1981. Steroid hormone regulation of ovalbumin and conalbumin in gene transcription, *J. Biol. Chem.* 256:7910–7916.

References

1. Baldacci, P., Royal, A., Bregegere, F., Abastado, J. P., Cami, B., Daniel, F., and Kourilsky, P. 1981. DNA organization in the chicken lysozyme gene region, *Nucleic Acids Res.* 9:3575–3588.

2. Benoist, C., O'Hare, K., Breathnach, R., and Chambon, P. 1980. The ovalbumin gene-sequence of putative control regions, *Nucleic Acids. Res.* 8:127–142.

3. Catterall, J. F., Stein, J. P., Kristo, P., Means, A. R., and O'Malley, B. W. 1980. Primary sequence of ovomucoid messenger RNA as determined from cloned complementary DNA, *J. Cell Biol.* 87:480–487.

4. Chambon, P., Benoist, C., Breathnach, R., Cochet, M., Gannon, F., Gerlinger, P., Krust, A., LeMeur, M., LePennec, J. P., Mandel, J. L., O'Hare, K., Perrin, F. 1979. Structural organization and expression of ovalbumin and related chicken genes, in: *From Gene to Protein: Information Transfer in Normal and Abnormal Cells* (T. R. Russell, K. Brew, H. Faber, J. Schultz, eds.), pp. 55–81, Academic Press, New York.

5. Chambon, P., Dierich, A., Gaub, M. P., Jakowlev, S., Jongstra, J., Krust, A., LePennec, J. P., Oudet, P., and Reudelhuber, T. 1984. Promoter elements of

genes coding for proteins and modulation of transcription by estrogens and progesterone, *Recent Progr. Horm. Res.* 40:1–42.

6. Chambon, P., Gaub, M. P., LePennec, J. P., Dierich, A., and Astinotti, D. 1984. Steroid hormones relieve repression of the ovalbumin gene promoter in chick oviduct tubular gland cells, in: *Endocrinology: Proceedings of the 7th International Congress of Endocrinology* (F. Labrie, L. Proulx, eds.), pp. 3–10, Excerpta Medica, Amsterdam.

7. Ciejek, E. M., Tsai, M.-J., and O'Malley, B. W. 1983. Actively transcribed genes are associated with the nuclear matrix, *Nature (London)* 306:607–609.

8. Cochet, M., Gannon, F., Hen, R., Maroteaux, L., Perrin, F., and Chambon, P. 1979. Organization and sequence studies of the 17-piece chicken conalbumin gene, *Nature (London)* 282:567–574.

9. Colbert, D. A., Knoll, B. J., Woo, S. L. C., Mace, M. L., Tsai, M.-J., and O'Malley, B. W. 1980. Differential hormonal responsiveness of the ovalbumin gene and its pseudogenes in the chick oviduct, *Biochemistry* 19:5586–5592.

10. Compton, J. G., Schrader, W. T., and O'Malley, B. W. 1983. DNA sequence preference of the progesterone receptor, *Proc. Nat. Acad. Sci. U.S.A.* 80:16–20.

11. DeFranco, D., Wrange, O., Merryweather, J., and Yamamoto, K. R. 1985. Biological activity of a glucocorticoid regulated enhancer: DNA sequence requirements and interactions with other transcriptional enhancers, *UCLA Symp. Mol. Cell. Biol. New Ser.* 20:305–321.

12. Dean, D. C., Knoll, B. J., Riser, M. E., and O'Malley, B. W. 1983. A 5′-flanking sequence essential for progesterone regulation of an ovalbumin fusion gene, *Nature (London)* 305:551–553.

13. Dean, D. C., Gope, R., Knoll, B. J., Riser, M. E., and O'Malley, B. W. 1984. A similar 5′-flanking region is required for estrogen and progesterone induction of ovalbumin gene expression, *J. Biol. Chem.* 259:9967–9970.

14. Doolittle, R. F. 1983. Angiotensinogen is related to the antitrypsin-antithrombin-ovalbumin family, *Science* 222:417–419.

15. Evans, M. E., and McKnight, G. S. 1984. Regulation of the ovalbumin gene: Effects of insulin, adenosine 3′, 5′-monophosphate, and estrogen, *Endocrinology* 115:368–377.

16. Evans, M. I., Hager, L. J., and McKnight, G. S. 1981. A somatomedin-like peptide hormone is required during the estrogen-mediated induction of ovalbumin gene transcription, *Cell* 25:187–193.

17. Fritton, H. P., Igo-Kemenes, T., Nowock, J., Strech-Jurk, U., Theisen, M., Sippel, A. E. 1984. Alternative sets of DNase I-hypersensitive sites characterize the various functional states of the chicken lysozyme gene, *Nature (London)* 311:163–165.

18. Gerlinger, P., Krust, A. LeMeur, M., Perrin, F., Cochet, M., Gannon, F., Dupret, D., and Chambon, P. 1982. Multiple initiation sites for the chicken ovomucoid transcription unit, *J. Mol. Biol.* 162:345–364.

19. Grez, M., Land, H., Giesecke, K., and Schutz, G. 1981. Multiple mRNAs are generated from the chicken lysozyme gene, *Cell* 25:743–752.

20. Heilig, R., Perrin, F., Gannon, F., Mandel, J. L., and Chambon, P. 1980. The ovalbumin gene family: Structure of the X gene and evolution of duplicated split genes, *Cell* 20:625–637.

21. Heilig, R., Muraskowsky, R., and Mandel, J.-L. 1982. The ovalbumin gene family: the 5' end region of the X and Y genes, *J. Mol. Biol.* 156:1–19.

22. Hughes, S. H., Stubblefield, E., Payvar, F., Engel, J. D., Dodgson, J. B., Spector, D., Cordell, B., Schimke, R. T., and Varmus, H. E. 1979. Gene localization by chromosome fractionation: Globin genes are on at least two chromosomes and three estrogen-inducible genes are on three chromosomes, *Proc. Nat. Acad. Sci., U.S.A.* 76:1348–1352.

23. Jeltsch, J. -M., and Chambon, P. 1982. The complete nucleotide sequence of the chicken ovotransferrin mRNA, *Eur. J. Biochem.* 122:291–295.

24. Jung, A., Sippel, A. E., Grez, M., and Schutz, G. 1980. Exons encode functional and structural units of chicken lysozyme, *Proc. Nat. Acad. Sci. U.S.A.* 77:5759–5763.

25. Kaye, J. S., Bellard, M., Dretzen, G., Bellard, F., and Chambon, P. 1984. A close association between sites of DNase I hypersensitivity and sites of enhanced cleavage by micrococcal nuclease in the 5'-flanking region of the actively transcribed ovalbumin gene, *EMBO J.* 3:1137–1144.

26. Lee, D. C., McKnight, G. S., and Palmiter, R. D. 1978. The action of estrogen and progesterone on the expression of the transferrin gene, *J. Biol. Chem.* 253:3494–3503.

27. Lee, D. C., McKnight, G. S., and Palmiter, R. D. 1980. The chicken transferrin gene: Restriction endonuclease analysis of gene sequences in liver and oviduct DNA, *J. Biol. Chem.* 255:1442–1450.

28. LeMeur, M., Glanville, N., Mandel, J. L., Gerlinger, P., Palmiter, R., and Chambon, P. 1981. The ovalbumin gene family: Hormonal control of X and Y transcription and mRNA accumulation, *Cell* 23:561–571.

29. Matthias, P. D., Renkawitz, R., Grez, M., and Schutz, G. 1982. Transient expression of the chicken lysozyme gene after transfer into human cells, *EMBO J.* 1:1207–1212.

30. McKnight, G. S. 1978. The induction of ovalbumin and conalbumin mRNA by estrogen and progesterone in chick oviduct explant cultures, *Cell* 14:403–413.

31. McKnight, G. S., and Palmiter, R. D. 1979. Transcriptional regulation of the ovalbumin and conalbumin genes by steroid hormones in chick oviduct, *J. Biol. Chem.* 254:9050–9058.

32. McKnight, G. S., Hager, L., and Palmiter, R. D. 1980. Butyrate and related inhibitors of histone deacetylation block the induction of egg white genes by steroid hormones, *Cell* 22:469–477.

33. McKnight, G. S., Lee, D. C. Hemmaplardh, D., Finch, C. A., and Palmiter, R. D. 1980. Transferrin gene expression: Effects of nutritional iron deficiency, *J. Biol. Chem.* 255:144–147.

34. McKnight, G. S., Hammer, R. E., Kuenzel, E. A., and Brinster, R. L. 1983. Expression of the chicken transferrin gene in transgenic mice, *Cell* 34:335–341.

35. McReynolds, L., O'Malley, B. W., Nisbet, A. D., Fothergill, J. E., Givol, D., Fields, S., Robertson, M., and Brownlee, G. G. 1978. Sequence of chicken ovalbumin mRNA, *Nature (London)* 273:723–728.

36. Meek, R. L., Walsh, K. A., and Palmiter, R. D. 1982. The signal sequence of ovalbumin is located near the NH_2 terminus, *J. Biol. Chem.* 257:12245–12251.

37. Miesfeld, R., Okret, S., Mikstrom, A.-C., Wrange, O., Gustafsson, J.-A., and

Yamamoto, K. R. 1984. Characterization of a steroid receptor gene and mRNA in wild type and mutant cells, *Nature (London)* 312:779–781.

38. Mirkovitch, J., Mirault, M.-E., and Laemmli, U. K. 1984. Organization of the higher-order chromatin loop: Specific DNA attachment sites on nuclear scaffold, *Cell* 39:223–232.

39. Mulvihill, E. R., LePennec, J.-P., and Chambon, P. 1982. Chicken oviduct progesterone receptor: Localization of specific regions of high-affinity binding in cloned DNA fragments of hormone-responsive genes, *Cell* 28:621–632.

40. Palmiter, R. D., Mulvihill, E. R., Shepherd, J. H. and, McKnight, G. S. 1981. Steroid hormone regulation of ovalbumin and conalbumin gene transcription, *J. Biol. Chem.* 256:7910–7916.

41. Pardoll, D.M., and Vogelstein, B. 1980. Sequence analysis of nuclear matrix associated DNA from rat liver, *Exp. Cell. Res.* 128:466–470.

42. Renkawitz, R., Beug, H., Graf, T., Matthias, P., Grez, M., and Schutz, G. 1982. Expression of a chicken lysozyme recombinant gene is regulated by progesterone and dexamethasone after microinjection into oviduct cells, *Cell* 31:167–176.

43. Renkawitz, R., Schutz, G., von der Ahe, D., and Beato, M. 1984. Sequences in the promoter region of the chicken lysozyme gene required for steroid regulation and receptor binding, *Cell* 37:503–510.

44. Robinson, S. I., Small, D., Idzerda, R., McKnight, G. S., and Vogelstein, B. 1983. The association of transcriptionally active genes with the nuclear matrix of the chicken oviduct, *Nucleic Acids Res.* 11:5113–5130.

45. Sanders, M. M., and McKnight, G. S. 1985. Chicken egg white genes: Multihormonal regulation in a primary cell culture system, *Endocrinology* 116:398–405.

46. Shepherd, J. H., Mulvihill, E. R., Thomas, P. S., and Palmiter, R. D. 1980. Commitment of chick oviduct tubular gland cells to produce ovalbumin mRNA during hormonal withdrawal and restimulation, *J. Cell Biol.* 87:142–151.

47. Stalder, J., Groudine, M., Dodgson, J. B., Engel, J. D., and Weintraub, H. 1980. HB switching in chickens, *Cell* 19:973–980.

48. Stein, J. P., Catterall, J. F., Kristo, P., Means, A. R., and O'Malley, B. W. 1980. Ovomucoid intervening sequences specify functional domains and generate protein polymorphism, *Cell* 21:681–687.

49. Stumph, W. E., Kristo, P., Tsai, M.-J., and O'Malley, B. W. 1981. A chicken middle-repetitive DNA sequence which shares homology with mammalian ubiquitous repeats, *Nucleic Acids Res.* 9:5383–5397.

50. Stumph, W. E., Baez, M., Beattie, W. G., Tsai, M.-J., and O'Malley, B. W. 1983. Characterization of deoxyribonucleic acid sequences at the 5' and 3' borders of the 100 kilobase pair ovalbumin gene domain, *Biochemistry* 22:306–315.

51. von der Ahe, D., Janich, S., Scheidereit, C., Renkawitz, R., Schutz, G., and Beato, M. 1985. Glucocorticoid and progesterone receptors bind to the same sites in two hormonally regulated promoters, *Nature (London)* 313:706–709.

52. Woo, S. L. C., Beattie, W. G., Catterall, J. F., Dugaiczyk, A., Staden, R., Brownlee, G. G., and O'Malley, B. W. 1981. Complete nucleotide sequence of the chicken chromosomal ovalbumin gene and its biological significance, *Biochemistry* 20:6437–6446.

53. Zaret, K. S., and Yamamoto, K. R. 1984. Reversible and persistent changes in chromatin structure accompanying activation of a glucocorticoid-dependent enhancer element, *Cell* 38:29–38.

54. Zarucki-Schulz, T., Tsai, S. Y., Itakura, K., Soberon, X., Wallace, R. B., Tsai, M.-J., Woo, S. L. C., and O'Malley, B. W. 1982. Point mutagenesis of the ovalbumin gene promoter sequence and its effect on *in vitro* transcription, *J. Biol. Chem.* 257:11070–11077.

55. Zarucki-Schulz, T., Kulomaa, M. S., Headon, D. R., Weigel, N. L., Baez, M., Edwards, D. P., McGuire, W. L., Schrader, W. T., and O'Malley, B. W. 1984. Molecular cloning of a cDNA for the chicken progesterone receptor B antigen, *Proc. Nat. Acad. Sci. U.S.A.* 81:6358–6362.

Questions for Discussion with the Editors

1. *How does withdrawal of steroid affect tubular gland cells? Does it lead to cell death and removal, cytological differentiation, or reduction in cell size?*

After withdrawal of estrogen *in vivo*, tubular gland cells are maintained without obvious morphological changes for the first 3 days. These cells can be restimulated by steroids to produce a rapid increase in ovalbumin mRNA. In addition, the ovalbumin gene retains its sensitivity to DNase I. Subsequently, the majority of tubular gland cells are lost from the oviduct although a small population remains. These remaining cells still induce ovalbumin mRNA rapidly in response to steroids. Thus, the oviduct does not return to its original, undifferentiated condition after removal of steroid.

2. *Which of the following features of hormone-inducible gene systems have analogous counterparts in other gene regulation systems: DNase I sensitivity; nuclear matrix interactions; binding of activar/receptor complexes?*

All of these features have counterparts in other gene regulation systems. A common feature of any gene undergoing developmental activation is the appearance of DNase I sensitive sites in the vicinity of the gene. Some of these sites may be associated with specific inducers such as those appearing in response to steroids while others signal the transcriptional capacity of the gene in that tissue.

Association of transcriptionally active genes with the nuclear matrix appears to be a general feature of matrices prepared by a high-salt extraction procedure, with no sequence-specific pattern of association. However, using a modified method, Mirkovitch and colleagues have demonstrated sequence-specific binding sites in the 5'-flanking region of the hsp 70 heat-shock genes and in the spacer region between histone gene clusters but not in the actively transcribed portion of these genes. They propose that the matrix facilitates transcription by bringing promoter and regulatory regions of active genes in close proximity to RNA polymerase and other regulatory factors on the matrix. At this time, the role of the nuclear matrix in gene transcription remains unresolved.

While a number of eukaryotic genes are thought to be regulated by the binding of activator or repressor molecules to specific DNA sequences, these interactions are not yet well understood. With the possible exception of steroid receptor complexes, proteins that regulate viral genes are the best characterized of the proteins that regulate genes transcribed by RNA polymerase II. For example, Robert Tjian and colleagues have identified a transcription factor, Sp_1, that binds to a specific sequence in promoter regions of the SV40 genome. They have demonstrated that Sp_1 interacts with three to four of these sequences and that these contacts fall on one strand of DNA and are arranged in the major groove of the helix. More recently, this sequence has been found in the promoter region of several cellular genes and thus may have a more general role in transcription. (For a review of sequence-specific DNA binding proteins, see Dyan and Tjian.[1])

[1]Dynan, W. S., and Tjian, R. 1985. Control of eukaryotic messenger RNA synthesis by sequence-specific DNA binding proteins, *Nature (London)* 316:774–778.

CHAPTER 9

Amplification and Expression of Modular cDNA Genes

Randal J. Kaufman

THE ISOLATION OF SPECIFIC GENES, their modification *in vitro*, and their introduction into mammalian cells has facilitated the understanding of gene regulation and has made possible the isolation of new variants of mammalian cells. These new variants have been used to study cellular, biochemical, and regulatory pathways as well as to produce large quantities of specific proteins that were previously difficult to obtain by other means. Unfortunately, many genes of interest are difficult to manipulate *in vitro* due to the presence of large or multiple introns. One solution to the expression of such genes is to introduce a cDNA copy of the mRNA into an expression vector that contains appropriate signals for mRNA production relevant to the questions to be studied. The relevant questions relate to both protein expression from the cDNA itself and to transcriptional and translational control signals. The purpose of this chapter is to summarize how cDNA genes can be used to study biological processes.

In general, cDNA genes can be used to study the control signals required for efficient or regulated expression of a particular gene, or to produce a specific protein. Heterologous DNA signals encoding transcriptional, RNA processing, or translational capabilities have been added to protein coding regions (cDNAs) in order to ask what DNA sequences are required to yield appropriate expression of the cDNA gene. This chapter will focus on the utility of dihydrofolate reductase (DHFR) cDNA genes to study gene expression.

The expression of heterologous cDNA genes in mammalian cells with the aim of producing a specific protein has had several important applications in modern biology and pharmaceutical development. First, cDNA clones isolated by a variety of means have been verified by their ability to direct the synthesis of a protein with the desired activity after introduction into mammalian cells. Probably the most dramatic example of the ability of heterologous mammalian cells to express a cDNA gene is the expression of active antihemophilic factor VIII, a large complex protein deficient in hemophilia A (103, 112). Second, the evaluation of the role of specific mutations introduced into cDNAs is readily possible by analyzing proteins expressed in mammalian cells. Third, expression of cDNA genes at high levels in mammalian cells allows an analysis of the effect of expression level of a specific protein on a particular property of the cell. It also allows the production of proteins of potential therapeutic value which were previously difficult to obtain by other means. Finally, it has been possible to directly isolate particular cDNAs by screening recipient mammalian cells for the production of biologically active proteins (35, 51, 111).

A variety of approaches have been used to express heterologous cDNAs in mammalian cells. First, DNA can be introduced typically by calcium phosphate (25) or DEAE-dextran mediated DNA transfection (97), or by protoplast fusion of bacteria to cells (89). These procedures allow for approximately 10 percent of the population to acquire and express DNA over a period of several days to several weeks (transiently). One particularly useful host-vector system is based on the use of recombinant bacterial plasmids that can be replicated and expressed in SV40 transformed monkey kidney cell lines (COS cells). In the absence of DNA replication, SV40 transforms monkey cells. The transformed monkey kidney cells containing integrated origin defective SV40 DNA, designated COS (24), produce high levels of the SV40 viral A gene product (T antigen) and are permissive for SV40 DNA replication. The cells efficiently support the replication of bacterial plasmids that contain the SV40 origin of replication (74).

This vector system has been used to identify transcriptional regulatory elements in eukaryotic genes (64), has simplified the investigation of SV40 T antigen, a regulatory protein involved in DNA replication, transcription, and neoplastic transformation (74, 83), and has permitted the propagation of pure SV40 recombinant virus stocks that are defective in early viral functions (15). In addition, this vector system has allowed for the convenient positive verification of cDNA clones by the ability to effiiently express the desired protein (51, 103, 111, 112). However, for many purposes transient expression in COS cells is not appropriate since the transfection process disrupts cellular metabolism and is cumbersome, and the resultant expression levels are not sufficient to obtain large quantities of proteins.

Another approach to the expression of cDNAs has been to use lytic DNA viral vectors by substitution into the early or late region of SV40 (22, 28, 70, 72) or into nonessential regions of human adenovirus (59, 113). These systems can express significant quantities of protein from the introduced gene (1 to 10 μg/10^6 cells) but are limited by the packaging constraints of the virus. SV40 can accommodate 2.5 kb where adenovirus may accommodate up to 7 kb of heterologous DNA. In addition, due to the viral life cycle, the cells lyse prior to the accumulation of high levels of the desired protein. Thus, it is not possible to study the effect of expression of a particular protein on cellular metabolism, and it is difficult to produce large quantities of the desired protein.

Other approaches to cDNA expression have involved the development of stable cell lines that constitutively express the heterologous gene. Two general approaches have involved the use of nonlytic viral vectors, such as retroviruses or bovine papilloma virus (BPV), and the use of DNA cotransformation with selectable markers. Particular advantages of retroviral vectors include the abilities (a) to transduce genes into a variety of cell types and into a variety of species, (b) to produce stable cell lines as a result of retrovirus integration into the host chromosome, (c) to transform nearly 100 percent of the host cells, due to the high infectivity of retroviruses, and (d) to introduce foreign genes into animals (68, 110). However, protein expression from retroviral-based vectors has been low due to problems with RNA splicing and mRNA translation. Furthermore, packaging constraints again limit the size of the inserted segment to approximately 6 to 7 kb. The prototype of DNA-based nonlytic viral vectors is BPV. BPV is a small DNA virus that morphologically transforms a variety of cells. Vectors containing the entire BPV genome or a 69 percent (5.5 kb) subgenomic transforming fragment are in many cases stable as multicopy (20 to 100 copies per cell) extrachomosomal elements in transformed cells (18, 19, 50, 87, 115). In other cases the vector sequences are maintained as oligomers integrated into the host chromosome in a head-to-tail tandem array. The multicopy nature of the DNA in BPV transformed cells is in part responsible for the high-level expression of the foreign gene contained on BPVc-based vectors. Derivatives of BPV vectors contain selectable markers, which obviates relying on morphological transformation to obtain cells harboring exogenous DNA (58). Many secreted proteins have been expressed in BPV-based vectors such as β-interferon (69, 115) and human growth hormone (78). Although there has been considerable success with BPV expression systems, the biology of BPV expression and requirements for its plasmid maintenance and regulation of copy number are poorly understood. Thus, in general, results have been variable. Potential problems include obtaining proper transcription of the inserted gene, the instability of plasmid maintenance, and DNA rearrangements that occur upon propagation. As more features of the biology of BPV become understood, BPV will become a more useful system for expression of a wide variety of heterologous genes.

Finally, investigators have used DNA cotransformation with selectable markers such as the bacterial neomycin phosphotransferase gene (93), which encodes resistance to the antibiotic G418, or the bacterial xanthine-guanine phosphoribosyltransferase gene for which expression can be selected by forcing the cells to use xanthine as a purine source (71). A particularly useful approach to obtain permanent cell lines expressing high levels of heterologous proteins is to cotransform and subsequently coamplify the copy number of the integrated DNA. Although a number of amplifiable markers are potentially available, the most characterized system is the amplification of the DHFR gene which occurs upon selection of cultured cells in stepwise increasing concentrations of methotrexate (MTX) (3, 90, 91). The folic acid analog MTX kills cells by stoichiometric inhibition of DHFR. Upon selection for growth of cells in increasing concentrations of MTX, cells are obtained that express higher levels of DHFR. Most frequently this increase results from an amplification of the DHFR gene.

MTX selection for amplification of heterologous genes has proved particularly successful in Chinese hamster ovary (CHO) cells which are deficient in DHFR (14). These cells can be transformed with a DHFR gene and any particular gene of interest. DHFR-positive (DHFR$^+$) transformants can be selected on the basis of growth in the absence of nucleosides. Subsequent selection in increasing MTX results in amplification of the transfected DNA. DNA that has been coamplified in this manner is also coexpressed. (43, 49). This approach has been used to obtain cell lines that constitutively express and secrete high levels of human interferons (30, 63, 88).

This chapter will first focus on the development of an efficient modular DHFR cDNA gene that can be used as a selectable and amplifiable marker in DHFR deficient CHO cells. The modular nature of this cDNA allows for easy removal and insertion of various DNA segments in order to study their function. DHFR is a useful marker for monitoring gene activity for several reasons. First, DHFR can be easily monitored by activity or immunologically. Second, the DHFR-deficient (DHFR$^-$) CHO cell line (14) can be used as a recipient for the introduction of DHFR cDNA genes, and DHFR expression can be measured by transformation to the DHFR$^+$ phenotype. Since very low levels of DHFR expression can elicit a DHFR$^+$ phenotype, the DHFR transformation assay provides a uniquely sensitive assay for expression from DHFR cDNA genes. Generally, the transformation efficiency of a particular DHFR cDNA gene, as measured by the number of colonies that appear in nucleoside-free media, is directly related to the ability of that cDNA gene to direct DHFR synthesis.

It is also possible to quantitate DHFR expression from DHFR cDNA genes by monitoring the plating efficiency of transfected cells as a function of increasing concentrations of MTX. The ability to amplify the copy number of transfected DNA by selection for growth in increasing concentrations of MTX has facilitated the analysis of the signals utilized for mRNA production as well as the analysis regulatory events, such as transcriptional activation by trans-acting factors. In addition, selection for decreased DHFR expression can be achieved using suicide selections based on [^{3}H]uridine incorporation (14).

Finally, it is possible to monitor and isolate cells on the basis of their intracellular level of DHFR by saturating DHFR with a fluorescent derivative of MTX and use of the fluorescence-activated cell sorter (36). Two examples of the use of DHFR cDNA genes to study gene regulation will be summarized: (a) transcriptional regulation of the adenovirus early region 2 promoter and (b) posttranscriptional growth-dependent regulation of DHFR mRNA levels.

The DHFR cDNA gene has also been used to introduce and subsequently amplify heterologous genes in cells to produce foreign proteins at high levels. Amplification of transfected DNA has allowed the study of the nature of the chromosomal events associated with DNA amplification in different regions of the host chromosome. The latter part of this chapter will focus on the development of efficient cDNA expression vectors for monkey COS cells that exploit adenovirus translational control signals to enhance mRNA translation. The use of these improved cDNA expression vectors has made possible the isolation of cDNA clones by directly screening for protein expression in transiently transfected COS cells.

Expression, Amplification, and Regulation of cDNA Genes in CHO Cells

Signals Required for Efficient Expression of a Modular DHFR cDNA Gene

Modular DHFR cDNA genes have been constructed to transform CHO DHFR$^-$ cells in order to determine what DNA signals are required for efficient DHFR expression (40). The transformation efficiency of CHO DHFR$^-$ cells to the DHFR$^+$ phenotype after calcium phosphate-mediated DNA transfection of the DHFR cDNA genes correlates with the ability of the cDNA genes to express DHFR. Transformation of DHFR$^-$ to DHFR$^+$ phenotype was not observed with the DHFR cDNA clone in pBR322 without any additional mRNA expression signals. With the introduction of the adenovirus major late promoter including the first late adenovirus leader with 5' splice site and a 3' splice site derived from an immunoglobulin gene, DHFR$^+$ transformants arose at a low frequency (10^{-7} to 2×10^{-6} in different experiments) (Fig. 1). Recombinants containing the entire early region of SV40 insert yielded DHFR$^+$ transformants at a higher frequency.

Analysis of the DHFR mRNA expressed in four individual transformants after MTX selection and gene amplification indicated that all clones utilized the adenovirus major late promoter for transcription initiation, and a hybrid intron formed by the 5' splice site of the adenovirus major late leader and the 3' splice from the immunoglobulin gene was properly excised. The mRNA was not efficiently polyadenylated at sequences in the 3' end of the DHFR cDNA but rather utilized polyadenylation sites downstream from the DHFR cDNA. Three transformants produced DHFR mRNAs which were polyadenylated at the SV40 late polyadenylation signal. The fourth transformant apparently used cellular sequences for polyadenylation that were downstream from the site of integration of the cDNA.

The analysis of the DHFR mRNA expressed in these four transformants indicated that SV40 might increase the transformation efficiency by providing a more efficient polyadenylation signal. Alternatively, SV40 may have provided an enhancer element to increase DHFR expression. These possibilities were tested by introduction of the SV40 early polyadenylation signal into the 3' end of the DHFR cDNA in pAdD26-1 [(5) in Fig. 1]. This plasmid transformed DHFR$^-$ cells to the DHFR$^+$ phenotype 10-fold more efficiently than similar plasmids containing the entire SV40 genome. Thus, the presence of an efficient polyadenylation signal significantly increased expression from the

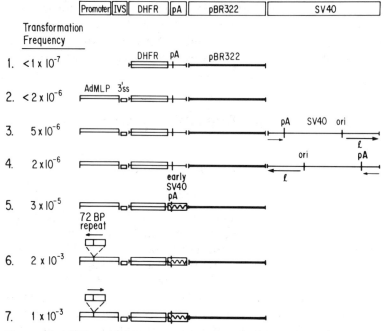

Figure 1 DHFR⁻ to DHFR⁺ transformation frequency from various cDNA genes. Calcium phosphate-mediated transfection of plasmid DNA (1 μg/10⁶ cells) in the absence of carrier DNA into CHO DHFR⁻ cells has been described (40). Transformation frequencies are determined from the number of DHFR⁺ transformants arising per number of cells plated into selective media. Recombinants indicated are (1) pDHFR26; (2) pAdD26-1; (3) pASD11; (4) pASD12 which contain the entire SV40 genome in either orientation cloned into the recombinants by their EcoRI sites (in the late region of SV40 and in pBR322 of pAdD26-1); (5) pAdD26SVp(A)3) which contains an early SV40 polyadenylation site; and (6) pCVSVE and (7) pCVSVL which contain the 72-bp repeat in either orientation. The arrows depict the late transcription unit of SV40 in (3) and (4) and the direction of the SV40 late promoter in (6) and (7). Also shown is the inefficient polyadenylation site in the 3' end of the DHFR cDNA and the location of polyadenylation sites supplied by SV40 sequences (pA).

cDNA gene. The introduction of the SV40 enhancer element in either orientation 250 bp upstream from the adenovirus major late promoter resulted in an additional 50- to 100-fold increase in the transformation efficiency [(6) and (7) in Fig. 1]. This indicated the SV40 enhancer can increase expression from the adenovirus major late promoter. With the expression vector that contains the SV40 enhancer, adenovirus major late promoter, 3' splice site, DHFR cDNA, and the SV40 early polyadenylation site, one in a thousand cells were transformed to the DHFR⁺ phenotype.

Amplification and Expression of Transfected cDNA Genes

COAMPLIFICATION OF TRANSFECTED GENES

The copy number of a wide variety of genes can be amplified as a consequence of applying appropriate selective pressures. However, direct selection methods are not available to obtain amplification for many genes. In these cases it has been possible to cotransfect the gene to be amplified with an amplifiable marker gene and subsequently select for amplification of the marker gene to generate cells that have amplified the desired gene. DHFR is the most widely used and understood amplifiable marker gene for this purpose. CHO cells deficient in DHFR have been used to express a variety of proteins by cotransfection and coamplification with a DHFR cDNA gene (30, 39, 43, 49, 63, 88).

An example of this latter approach is the development of a CHO cell line that produces high levels of human tissue-type plasminogen activator (t-PA), a potential thrombolytic agent (39). The DHFR cDNA gene utilized for these experiments [pAdD26SVp(A)3; Fig. 1] lacks an enhancer element and thus transforms CHO DHFR⁻ cells inefficiently. However, cotransfection of this DHFR cDNA gene with a t-PA cDNA gene containing an enhancer increased the efficiency of transformation. This increase occurred by ligation of the two separate cDNA genes resulting in enhancement of DHFR expression by the t-PA associated enhancer element. Efficient DHFR expression was thus dependent on continued association with the enhancer from the t-PA cDNA gene. As a consequence, when cells were selected for increasing MTX resistance, amplification of both transcription units occurred and cells were obtained that expressed high levels of human t-PA. Figure 2 shows t-PA synthesis in amplified CHO cells analyzed by [^{35}S]methionine labeling of CHO cells in culture for either 30 minutes to examine intracellular protein or for 4 hours to examine secreted protein. Gel electrophoresis of total intracellular protein demonstrated the rate of t-PA synthesis correlated with increased MTX resistance (Fig. 2) and DHFR and t-PA gene copy number. tPA synthesis in cells resistant to 20 μm MTX represented 5 to 10 percent of the total protein synthesis (several-fold higher than actin, a major cellular protein). Secreted t-PA represented more than 50 percent of the total secreted protein (lane 3), was immune-precipitated with a polyclonal antibody directed toward human t-PA (lane 2), and comigrated with t-PA synthesized by a human melanoma cell line (lane 4). The diffuse band for secreted t-PA is attributed to glycosylation patterns. Analysis of the t-PA synthesized by CHO cells indicated that it has approximately the same specific activity as native human t-PA, is glycosylated in a similar but not identical manner, and has an identical amino-terminal sequence (data not shown).

The linearity observed between MTX resistance, DHFR expression, and t-PA expression indicated that CHO cells have the capacity to accommodate the synthesis, glycosylation, and secretion of significantly large quantities of heterologous proteins. Cell lines have presently been established that express high levels of γ-interferon (30, 88), β-interferon (63), Herpes simplex virus

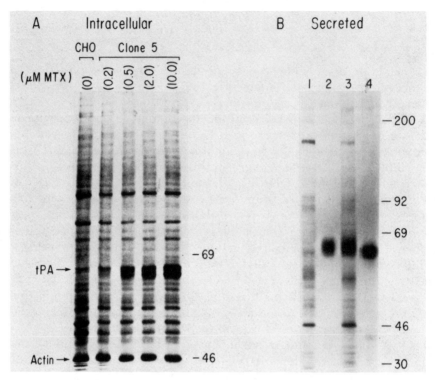

Figure 2 Polyacrylamide gel electrophoresis of intracellular and secreted proteins from transformed CHO cells. Cells were labeled with [^{35}S] methionine for (A) 30 minutes or (B) 4 hours, and cell extracts and conditioned media were prepared and analyzed as described in the text. (A) represents approximately equal numbers of counts from cell extracts isolated from the original DHFR CHO cell line, and from a DHFR t-PA-producing clone selected in 0.2, 0.5, 2.0, and 10.0 μM MTX. (B) represents total labeled secreted protein from the original DHFR⁻ CHO cells (lane 1) and a t-PA producing clone propagated 0.1 μM-MTX (lane 3). Also shown are results from conditioned media immunoprecipitated with a rabbit antihuman t-PA polyclonal antibody for the CHO clone in 0.1 μM MTX and for the original Bowes human melanoma cell line (lanes 2 and 4, respectively). [See (39) for details.]

glycoprotein D (8), and human t-PA (39). Coamplification in CHO cells is generally applicable to the development of cell lines to produce high levels of proteins that may prove therapeutically useful but have been previously difficult to obtain in sufficient quantities by other means.

In addition to the use of coamplification for the development of cell lines to facilitate purification of desired proteins, cell lines can be engineered to study the effect of increase in gene dosage on cell metabolism or gene regulation. A good example of the utility of this approach is the expression of antisense mRNA in order to block expression of specific genes (32). This approach allows for the development of cell lines that harbor defects in expression of specific genes that were previously difficult to obtain by classical genetic selection. In order to coamplify a particular DNA sequence, it is important that the DNA sequence not interfere with amplification or be toxic to the cell. Since a minor subset of cells from the total population survives during each step of the MTX selection and amplification process, it is possible to select for mutations in the coamplified gene that allow cell viability. Thus, the protein expressed at high level may be different from the native protein. One example of mutation occurring upon amplification has been the high-level expression of SV40 small t antigen (43). CHO cells selected to 1 μm MTX resistance after cotransfection and amplification with DHFR express a normal t antigen monitored by polyacrylamide gel electrophoresis. Further selection to resistance of 40 μm MTX rendered cells in which small t-antigen expression increased 5- to 10-fold. However, the t antigen from these latter cells migrated at a molecular weight 2000 daltons less than the wild-type t antigen. While the basis of this change is not understood, it demonstrates the potential difficulty in assuring the fidelity of proteins produced after selection for gene amplification.

TRANSFECTED cDNA GENES AND THE ANALYSIS OF GENE AMPLIFICATION

Amplification of transfected DNA has been compared to amplification of endogenous genes. Since transfected DNA integrates in different chromosomal locations, it is possible to compare and contrast amplification of DNA sequences in different chromosomal locations. Results indicate that DNA sequences in different chromosomal locations have dramatically different potential for amplification (43, 107). It is not known if this reflects proximity to some specific sequence or to its general location in the chromosome. Transfected sequences that are more unstably associated with the host DNA appear to be amplified preferentially to those that are stably associated (43). For any one transformant, either a subset or all of the transfected sequences may be amplified. This latter result is consistent with a hypothesis that a gradient of DNA amplification exists (84). Transfected sequences at the center of the gradient may be amplified frequently whereas those more distal would be less likely to be amplified. It is not known what sequences determine the center of the gradient, but they might function as an origin of replication.

The chromosomal aberrations associated with amplification of transfected DNA are similar to those found with amplification of endogenous genes. Amplification of DHFR cDNA genes transfected into murine 3T3 cells are localized to double minute chromosomes (73), similar to amplified endogenous DHFR genes in murine 3T cells (38). In contrast, amplification of DHFR cDNA genes transfected into CHO cells results in expanded chromosomes (37, 39, 49) originally designated homogeneously staining regions (HSRs) due to their appearance after Giemsa-trypsin banding procedures (9). These HSRs are similar to the frequent karyotypic alterations seen in CHO cells after amplification of the endogenous DHFR gene (77). In a few cases, the amplified transfected DNA may be present with little cytological perturbation (37). This may reflect a smaller size of the amplified unit. Further analysis of the HSRs in CHO cells after DHFR cDNA gene transfection and amplification demonstrated that the amplified region may be either AT or GC rich (37). Thus, there may be qualitative differences in the composition of the DNA surrounding the amplified sequences in the resulting HSRs. In addition, all transfected and amplified sequences studied to date have been found to replicate early in the S phase of the cell cycle (37). This is again similar to amplified endogenous genes (29, 44). Whether this reflects a requirement for expression of the transfected DNA (i.e., most DNA that is expressed is replicated early in the S phase) or whether it reflects a requirement for DNA amplification, is not known.

Frequently transfected and amplified DNA is found in HSRs associated with dicentric chromosomes (Fig. 3). This finding has led to the proposal that bridge breakage-fusion cycles (17, 37, 62) may be involved in some gene amplifications. The loss of telomeric function with subsequent bridge breakage-fusion cycles could be important for generating chromosome aberrations and perhaps duplications and instability in DNA transfer experiments. Recombination of exogenous DNA with a chromosome might displace or disturb function of a telomere, thus rendering the chromosome susceptible to instability (Fig. 3). After chromosome replication both sister chromatids would contain abnormal termini which might spontaneously fuse to form chromosomal loops or circles, and eventually dicentric chromosomes. Repeated cycles would generate duplicate inverted chromosomal regions and stabilization could result from acquisition of telomere function, perhaps by chromosome translocation. Chromosomal inversions and translocations have been observed in DNA transfection experiments (5, 37, 85). Some chromosomal amplification in situations not involving DNA transfer could be generated by bridge breakage-fusion events. Examples of amplified endogenous DNA being associated with chromosomal structures that may have resulted from bridge breakage-fusion cycles are chromosome circles (11), chromosomal inversions (65), and dicentric chromosomes (67, 108).

Regulation of Molecular DHFR cDNA Genes

Transcriptional Regulation of DHFR cDNA Genes

The first example of the use of DHFR cDNA genes to study gene regulation was the addition of the mouse mammary tumor virus (MMTV) promoter to a DHFR cDNA gene (52). CHO DHFR$^-$ cells selected for the DHFR$^+$ phenotype after DNA transfection expressed DHFR from the MMTV promoter. Addition of dexamethasone resulted in a 3- to 5-fold increase in DHFR synthesis due to an increase in the DHFR MRNA level. This approach provides a unique method to study the genes and proteins that mediate the dexamethasone-induced expression from the MMTV-DHFR hybrid gene. The

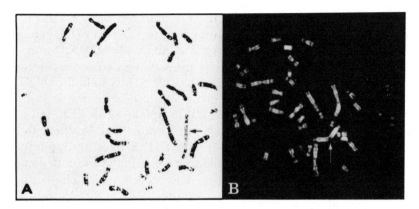

C   DNA Incorporation and Bridge–Breakage–Fusion Cycles

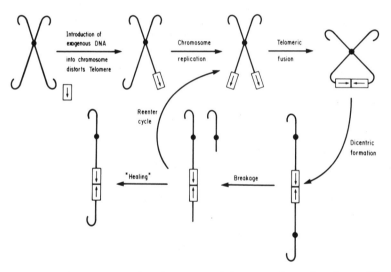

Figure 3 Chromosomal regions containing transfected and amplified DNA. (A) Metaphase with an HSR in a dicentric chromosome (arrow) stained with Giemsa-trypsin. (B) Metaphase spread stained with quinicrine that contains a chromatid break in a dicentric chromosome. (C) Hypothetical scheme for DNA incorporation and bridge breakage-fusion cycles [see (37)].

ability to isolate variants based on differing DHFR levels in response to dexamethasone by fluorescent MTX staining and fluorescence-activated cell sorting (36) will facilitate analysis of the mechanism of steroid hormone induction of gene expression.

Another approach utilizing modular DHFR cDNA genes to study transcriptional activation of DHFR has been presented by Kingston et al. (45). In this study a DHFR cDNA was attached to the adenovirus 2 (Ad 2) early region 2 promoter. This promoter is susceptible to trans-activation by the adenovirus E1a gene product. DHFR$^-$ cells were selected for the DHFR$^+$ phenotype after transfection of early region 2 promoter–DHFR cDNA gene chimeras. The introduction of an E1a transcription unit into the DHFR$^+$ transformants was monitored by the ability of cells to grow in increasing concentrations of MTX. Cells transfected with E1a grew more efficiently in higher MTX concentrations than control transfected cells. The ability of these cells to grow in MTX was shown to result from stable uptake of the E1a gene. Although three species of transcripts are derived from the E1a gene, the 9S, 12S, and the 13S mRNAs, the ability to induce DHFR transcription from the Ad 2 early region 2 promoter was shown to involve only the 13S E1a gene transcript. This transcript encodes a 289 amino acid protein which elicits immortalization of cells upon adenovirus transformation. Analysis of transcription from episomal DNA after transient transfection has indicated the 289 amino acid E1a gene product acts to increase transcription generally for promoters that require enhancers (26, 104). The latter studies have analyzed expression of episomal DNA upon transient transfection. Kingston et al. have shown that a chromosomally integrated E2 promoter also can be regulated in a similar manner to episomal DNA (45).

A similar approach has been taken to show that both the E1a and c-myc gene products can increase expression from a *Drosophila* heat shock promoter–DHFR cDNA chimeric gene (46). Presently, this is the only observation that indicates that the c-myc gene product can act as a trans-acting transcriptional activator. This is provocative due to the limited amino acid homology between c-myc and E1a (80) and the finding that both E1a and c-myc can complement ras in transformation of primary cells (48, 86). However, deletion analysis has indicated that adenovirus E2 promoter induction by E1a does not involve any specific DNA sequence and appears to overlap with transcription initiation signals whereas the c-myc-inducible element of the *Drosophila* heat shock promoter lies greater than 200 bp upstream from the mRNA cap site, separate from transcription initiation signals. [See (47) for review.]

The interaction of trans-acting factors with promoters may well be studied with DHFR cDNA genes. Recent studies have demonstrated that E1a may act not only as an activator of certain genes, but also as an inhibitor of enhancer activity (10, 106). A greater understanding of the cis- and trans-acting elements may be provided by studying the effect of trans-acting elements on the expression of integrated chimeric cDNA genes. As a hypothetical example, it should be possible to directly isolate the gene(s) that are responsible for transcriptional activation by their transfection into cells containing hybrid DHFR cDNA genes expressed from promoters and enhancers that are responsive to the transcriptional activator. Transfected cells capable of growth in higher levels of MTX should have acquired DNA encoding functions necessary to enhance expression from the hybrid DHFR cDNA genes. This assay should allow for the identification and cloning of genes necessary for activation of the promoter in question. However, this approach does have several potential difficulties. First, transfected DNA at different locations in the chromosome may respond differently to trans-acting factors. This differential response may be overcome by the analysis of several independent clones or by analyzing pools of initial transformants. Second, it has recently become clear that transcriptional control signals may involve more than sequences 5′ from the cap site for mRNA biogenesis. One well-studied example is the expression of the immunoglobulin mu heavy-chain gene (6, 23, 27, 79) which requires signals downstream from the mRNA cap site for mRNA transcription. Thus, when constructing hybrid transcription units, one must be concerned that all elements are present that are responsible for regulated transcription of the gene in question. Finally, there may be a number of posttranscriptional events that also mediate alterations in gene expression.

Posttranscriptional Regulation of Amplified DHFR cDNA Genes

Many enzymes involved in DNA precursor biogenesis are potently expressed during the replication (S) phase of the cell cycle (11, 31, 56, 75, 100, 105, 109). The levels of these proteins also decline as cells are arrested in growth or withdraw from the cell cycle during terminal differentiation (66). DHFR is one of these enzymes that has been extensively studied because of the availability of cells that produce high levels of DHFR as a result of gene amplification. In this case, gene amplification has facilitated a molecular analysis of the events involved in growth-dependent gene expression.

Analysis of amplified endogenous DHFR genes has demonstrated that regulation of DHFR mRNA levels occurs posttranscriptionally. Agents that promote cell growth, such as insulin, dibutryl cyclic GMP, and serum, all act to increase DHFR mRNA levels in MTX-resistant cells that harbor a high degree of amplification of the DHFR gene. All the amplified copies appear to correctly respond to regulatory stimuli that affect a single copy gene. Growth stimulation does not appear to affect the transcription of the DHFR gene but rather affects the efficiency of maturation and/or stability of DHFR transcripts (54, 55).

To examine this regulation in more detail, DHFR cDNA genes have been introduced into CHO DHFR⁻ cells and amplified by selection of cells resistant to MTX. Their regulation has been studied by monitoring changes on growth stimulation in confluent cells that have been subcultured into fresh media (42). A variety of transcription units have been constructed and analyzed after amplification in CHO cells as described in Fig. 4. All transcription units utilize the adenovirus major late promoter for transcription initiation, have 5′ and 3′ splice sites for RNA processing, contain the intact DHFR coding region, and have different sequences 3′ of the DHFR coding region. Several cell lines utilize the SV40 late polyadenylation signal as well as the DHFR polyadenylation signals within the 3′ end of the DHFR cDNA. Other cell lines utilize the SV40 early polyadenylation signal as well as signals present in the 3′ end of the DHFR cDNA. Finally, one line has recombined the transfected DNA so it utilizes a polyadenylation site downstream from the

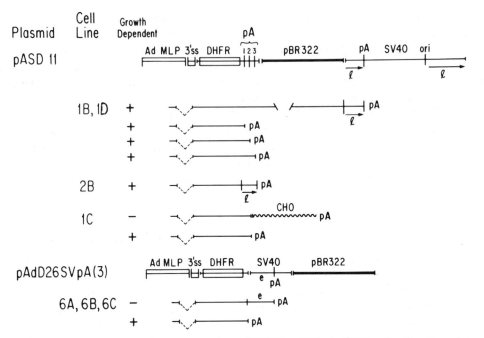

Figure 4 Growth-dependent expression of DHFR mRNAs in CHO cells after transfection and amplification of DHFR cDNA genes. The major DHFR mRNA in independent transformants isolated by transformation with pASD11 (see Fig. 1) are indicated. In all transformants, the adenovirus major late leader is properly spliced to the 3′ splice site. In CHO transformants 1B and 1D, the major DHFR mRNA contains an additional segment of RNA (approximately 1 kb) which is removed, presumably by RNA splicing of sequences in pBR322. The mRNA is polyadenylated at the SV40 late polyadenylation signal. Transformant 2B has a rearrangement of the DNA sequences so that the 3′ end of the SV40 late transcription unit and the polyadenylation signal are positioned 3′ to the DHFR segment. The 3′ end of DHFR mRNA in transformant 1C is derived from cellular DNA sequences that are adjoined to the plasmid DNA sequences. Transformants 1B, 1C, and 1D also utilize, at a low frequency, polyadenylation signals (1, 2, 3) in the 3′ end of the DHFR cDNA. The bottom depicts the structure of pAdD26sVp(A)3 with the two potential DHFR mRNA species derived by polyadenylation at a DHFR polyadenylation signal (the lower species) or at the SV40 early polyadenylation signal (epA). The column marked "growth dependent" shows whether the particular mRNA in that cell line is expressed in a growth-dependent manner (+) or is expressed constitutively (−). For details see (42).

integration site of the transfected DNA as well as a site in the 3′ end of the DHFR cDNA. RNA was harvested from growth-arrested confluent cells and from similar cells subcultured into fresh media. By hybridization of RNA to 3′-end labeled probes, and analysis of S1 nuclease digestion products by gel electrophoresis, the abundance of different 3′ ends was determined. The results are summarized in Fig. 4.

There is growth-dependent regulation of the level of those mRNAs that contain polyadenylation sites in the 3' end of the DHFR cDNA or within the SV40 late polyadenylation signal. In contrast, DHFR mRNAs polyadenylated at the SV40 early site or at a random site in CHO DNA exhibit constitutive mRNA expression. Cells that have two mRNA species exhibit a shift from the regulated species (+) to the constitutively produced species (−) upon growth arrest. Nuclear run-off experiments that measure the rate of transcription of the amplified DHFR cDNA demonstrated that the difference in mRNA levels (up to 20-fold in some cell lines) during growth stimulation was not due to a difference in the transcription rate of the DHFR cDNA gene (data now shown). Thus, posttranscriptional processing likely rendered these genes susceptible to growth regulation. This form of regulation has been observed for a variety of genes (13, 16, 55).

These results suggest that the rates of synthesis and degradation of mRNA with polyadenylation sites in DHFR and in the SV40 late region might be growth dependent. The growth dependence could be explained if the physiological state of cells in S phase favor accumulation of these mRNAs. The early functions of SV40 stimulate cells to enter S phase, before late mRNAs accumulate. The restricted accumulation of SV40 late mRNAs at early times after infection may be posttranscriptional. This would be consistent with the detection of transcription from the late region at early times after infection (22). These experiments also suggest that sequences at the 3' end of the transcription unit can have profound effects on the stability of mRNAs produced in growing and resting cells.

There is other evidence that alternate utilization of polyadenylation signals may modulate gene expression. Mitogenic stimulation of resting B-cell lymphocytes results in an upstream shift in the utilization of polyadenylation sites specifying an mRNA for a secreted mu heavy-chain immunoglobulin as opposed to a membrane-bound mu heavy chain (2, 20). Shifts in utilization of polyadenylation sites have also been observed in the maturation of transcripts from the adenovirus major late promoter (1, 76, 92) and also in the tissue-specific expression of the calcitonin gene (4).

Efficient cDNA Expression Vectors for Transient Expression in COS Cells

Translational Control Signals from Human Adenovirus Augment Expression from cDNA Genes

The development of stably transformed cell lines to study gene expression and to produce specific proteins is time consuming and requires a significant effort. An alternative approach uses the transient expression of heterologous genes. A useful transient transfection system utilizes recombinant plasmids containing a functional promoter fused to the coding region of interest and an SV40 origin of replication for transfection of monkey COS cells which express

SV40T antigen. As a result of SV40-mediated DNA replication, the plasmid copy number increases to greater than 10,000 copies in transfected COS cells, and mRNA expression is easily detected. However, relative to the mRNA levels, the levels of protein expressed in COS cells have been low. It is possible to overcome this translational inefficiency by introducing adenovirus translational control signals into COS cell cDNA expression vectors. Resultant expression vectors yield 10- to 20-fold greater levels of protein for the same level of mRNA.

Early after adenovirus infection, mRNA is expressed from at least seven regions of the adenovirus genome. At late times after infection expression from the early region is repressed and the majority of adenovirus-specific mRNAs comprise a set of transcripts expressed from the adenovirus major late promoter. These mRNAs contain a common tripartite 5′ untranslated leader that is spliced together from three small exons (7). In addition, late infected cells contain an abundance of small polymerase III transcripts, the virus-associated RNA (VA RNA) I and II (60, 82, 96). The VA RNAs are transcribed from the adenovirus genome and are approximately 160 nucleotides in length. A small proportion of VA RNA is found associated in a ribonucleoprotein particle with the cellular lupus antigen, La (53). Involvement of the VA RNAs in translation was first suggested by studies with mutant adenoviruses that contained deletions in the VA I gene (101). Infection of cells with this mutant virus proceeds normally through the early phase of infection but is defective in the translation of late viral proteins. The late mRNAs are normal in structure and can be translated *in vitro* but are inefficiently translated *in vivo*. Further studies indicated that the VA RNA may function in translation to facilitate mRNA interaction with a 43S ribosomal preinitiation complex to form a 48S species (94). Other studies with recombinant adenovirus genomes demonstrated that the full adenovirus tripartite leader is required for efficient translation of the late viral mRNAs (57, 102). These studies suggest that the VA RNA plus the tripartite leader may be critical for highly efficient translation of adenovirus late mRNAs, but are complicated due to the multitude of changes that occur in the cell at late times of the adenovirus infectious cycle.

In order to study the translational control signals from adenovirus, a transient expression system was devised to dissect the roles of the tripartite leader and the VA genes in mRNA translation (40). Hybrid transcription units containing the adenovirus major late promoter fused to various coding regions were introduced into COS monkey cells in the presence or absence of VA RNA provided either by adenovirus infection of transfected cells or by co-transfection with plasmids containing the VA genes. Figure 5A shows four plasmids designed to express human γ-interferon. pγIF-6 contains only the adenovirus first late leader. The three other plasmids contain a cDNA copy of the adenovirus tripartite leader. All plasmids contain the SV40 origin of replication and the SV40 enhancer element upstream from the adenovirus major late promoter, a small intron, the γ-interferon coding region, a 3′ untranslated region derived from DHFR, and the SV40 early polyadenylation site. pQ2 and pQ3 contain the VA I and VA II genes inserted in either orientation.

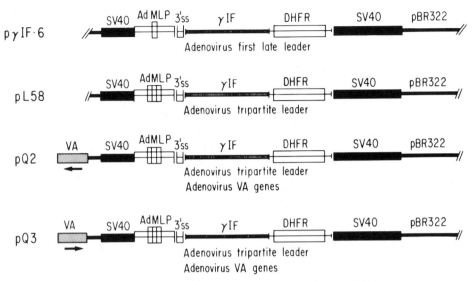

Figure 5A Adenovirus translational controls in γ-interferon cDNA expression vectors. Plasmids that contain the SV40 origin of replication and enhancer element, the adenovirus major late promoter, a 3′ splice site, the human γ-interferon coding region, a 3′ segment from murine DHFR cDNA, the SV40 early polyadenylation signal, in a derivative of pBR322 are depicted (41). pγIF-6 contains only the first leader from the adenovirus major late transcripts, whereas pL58, pQ2, and pQ3 contain in addition the spliced copy (derived from a cDNA clone) of the second and majority of the third leaders from adenovirus late mRNAs. pQ2 and pQ3 contain the adenovirus VA genes I and II in the orientation shown. The arrows indicate the transcriptional orientation of the VA genes.

Results from transfection of these plasmids into COS cells and [^{35}S]methonine labeling with subsequent immune-precipitation of the conditioned media are shown in Fig. 5B. Very little difference in the synthesis of γ-interferon is detected between pXIF-6 and pL58. However, both pQ2 and pQ3 express up to 20-fold greater levels of γ-interferon. Northern analysis indicated that the γ-interferon mRNA level was similar from all samples (41). Thus, the difference in expression can be attributed to an increased translational efficiency of the γ-interferon mRNA.

Other experiments indicated that the presence of VA RNA only elicited a 2-fold increase in γ-interferon expression from the plasmid containing a single first late leader (pγ IF-6). In addition, when the orientation of the second and third late leader was altered, the translational stimulation by VA RNA was dramatically reduced.

These experiments demonstrated that adenovirus VA RNA can enhance translation in the absence of any other adenovirus gene product. Although the mechanism by which VA RNA enhances translation is presently not understood, recent experiments have yielded significant insights (81). When *in vitro* translation extracts from HeLa cells infected with either wild-type or VA-mu-

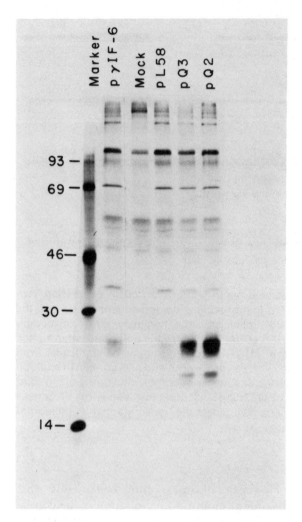

Figure 5B Human γ-interferon synthesis in transfected COS cells. COS cells were transfected with the indicated plasmid DNAs and 56 hours later cells were labeled with [³⁵S]methionine for 2 hours. Equal aliquots of conditioned labeled media were immunoprecipitated with a mouse antihuman γ-IF monoclonal antibody and electrophoresed on an SDA-polyacrylamide gel. Mock refers to COS cells that received no DNA. Sizes on the left are shown as molecular weight in kilodaltons [see (41) for details].

tant adenoviruses were compared, it was found that only the VA mutant extract was deficient in translation and that addition of VA RNA could not restore translatability. However, addition of eukaryotic initiation factor eIF-2 could restore translation. By analogy to the translation block in reticulocytes in which conditions such as haem deficiency and the presence of double-stranded RNA. Indeed, recently it has been demonstrated that VA I RNA subunit of eIF-2 (61), it has been proposed that upon adenovirus infection, virally derived double-stranded RNA can activate the double-stranded RNA-dependent protein kinase which phosphorylates the α subunit of eIF-2 causing its inactivation. Since VA RNA lacks extended regions of perfect duplex structure necessary to activate the protein kinase, it has been proposed that VA RNA may bind and actually inhibit activation of the kinase by double-stranded RNA (81). Indeed, recently it has been demonstrated that VA I RNA can prevent phosphorylation of the α subunit of eIF-2 (95). This mechanism may also account for the effect of VA RNA on translation in COS cells. Due to the plasmid DNA replication in COS cells, transcription can occur on both DNA strands of the plasmid leading to the formation of double-stranded RNA to similarly activate the double-stranded RNA-dependent protein kinase. Thus, as in adenovirus-infected cells, VA RNA may be expected to restore translatability in transfected COS cells.

Another unanswered question related to translational control mediated by VA RNA is whether there is any mRNA preference to the translational stimulation. Experiments have demonstrated that translation of mRNAs containing either the adenovirus tripartite leader or the SV40 early leader can be increased by VA RNA, and that major alterations in the tripartite leader decrease the translational enhancement (41). Thimmappaya (unpublished) has recently found that translation of mRNAs containing the SV40 early leader but not the SV40 late leader can be enhanced by VA RNA. It is interesting to note that the SV40 early leader exhibits some homology to the second and third leaders of the adenovirus tripartite leader (41). Although this sequence may be relevant to the translational stimulation by VA RNA, the mouse β-globin leader, which does not contain sequence homology to the adenovirus tripartite leader, also exhibits increased translation by the presence of VA RNA (99). Thus, it appears that there is no absolute sequence requirement for translational enhancement by VA RNA. Certainly, alteration in the level of translation initiation factors might result in a differential shift in mRNA translation due to mRNA competition for these factors.

The translational stimulation mediated by VA RNA is not limited to COS cells. Monkey CV1, HeLa, and adenovirus transformed human 293 cells all exhibit a VA stimulation of translation (41, 98). The translation of both cytoplasmic (DHFR and chloramphenicol acetyltransferase) and secreted (γ-interferon, human interleukin II, and human proinsulin) proteins can be stimulated by VA RNA. COS cells transfected with plasmid pQ2 produce over 1 μg of γ-interferon per 10^6 cells 72 hours after transfection compared to 0.1 μg after transfection with pL58. Thus, the increased translational efficiency of heterologous mRNAs in COS cells mediated by VA RNA provides an improved means of expressing heterologous cDNAs.

Isolation of cDNA Clones by Screening for Expression in COS Cells

When the γ-interferon expression plasmid (pQ2) was diluted 1000-fold with other plasmid DNA and transfected into COS cells, γ-interferon activity could be assayed in the conditioned media of transfected cells 72 hours posttransfection. Thus, it appeared possible to directly screen and isolate cDNA clones contained in SV40 derived plasmids on the basis of their expression in COS cells, provided there is a sensitive assay available for detection of the desired protein. This approach (diagrammed in Fig. 6) has been applied to the cloning of human granulocyte–macrophage colony-stimulating factor (GM-CSF) cDNA (111). Many factors (known as colony-stimulating factors, CSFs) are necessary for the proliferation and differentiation of hematopoetic

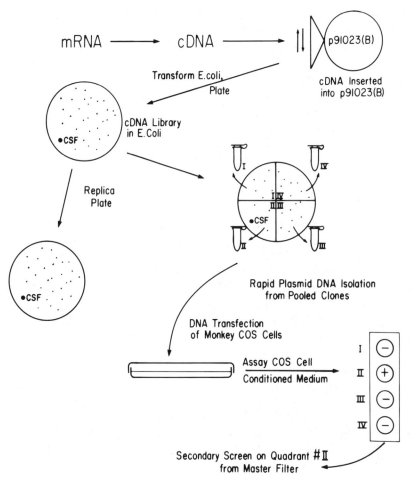

Figure 6 Scheme for cDNA cloning by transient expression in COS cells. A cDNA library is introduced into p91023(B), a plasmid identical to pQ2 (see Fig. 5) except that it lacks the γ-interferon sequence. Multiple DNA preparations from pools of recombinants grown in *E. coli* are transfected into COS cells, and samples of the conditioned medium are taken for assay. The master filters for which the assay results are positive are then subdivided and analyzed in a secondary screen.

progenitor cells and their activity is measured by their ability to stimulate progenitor cells to form colonies in a semisolid medium. The CSFs are classified by the type of mature blood cells found in the resulting colonies. GM-CSF stimulates the proliferation of the granulocyte and macrophage lineages. In order to clone GM-CSF, mRNA from a naturally occurring HTLV transformed T lymphoblast cell line which expressed GM-CSF was isolated and cDNA libraries were constructed in the expression vector p91023(B) (Fig. 6). DNA from 200 pools of approximately 500 recombinants each were prepared and tested for their ability to direct the synthesis of GM-CSF after transformed T lymphoblast cell line which expresses GM-CSF was isolated GM-CSF activity. One single CSF clone in each of three of the original six positives was identified by sibling selection. The three independent clones contained the same cDNA sequence and, by colony hybridization, the three clones in the remaining pools were also identified to be the same. Impressively, the abundance of the GM-CSF mRNA was estimated to be only 0.01 percent of total mRNA in the T-cell line.

The GM-CSF expression plasmid was used to transfect 4- to 5-L batches of COS cells to produce serum-free conditioned medium with GM-CSF activity 30-fold higher than that in the original T-cell conditioned medium. One milligram GM-CSF was purified to apparent homogeneity and compared to GM-CSF isolated from T-cell conditioned media. Amino terminal sequencing indicated the 17 amino acid signal peptide was properly removed in COS cells. Gel electrophoresis demonstrated both species to be similarly heterogeneous [18 to 24 kilodaltons (kd)], presumably due to heterogenety in glycosylation. The specific activity of the GM-CSF was indistinguishable from the native protein isolated from the T-cell line. The recombinant protein also had neutrophil migration inhibitory activity. Thus, a single hematopoeitin can not only stimulate proliferation and differentiation of progenitor cells in bone marrow but can also activate the biological function of the resulting circulating mature cells. These experiments have demonstrated the utility of an efficient cDNA expression vector both for the isolation of a full length, biologically active cDNA clone, and for the biological and biochemical characterization of protein prepared from cells transfected with that clone (111).

Conclusions

The utility of cDNA gene expression has been reviewed with particular emphasis given to how DHFR cDNA genes can be applied to study gene regulation and gene amplification as well as to express heterologous genes at high levels in mammalian cells. Although cDNA genes have applications to a wide variety of different vector systems for mammalian cells, this chapter has focused on the use of DHFR cDNA genes in DNA transfer experiments involving DHFR-deficient CHO cells and also on the utility of efficient cDNA expression systems for monkey COS cells. As the ability to manipulate DNA becomes more convenient, these two approaches will have considerable importance in both understanding gene regulation and in the characterization of proteins that harbor specific alterations.

General References

GLUZMAN, Y. (ed.) 1982. *Eukaryotic Viral Vectors,* Cold Spring Harbor Laboratory, Cold Spring Harbor, N. Y.

STARK, G. R., and WAHL, G. M. 1984. Gene Amplification, *Annu. Rev. Biochem.* 53:447–491.

References

1. Akusjarvi, G., and Persson, H. 1980. Controls of RNA splicing and termination in the major late adenovirus transcription unit, *Nature (London)* 292:420–426.

2. Alt, F. W., Bothwell, A. L. M., Knapp, M., Siden, E., Mather, E., Koshland, M., and Baltimore, D. 1980. Synthesis of secreted membrane-bound immunoglobulin mu heavy chains is directed by mRNAs that differ at their 3' ends, *Cell* 20:293–301.

3. Alt, F. W., Kellems, R. E., Bertino, J. R., and Schimke, R. T. 1978. Selective amplification of dihydrofolate reductase genes in methotrexate-resistant variants of cultured murine cells, *J. Biol. Chem.* 253:1357–1370.

4. Amara, S. G., Jonas, V., Rosenfeld, M. G., Ong, E. S., and Evans, R. M. 1982. Alternative RNA processing in calcitonin gene expression generates mRNAs encoding different polypeptide products, *Nature (London)* 292:420–426.

5. Andrulis, J. L., and Siminovitch, L. 1982. Amplification of the gene for asparagine synthetase, in: *Gene Amplification* (R. T. Schimke, ed.) pp. 75–80, Cold Spring Harbor Laboratory, Cold Spring Harbor, N. Y.

6. Banerji J., Olson, L., and Schaffner, W. 1983. A lymphocyte-specific cellular enhancer is located downstream of the joining region in immunoglobulin heavy chain genes, *Cell* 33:729–740.

7. Berget, S. M., Moore, C., and Sharp, P. A. 1977. Spliced segments at the 5' terminus of adenovirus 2 late mRNA, *Proc. Nat. Acad. Sci. U.S.A.* 74:3171–3175.

8. Berman, P., Dowbenko, D., Lasky, L., and Simonsen, C. 1983. Detection of antibodies to Herpes Simplex Virus with a continuous cell line expression cloned glycoprotein D, *Science* 222:524–527.

9. Biedler, J. L., and Spengler, B. A. 1976. Metaphase chromosome anomaly: association with drug resistance and cell specific products. *Science* 191:185–187.

10. Borrelli, E., Hen, R., and Chambon, P. 1984. Adenovirus-2 E1A products repress enhancer-induced stimulation of transcription, *Nature (London)* 312:608–612.

11. Bostock, C. J., and Tyler-Smith, C. 1981. Gene Amplification in methotrexate resistant mouse cells. II. Rearrangement and amplification of non-dihydrofolate reductase gene sequences accompany chromosomal changes, *J. Mol. Biol.* 153:219–236.

12. Brent, T. 1971. Periodicity of DNA synthetic enzymes during the HeLa cell cycle, *Cell Tissue Kinet.* 4:297–305.

13. Carneiro, M., and Schibler, V. 1984. Accumulation of rare and moderately abundant mRNAs in mouse L-cells is mainly post-transcriptionally regulated, *J. Mol. Biol.* 178:869–880.

14. Chasin, L. A., and Urlaub, G. 1980. Isolation of Chinese hamster cell mutants deficient in dihydrofolate reductase activity, *Proc. Nat. Acad. Sci. U.S.A.* 77:4216–4220.

15. Clark, R., Peden, K., Pipas, J. M., Nathans, D., and Tjian, R. 1983. Biochemical activities of T-antigen proteins encoded by simian virus 40 A gene deletion mutants, *Mol. Cell. Biol.* 3:220–228.

16. Collins, M. L., Wu, J. R., Santiago, C. L., Hendrickson, S. L., and Johnson, L. F. 1983. Delayed processing of dihydrofolate reductase heterogeneous nuclear RNA 6. *Mol. Cell. Biol.* 3:1792–1802.

17. Cowell, J. K., and Miller, O. J. 1983. Occurrence and evolution of homogeneously staining regions may be due to breakage-fusion-bridge cycles following telomere loss, *Chromosoma* 88:216–221.

18. DiMaio, D., Treisman, R. H., and Maniatis, T. 1982. Bovine papillomavirus vector that propagates as a plasmid in both mouse and bacterial cells, *Proc. Nat. Acad. Sci. U.S.A.* 79:4030–4034.

19. DiMaio, D., Corbin, V., Sibley, E., Maniatis, T. 1984. High-level expression of a cloned HLA heavy chain gene introduced into mouse cells on a bovine *papillomavirus* vector, *Mol. Cell. Biol.* 4:340–350.

20. Early, P., Rogers, J., Davis, M., Calame, K., Bond, M., Wall, R., and Hood, L. 1980. Two mRNAs with different 3′ ends encode membrane-bound and secreted forms of immunoglobulin chains, *Cell* 20:313–319.

21. Ferdinand, F., Brown, M., and Khoury, G. 1977. Characterization of early simian virus 40 transcriptional complexes: Late transcription in the absence of detectable DNA replication, *Proc. Nat. Acad. Sci. U.S.A.* 74:5443–5447.

22. Gething, M. J., Sambrook, J. 1982. Construction of influenza haemagglutinin genes that code for intracellular and secreted forms of the protein, *Nature (London)* 300:598–603.

23. Gillies, S. D., Morrison, S. L., Oi, V. T., and Tonegawa, S. 1983. A tissue-specific transcription enhancer element is located in the major intron of a rearranged immunoglobulin heavy chain gene, *Cell* 33:717–728.

24. Gluzman, Y. 1981. SV40-transformed simian cells support the replication of early SV40 mutants, *Cell* 23:175–192.

25. Graham, F., and Van der Eb, H. 1973. A new technique for the assay of infectivity of human adenovirus 5 DNA, *Virology* 52:456–467.

26. Green, M. R., Treisman, R., and Maniatis, T. 1983. Transcriptional activation of cloned human B-globin genes by viral immediate-early gene products, *Cell* 35:137–148.

27. Grosschedl, R., and Baltimore, D. 1985. Cell-type specificity of immunoglobulin gene expression is regulated by at least three DNA sequence elements, *Cell* 41:885–897.

28. Hamer D. H., Leder, P. 1979. SV40 recombinants carrying a functional RNA splice junction and polyadenylation site from the chromosomal mouse beta major globin gene, *Cell* 17:737–747.

29. Hamlin, J. L., and Biedler, J. L. 1981. Replication pattern of a large homogen-

eously staining region in antifolate-resistant Chinese hamster cell lines, *Cellular Physiology,* 107:101–114.

30. Haynes, J., and Weissman, C. 1983. Constitutive, long term production of human interferons by hamster cells containing multiple copies of a cloned interferon gene, *Nucleic Acids Res.* 11:687–706.

31. Howard, D., Hay, J., Melvin, W., and Durham, J. 1974. Changes in DNA and RNA synthesis and associated enzyme activities after the stimulation of serum-depleted BHK 21/c13 cells by the addition of serum, *Exp. Cell Res.* 86:31–42.

32. Izant, J. G., and Weintraub, H. 1985. Constitutive and conditional suppression of exogenous and endogenous genes by anti-sense RNA. *Science* 229:345–352.

33. Jackson, R. J. 1982. Cytoplasmic control of protein synthesis, in: *Protein Biosynthesis in Eukaryotes* (R. Perez-Bercoff, ed.) pp. 363–418, Plenum, New York.

34. Jagus, R., Anderson, W. F., and Safer, B. 1981. The regulation of initiation of mammalian protein synthesis, *Prog. Nucleic Acid Res. Mol. Biol.* 25:127–185.

35. Jolly, D. J., Okayama, H., Berg, P., Esty, A. C., Filpula, D., Bohlen, P., Johnson, G. G., Shively, J. E., Hunkapillar, T., and Friedman, T. 1983. Isolation and characterization of a full-length expressible cDNA for human hypoxanthine phosphoribosyltransferase, *Proc. Nat. Acad. Sci. U.S.A.* 80:477–481.

36. Kaufman, R. J., Bertino, J. R., and Schimke, R. T. 1978. Quantitation of dihydrofolate reductase in individual parental and methotrexate-resistant murine cells. *J. Biol. Chem.* 253:5852–5860.

37. Kaufman, R. J., Sharp, P. A., and Latt, S. A. 1983. Evolution of chromosomal regions containing transfected and amplified dihydrofolate reductase sequences. *Mol. Cell. Biol.* 3:699–711.

38. Kaufman, R. J., Brown, P. C., and Schimke, R. T. 1979. Amplified dihydrofolate reductase genes in unstably methotrexate-resistant cells are associated with double minute chromosomes. *Proc. Nat. Acad. Sci. U.S.A.* 76:5669–5673.

39. Kaufman, R., Wasley, L., Spilotes, L., Gossels, S., Latt S., Larsen G., and Kay, R. 1985. Coamplification and coexpression of human tissue-type plasminogen activator and murine dihydrofolate reductase sequences in Chinese hamster ovary cells, *Mol. Cell. Biol.* 5:1750–1759.

40. Kaufman, R. J., and Sharp, P. A. 1982. Construction of a modular dihydrofolate reductase cDNA gene: Analysis of signals utilized for efficient expression, *Mol. Cell. Biol.* 2:1304–1319.

41. Kaufman, R. J. 1985. Identification of the components necessary for adenovirus translational control and their utilization in cDNA expression vectors, *Proc. Nat. Acad. Sci. U.S.A.* 82:689–693.

42. Kaufman, R. J., and Sharp, P. A. 1983. Growth-dependent expression of dihydrofolate reductase mRNA from modular cDNA genes, *Mol. Cell. Biol.* 3:1598–1608.

43. Kaufman, R. J., and Sharp, P. A. 1982. Amplification and expression of sequences cotransfected with a modular dihydrofolate reductase cDNA gene, *Mol. Cell. Biol.* 159:601–621.

44. Kellems, R. E., Leys, E. J., Harper, M. E., and Smith, L. E. 1982. Control of dihydrofolate reductase DNA replication and mRNA production in: *Gene Amplification,* R. T. Schimke, ed. pp. 81–87, Cold Spring Harbor Laboratory, Cold Spring Harbor, N. Y.

45. Kingston, R. E., Kaufman, R. J., and Sharp, P. A. 1984. Regulation of transcription of the adenovirus EII promoter by EIa gene products: Absence of sequence specificity, *Mol. Cell. Biol.* 4:1970–1977.

46. Kingston, R. E., Baldwin, A. S., Jr., and Sharp, P. A. 1984. Regulation of heat shock protein 70 gene expression by c-myc, *Nature (London)* 312:280–282.

47. Kingston, R. E., Baldwin, A. S., and Sharp, P. A. 1985. Transcription control by oncogenes, *Cell* 41:3–5.

48. Land, H., Parada, L. F., and Weinberg, R. A. 1983. Tumorigenic conversion of primary embryo fibroblasts requires at least two cooperating oncogenes, *Nature (London)* 304:596–602.

49. Lau, Y. F., Lin, C. C., and Kan, Y. W. 1984. Amplification and expression of human α-globins genes in Chinese hamster ovary cells, *Mol. Cell. Biol.* 4:1469–1475.

50. Law, M. F., Lowy, D. R., Dvoretzky, I., and Howley, P. M. 1981. Bovine *papillomavirus* DNA is maintained as an extrachromosomal DNA element in transformed mouse cells, *Proc. Nat. Acad. Sci. U.S.A.* 78:2727–2731.

51. Lee, F., Yokota, T., Otsuka, T., Gemmell, L., Larson, N., Luh, J., Arai, K., and Rennick, D. 1985. Isolation of cDNA for a human granulocyte-macrophage colony-stimulating factor by functional expression in mammalian cells, *Proc. Nat. Acad. Sci. U.S.A.* 82:4360–4364.

52. Lee, F., Mulligan, R., Berg, P., Ringold, G. 1981. Glucocorticoids regulate expression of dihydrofolate reductase cDNA in mouse mammary tumor virus chimaeric plasmids, *Nature (London)* 294:228–232.

53. Lerner, M. R., Boyle, J. A., Hardin, J. A., and Steitz, J. A. 1981. Two novel classes of small ribonucleoproteins detected by antibodies associated with lupus ertyematosus, *Science* 211:400–402.

54. Leys, E. J., and Kellems, R. E. 1981. Control of dihydrofolate reductase messenger ribunucleic acid production, *Mol. Cell. Biol.* 1:961–971.

55. Leys, E. J., Crouse, G. F., and Kellems, R. E. 1983. Dihydrofolate reductase gene expression in cultured mouse cells is regulated by transcript stabilization in the nucleus, *J. Cell. Biol.* 99:180–187.

56. Littlefield, J. 1966. The periodic synthesis of thymidine kinase in mouse fibroblasts, *Biochim. Biophys. Acta* 114:398–403.

57. Logan, J., and Shenk, T. 1984. Adenovirus tripartite leader sequence enhances translation of mRNAs late after infection, *Proc. Nat. Acad. Sci. U.S.A.* 81:3655–3659.

58. Lusky, M., and Botchan, M. R. 1984. Characterization of the bovine papilloma virus plasmid maintenance sequences, *Cell* 36:391–401.

59. Mansour, S. L., Grodzicker, T., and Tjian, R. 1985. An adenovirus vector system used to express polyoma virus tumor antigens, *Proc. Nat. Acad. Sci. U.S.A.* 82:1359–1363.

60. Mathews, M. B. 1975. Genes for VA-RNA in adenovirus 2, *Cell* 6:223–229.

61. Mathews, M. B., and Francoeur, A. M. 1984. La antigen recognizes and binds to the 3′-oligouridylate tail at a small RNA, *Mol. Cell. Biol.* 4:1134–1140.

62. McClintock, B. 1941. The stability of broken ends of chromosomes in *Zea mays*, *Genetics* 26:234–282.

63. McCormick, F., Trahey, M., Innis, M., Dieckmann, B., and Ringold, G. 1984. In-

ducible expression of amplified human beta interferon genes in CHO cells, *Mol. Cell. Biol.* 4:166–172.

64. Mellon, P., Parker, V., Gluzman, Y., and Maniatis, T. 1981. Identification of DNA sequences required for transcription of the human α-globin gene in a new SV40-host vector system, *Cell* 27:279–288.

65. Melton, D. W., Konecki, D. S., Ledbetter, D. H., Hejtmancik, J. F., and Caskey, C. T. 1981. *In vitro* translation of HPRT mRNA: characterization of a mouse neuroblastoma cell line with elevated HPRT protein levels, *Proc. Nat. Acad. Sci. U.S.A.* 78:6977–6980.

66. Merrill, G. F., Hauschka, S. D., McKnight, S. L. 1984. Enzyme expression in differentiating muscle cells is regulated through an internal segment of the cellular tk gene, *Mol. Cell. Biol.* 4:1777–1784.

67. Milbrandt, J. D., Heintz, N. H., White, W. C., Rothman, S. M., and Hamlin, J. L. 1981. Methotrexate-resistant Chinese hamster ovary cells have amplified a 135 kb sequence which includes the gene for dihydrofolate reductase, *Proc. Nat. Acad. Sci. U.S.A.* 78:6043–6047.

68. Miller, A. D., Eckner, R. J., Jolly, D. J., Friendman, T., and Verma, I. M. 1984. Expression of a retrovirus encoding human HPRT in mice, *Science* 225:630–632.

69. Mitrani-Rosenbaum, S., Maroteaux, L., Mory, Y., Revel, M., Howley, P. M. 1983. Inducible expression of the human interferon B1 gene linked to a bovine papilloma virus DNA vector maintained extrachromosomally in mouse cells, *Mol. Cell. Biol.* 3:233–240.

70. Moriarty, A. M., Hoyer, B. H., Shih, J. W., Gerin, J. L., Hamer, D. H. 1981. Expression of the hepatitus B virus surface antigen gene in cell culture by using simian virus 40 vector, *Proc. Nat. Acad. Sci. U.S.A.* 78:2606–2610.

71. Mulligan, R. C., and Berg, P. 1980. Expression of a bacterial gene in mammalian cells, *Science* 209:1422–1427.

72. Mulligan, R. C., Howard, B. H., and Berg, P. 1979. Synthesis of rabbit beta-globin in cultured monkey kidney cells following infection with a SV40 beta-globin recombinant genome, *Nature (London)* 277:108–114.

73. Murray, M. J., Kaufman, R. J., Latt, S. A., and Weinberg, R. A. 1983. Construction and use of a dominant, selectable marker: A Harvey sarcoma virus-dihydrofolate reductase chimera, *Mol. Cell. Biol.* 3:32–43.

74. Myers, R., and Tjian, R. 1980. Construction and analysis of simian virus 40 origins defective in tumor antigen binding and DNA replication, *Proc. Nat. Acad. Sci. U.S.A.* 77:6419–6495.

75. Navalgund, L. G., Rossana, C., Meunch, A. J., and Johnson, L. F. 1980. Cell cycle regulation of thymidylate synthetase gene expression in cultured mouse fibroblasts, *J. Biol. Chem.* 255:7386–7395.

76. Nevins, J. R., and Darnell, J. E. 1978. Steps in the processing of Ad2 mRNA: Poly(A)+ nuclear sequences are conserved and poly(A) addition precedes splicing, *Cell* 15:1477–1493.

77. Nunberg, J. H., Kaufman, R. J., Schimke, R. T., Urlaub, G., and Chasin, L. A. 1978. Amplified dihydrofolate reductase genes are localized to a homogenously staining region of a single chromosome in a methotrexate-resistant Chinese hamster ovary cell line, *Proc. Nat. Acad. Sci. U.S.A.* 75:5553–5556.

78. Pavlakis, G. N., and Hamer, D. H. 1983. Regulation of a metallothionine-growth

hormone hybrid gene in bovine papilloma virus, *Proc. Nat. Acad. Sci. U.S.A.* 80:397–401.

79. Queen, C., and Baltimore, D. 1983. Immunoglobulin gene transcription is activated by downstream sequence elements, *Cell* 33:741–748.

80. Ralston, R., and Bishop, J. M. 1984. The protein products of the *myc* and *myb* oncogenes and adenovirus E1a are structurally related, *Nature (London)* 306:803–806.

81. Reichel, P. A., Merrick, W. C., Siekierka, J., and Mathews, M. B. 1985. Regulation of a protein synthesis initiation factor by adenovirus virus-associated RNA I, *Nature (London)* 313:196–200.

82. Reich, P. R., Forget, B. G., Weissman, S. H., and Rose, J. A. 1966. RNA of low molecular weight in KB cells infected with adenovirus type 2. *J. Mol. Biol.* 17:428–439.

83. Rio, D. C., and Tjian, R. 1983. SV40 antigen binding site mutations that affect autoregulation, *Cell* 32:1227–1240.

84. Roberts, J. M., Buck, L. B., and Axel, R. 1983. A structure for amplified DNA, *Cell* 33:53–63.

85. Robins, D. M., Axel, R., and Henderson, A. S. 1981. Chromosomal structure and DNA sequence alterations associated with mutation of a transformed gene, *J. Mol. Appl. Genet.* 1:91–123.

86. Ruley, H. E. 1983. Adenovirus early region 1A enables viral and cellular transforming genes to transform primary cells in culture, *Nature (London)* 304:602–606.

87. Sarver, N., Gruss, P., Law, M. F., Khoury, G., and Howley, P. M. 1981. Bovine papilloma virus deoxyribonucleic acid: A novel eucaryotic cloning vector, *Mol. Cell. Biol.* 1:486–496.

88. Scahill, S. J., Devos, R., Heyden, J. V., and Fiers, W. 1983. Expression and characterization of the product of a human interferon cDNA gene in Chinese hamster ovary cells, *Proc. Nat. Acad. Sci. U.S.A.* 80:4654–4658.

89. Schaffner, W. 1980. Direct transfer of cloned genes from bacteria to mammalian cells, *Proc. Nat. Acad. Sci. U.S.A.* 77:2163–2167.

90. Schimke, R. T. 1984. Gene amplification in cultured animal cells, *Cell* 37:705–713.

91. Schimke, R. T. (ed.). 1982. *Gene Amplification,* Cold Spring Harbor Laboratory, Cold Spring Harbor, N. Y.

92. Shaw, A., and Ziff, E. 1980. Transcripts from the adenovirus-2 major late promoter yield a single family of 3' coterminal mRNAs during early infection and five families at late times, *Cell* 22:905–916.

93. Southern, P. J., and Berg, P. 1982. Transformation of mammalian cells to antibiotic resistance with a bacterial gene under control of the SV40 early promoter region, *J. Mol. Appl. Genet.* 1:327–341.

94. Schneider, R. J., Weinberger, C., and Shenk, T. 1984. Adenovirus VA1 RNA facilitates the initiation of translation in virus-infected cells, *Cell* 37:291–298.

95. Schneider, R. J., Safer, B., Munemitsu, S. M., Samuel, C. E., and Shenk, T. 1985. Adenovirus VAI RNA prevents phosphorylation of the eukaryotic initiation factor 2α subunit subsequent to infection, *Proc. Nat. Acad. Sci. U.S.A.* 82:4321–4325.

96. Soderlund, H., Pettersson, U., Vennstrom, B., Philipson, L., and Mathews, M. B.

1976. A new species of virus-coded low molecular weight RNA from cells infected with adenovirus type 2, *Cell* 7:585–593.

97. Sompayrac, L. M., and Dana, K. J. 1981. Efficient infection of monkey cells with DNA of simian virus 40, *Proc. Nat. Acad. Sci. U.S.A.* 78:7575–7578.

98. Svensson, C., and Akusjarvi, G. 1984. Adenovirus VA RNAI: A Positive Regulator of mRNA Translation, *Mol. Cell. Biol.* 4:736–742.

99. Svensson, C., and Akusjarvi, G. 1985. Adenovirus VA RNAI mediates a translational stimulation which is not restricted to the viral mRNAs, *EMBO J.* 4:957–964.

100. Stubblefield, E., and Murphree, S. 1968. Synchronized mammalian cell cultures. II. Thymidine kinase activity in colcemid synchronized fibroblasts, *Exp. Cell Res.* 48:652–660.

101. Thimmappaya, B., Weinberger, C., Schneider, R. J., and Shenk, T. 1982. Adenovirus VAI RNA is required for efficient translation of viral mRNAs at late times after infection, *Cell* 31:543–551.

102. Thummell, C., Tjian, R., Hu, S.-L., and Godzicker, T. 1983. Translational control of SV40 T antigen expressed from the adenovirus late promoter, *Cell* 33:455–464.

103. Toole, J. J., Knopf, J. L., Wozney, J. M., Sultzman, L. A., Bucker, T. L., Pittman, D. D., Kaufman, R. J. Brown, E., Shoemaker, C., Orr, E. C., Amphlett, G. W. Foster W. B., Coe M. L., Knutson, G. J., Fass, D. N., and Hewick R. M. 1984. Molecular cloning of a cDNA encoding human antihaemophilic factor, *Nature (London)* 312:342–347.

104. Treisman, R., Green, M. R., and Maniatis, T. 1983. Cis and trans activation of globin gene transcription in transient assays, *Proc. Nat. Acad. Sci. U.S.A.* 80:7428–7432.

105. Turner, M., Abrams, R., and Lieberman, I. 1968. Levels of ribonucleotide reductase activity during the division cycle of the L cell, *J. Biol. Chem.* 243:3725–3728.

106. Velcich, A., and Ziff, E. 1985. Adenovirus E1a proteins repress transcription from the SV40 early promoter, *Cell* 40:705–716.

107. Wahl, G. M., Vincent, B. R. S., and DeRose, M. L. 1984. Effect of chromosomal position on amplification of transfected genes in animal cells, *Nature (London)* 307:516–520.

108. Wahl, G. M., Vitto, L., Padgett, R. A., and Stark, G. R. 1982. Single-copy and amplified CA genes in Syrian hamster chromosomes localized by a highly sensitive method for in situ hybridization, *Mol. Cell. Biol.* 2:308–319.

109. Weidemann, L. M., and Johnson, L. F. 1979. Regulation of dihydrofolate reductase synthesis in an overproducing 3T6 cell line during transition from resting to growing state, *Proc. Nat. Acad. Sci. U.S.A.* 76:2818–2822.

110. Williams, D. A., Lemischka, I. R., Nathan, D. G., and Mulligan, R. C. 1984. Introduction of new genetic material into pluripotent haematopoietic stem cells of the mouse, *Nature (London)* 310:476–480.

111. Wong, G. G., Witek, J. S., Temple, P. A., Wilkens, K. M., Leary, A. C., Luxenberg, D. P., Jones, S. S., Brown, E. L., Kay, R. M., Orr, E. C., Shoemaker, C. S., Golde, D. W., Kaufman, R. J., Hewick, R. M., Wang, E. A., and Clark, S. C. 1985. Human GM-CSF: Molecular cloning of the complementary DNA and purification of the natural and recombinant proteins, *Science* 228:810–815.

112. Wood, W. I., Capon, D. J., Simonsen, C. C., Eaton, D. L., Gitschier, J., Keyt, B., Seeburg, P. H., Smith, D. H., Hollingshead, P., Wion, K. L., Delwart, E., Tud-

denham, E, G. D., Vehar, G. A., and Lawn, R. M. 1984. Expression of active human factor VIII from recombinant DNA clones, *Nature (London)* 312:330–337.

113. Yamada, M., Lewis, J. A., and Gradzicker, T. 1985. Overproduction of the protein product of a nonselectable foreign gene carried by an adenovirus vector, *Proc. Nat. Acad. Sci. U.S.A.* 82:3567–3571.

114. Zain, S., Sambrook, J., Roberts, R. J., Keller, W., Fried, M., and Dunn, A. R. 1979. Nucleotide sequence analysis of the leader segments in a cloned copy of adenovirus 2 fiber mRNA, *Cell* 16:851–861.

115. Zinn, K., Mellon, P., Ptashne, M., and Maniatis, T. 1982. Regulated expression of an extrachromosomal human beta interferon gene in mouse cells, *Proc. Nat. Acad. Sci. U.S.A.* 79:4897–4901.

Questions for Discussion with the Editors

1. *Clearly, several heterologous proteins accumulate in large quantity in transfected cells. Is there any evidence that mammalian cells recognize foreign translation products and degrade them?*

A considerable amount of success has been achieved with genes for which the coding region has not been altered. For some cases the secreted protein may not have the desired activity due to the lack of a specific posttranslational modification which may not be efficient in the host cell. Examples include the removal of the C-peptide from proinsulin and the vitamin K-dependent γ-carboxylation of amino terminal glutamic acid residues in the clotting factor IX. When the coding regions of expressed genes have been altered (as in the removal of secretory leader segments, deletion of glycosylation sites, or the creation of fusion proteins), the protein may not be transported properly and may accumulate in either the endoplasmic reticulum or the Golgi apparatus (for review see Gething[1]). Alternatively, when the protein expressed is toxic to the cell, it is possible to select variants which express altered forms of the protein as mentioned in this chapter for SV40 small t antigen. Another unique example is tissue plasminogen activator which acts to convert the zymogen plasminogen, which is present in serum, to plasmin. The plasmin acts to proteolyze cells and also converts single-chain tissue plasminogen activator to a two-chain form.

2. *You have nicely reviewed some of the brilliant successes achieved by introducing cDNA constructs into mammalian cells. Please mention some of the more major disappointments experienced with this approach. Can they eventually be overcome?*

To date, no major disappointments have been documented. Certainly, there may be a number of different difficulties to overcome with different proteins. One can imagine problems at the DNA (replication, stability, and transcription), the RNA (transcription, processing, transport, and stability), and the protein (translation, posttranslational modification, transport, and stability) levels. The difficulties may

---

[1]Gething, M. J. (ed.). 1985. Current Communications, in: *Molecular Biology: Protein Transport and Secretion*, Cold Spring Harbor Laboratory, Cold Spring Harbor, N. Y.

or may not be overcome depending on the nature and severity of the problem. As an example, two observations to date indicate that coamplification of the bacterial xanthine-guanine phosphoribosyltransferase gene[2] and the human γ-interferon gene[3] results in 5- to 20-fold lower levels of mRNA compared to the selected DHFR gene. In order to increase the abundance of these mRNAs it is necessary to determine why their levels are low (i.e., does it result from transcription initiation, RNA processing, or RNA stability) and then make proper changes to correct the deficiencies.

[2]Ringold, G., Dieckmann, B., and Lee, F. 1981. Co-expresssion and amplification of dihydrofolate reductase cDNA and the *Escherichia coli* XGPRT gene in Chinese hamster ovary cells, *J. Mol. Appl. Genet.* 1:165–175.

[3]Haynes, J., and Weissmann, C. 1983. Constitutive long-term production of human interferons by hamster cells containing multiple copies of a cloned interferon gene, *Nucleic Acids Res.* 11:687–706.

CHAPTER **10**

Regulation of Heterologous Genes Injected into Oocytes and Eggs of *Xenopus laevis*

Laurence D. Etkin, Bradley Pearman, and Susana Balcells

THE MICROINJECTION of cloned genes into amphibian eggs and oocytes provides a system whereby one may analyze several questions regarding the regulation of gene expression during development and cellular differentiation. These include the role of specific DNA sequences in transcription, the effect of external trans-acting factors or effector molecules on the regulation of gene expression and effects of the changing cellular milieu during development on the expression of genes. In this chapter we will consider the use of the *Xenopus* system in analyzing such questions.

Amphibian oogenesis can last from several months to as long as a $1\frac{1}{2}$ years depending on the conditions under which the animals are kept. It is a period of intense metabolic activity. The oocyte sequesters yolk precursor protein from the liver and converts this into yolk during the process of vitellogenesis. The fully grown oocyte nucleus is a large vesicle approximately 400 mμ in diameter, and is referred to as the germinal vesicle (GV). The GV is a storehouse of many components including polymerases, histones, and nonhistone nuclear proteins, as well as several putative regulatory proteins such as the o$^+$ factor in the Mexican axolotl, all of which were synthesized during the long oogenetic process (7, 10).

During the early oogenetic period (pachytene) there is amplification of the ribosomal DNA. This amplified DNA resides in about 1500 extrachromosomal nucleoli. Other than DNA synthesis during meiotic S phase and ribosomal gene amplification, most synthetic activity during oogenesis is directed toward transcription. This includes synthesis of all major classes of RNA, some of which contribute to the large maternal store and some of which is translated into protein. It is not known whether these two classes are functionally segregated into different cellular compartments. The store of protein and RNA will be used by the embryo during the early developmental stages following fertilization. The transcriptional activity of the oocyte is the feature that makes it ideal as an *in vivo* transcription system to examine the expression of injected DNA sequences, since at the diplotene stage there is no DNA replication occurring. When the oocyte is fully grown it is approximately 1.5 mm in diameter, contains 4 μg of RNA, 12 pg of genomic DNA (tetraploid complement), and 1 mg of protein. Due to their size, macromolecules can be microinjected into either the cytoplasm or nucleus of the oocyte with a fine capillary micropipette (22).

Upon hormonal stimulation the transcriptional activity of the oocyte ceases. The GV breaks down and spews its contents into the cytoplasm. Meiosis continues with the release of the first polar body and then arrests in the metaphase of the second meiotic division. Protein synthesis at maturation is stimulated approximately 2-fold and takes place on stored maternal transcripts. The cell is now inert in terms of nucleic acid synthesis, but upon activation by either fertilization or pricking with a glass needle DNA synthesis begins (Fig. 1). Somatic cell nuclei injected into the oocyte prior to maturation become transcriptionally active without synthesizing DNA. Upon injection into the egg, however, they cease RNA synthesis and begin to

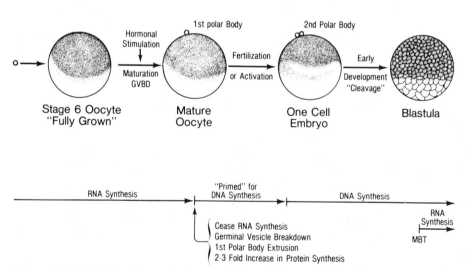

Figure 1 Biochemical events of late oogenesis and early development. Oogenesis (growth from a previtellogenic oocyte to a fully grown oocyte) is mainly characterized by transcription (RNA synthesis). Upon maturation of the oocyte (release of the first polar body) no nucleic acid is synthesized but protein synthesis is stimulated about 3-fold. DNA synthesis begins at fertilization (or at activation) and proceeds during early development. It is not until mid-blastula that transcription is turned on again.

replicate DNA. Thus the products of injected nuclei conform to those being synthesized by the host cell, indicative of the influence of cytoplasmic components of the oocyte and egg on injected nuclei. Specific gene products synthesized by somatic cell nuclei injected into oocytes also correspond to those being synthesized by the oocyte. The oocyte, therefore, is a transcriptionally active cell that is able to regulate the expression of genes of injected nuclei (Fig. 2) (11, 12, 23).

Cloned segments of DNA and isolated chromatin are also transcribed following injection into *Xenopus* oocyte nuclei [reviewed in (11, 12, 23, 24)]. This has provided an exceptional means by which to assay for (a) the role of specific DNA sequences and the modification of these sequences on the transcriptional regulation of gene expression, and (b) the effect of external factors, such as proteins and RNAs on the accuracy of initiation and termination, as well as the modulation of the rate of transcription. The *Xenopus* oocyte may be considered a cosmopolitan cell in that it is able to accurately transcribe, process, and translate RNAs synthesized under the direction of a wide variety of injected heterologous genes. An advantage of this system is that it provides a unique opportunity to introduce a test gene or test molecule into an *in vivo* environment complete with all the components necessary to sustain molecu-

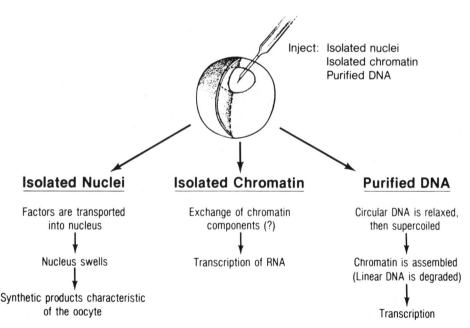

Figure 2 Fate and synthetic activity of genetic material injected into oocytes. Somatic nuclei isolated and injected into the GV of oocytes will take up factors, swell, and transcribe genes characteristic of the oocyte. Isolated chromatin and purified circular DNA will first reassemble into minichromosomes and then will also be transcribed.

lar processes such as chromatin assembly, transcription, RNA processing and modification, and translation (Fig. 2).

Following fertilization or activation, the cell cycle in the early developing embryo lasts from 10 to 20 minutes during which time DNA is replicated, but little or no transcription occurs. Transcription begins at the mid-blastula stage of development. It is also possible to introduce cloned genes into fertilized or unfertilized eggs. The early embryo may be used as a transient expression system to examine replication of exogenous DNA, as well as factors that influence transcription. Some copies of the injected DNA may be integrated into the frog genome and persist in adult frogs. These sequences may be transmitted to subsequent generations by nuclear transplantation or germ line transmission [14, and Etkin et al., unpublished observations]. The analysis of the expression of these genes in different tissues provides insights into the mechanism of tissue-specific gene regulation. Thus, the *Xenopus* system may be used to gather information regarding the mechanisms involved in the regulation of gene expression during development and cell differentiation which will be important in the establishment of fundamental concepts applicable to all organisms.

Injection of Cloned Genes into Oocytes

Fate of Injected DNA

The survival and expression of the injected DNA depends on its configuration and subcellular localization. Circular DNA molecules are stable when injected into the GV. This may be due to the presence of a DNA stabilizing factor present in the nucleus. Both linear and circular molecules injected into the oocyte cytoplasm are degraded over a period of time. Therefore, it is crucial to deposit the DNA into the GV for an accurate assessment of its transcriptional activity. This may be insured by gentle centrifugation of the oocytes making the area in which the GV is located visible, or by careful injection directly into the animal pole region of the oocyte. In the majority of cases, one normally injects between 1 to 10 ng DNA into the GV which represents 10^7 to 10^9 copies of the gene. Injected circular DNA that contain nicks are repaired. A complete second strand may be synthesized on single-stranded molecules following injection into the GV. Thus single-stranded M13 bacteriophage injected into the oocyte nucleus will have a second strand synthesized and ligated. This is probably a result of repair-type DNA synthesis.

Injected supercoiled DNA molecules are converted to form I (relaxed) or form II (nicked) molecules within the first 5 to 10 minutes following injection into oocyte nuclei. By 6 hours following injection most of the injected DNA molecules are in the supercoiled configuration. Micrococcal nuclease analysis of these minichromosomes shows a regular ladder of DNA fragments which are multiples of the typical 185 bp nucleosome unit. DNase I digestion produces single-strand fragments which are multiples of 10.4 bases. Therefore, it may be concluded that almost all of the injected DNA is assembled into nucleosomal particles with normal periodicity (18, 35). Linear DNAs microinjected into the oocyte nucleus, on the other hand, do not form chromatin with regularly spaced nucleosomes and are gradually degraded. Each oocyte contains enough histones to assemble 40 ng DNA into chromatin, yet the increasing amount of DNA is not transcribed in a linear fashion. It is quite possible that other factors needed for transcription and/or chromatin assembly may be in limiting quantities. Thus, when large quantities of DNA are injected, only a fraction may be reconstituted into proper chromatin structures and transcribed accurately.

Transcription complexes of injected DNA may be visualized under the electron microscope. Injected circular molecules form minichromosomes with nucleosomal structures. Results indicate that only a small percentage of injected molecules are transcribing RNA while the remaining fraction appears to be inactive. It is possible that actively transcribing molecules are associated with specific proteins in the oocyte that are required for transcription. In a recent analysis of chromatin structure of injected 5s DNA, Worcell's group suggested that injected DNA forms two types of chromatin, the so-called "static" chromatin, which is transcriptionally inactive but contains nucleosomal structures, and the "dynamic" chromatin, which is transcriptionally active (35). The difference between these two classes may be the presence of specific proteins and an ATP-driven DNA gyration (energy-requiring mechanism) associated with transcriptionally active dynamic chromatin.

Transcription of Injected Genes

One of the paradoxes relating to the transcription of injected genes is that some genes are transcribed relatively efficiently (10 to 20 transcripts per gene per hour), while others are transcribed less efficiently (0.01 to 1 transcripts per gene per hour) or not at all. Genes transcribed under the direction of RNA polymerase III (i.e., 5S, tRNA genes) are transcribed with the highest efficiency, while those transcribed under the direction of RNA polymerases I and II are transcribed to a lesser extent. There are, however, differences even within the group of genes transcribed by RNA polymerase II. Genes containing a viral promoter, such as SV40 and adenovirus, are transcribed efficiently and accurately with proper 5' and 3' ends and efficient splicing of the mRNA. Genes such as the early and late sea urchin histone genes, thymidine kinase, *Drosophila* alcohol dehydrogenase, and human ζ-globin (34) are also transcribed efficiently. However, chicken ovalbumin and *Xenopus* and rabbit globin appear to be poorly transcribed with little fidelity. Cloned *Xenopus* heat shock genes are constitutively transcribed in injected oocytes (5). An interesting observation is that *Xenopus laevis* α^1, and β^1-globin genes are actively transcribed in oocytes, but the majority of transcripts are inaccurately initiated. However, when injected into unfertilized eggs only accurately initiated transcripts were detected at low levels, but the level of accurate transcripts in the oocyte and egg were similar (2).

Some of the differences in transcription of various genes may be due to effects of vectors since these may interfere with proper promoter function and accuracy of initiation or termination. It is possible that DNA sequences necessary for proper function are not present on the segment of DNA that is being tested, leading to poor transcriptional activity. This must be considered, since sequences that are located at substantial distances from the site of initiation may influence the transcription of a gene. Also there is a great deal of variation between oocytes in different females with regard to their ability to transcribe injected DNA (2, 26, 27b). Thus, one cannot generalize at present regarding the reasons for efficient or inefficient transcriptional activity. It is necessary to test each gene in oocytes from different frogs in order to assess its transcriptional ability. In the next section we will consider the usefulness of the *Xenopus* oocyte system in the functional mapping of regulatory sequences on the DNA.

Mapping of the Function of Putative Regulatory Sequences (Cis-Acting Factors)

The organization of prokaryotic transcriptional units is well understood. There has been much speculation regarding the possible functional organization of eukaryotic genes. The logical assumption was that regulatory regions would surround the coding portion of a gene. Specifically, promoters or other regulatory sequences were envisioned as probably being located upstream or 5' to the coding portion. These regions were presumed to influence the expression of the adjacent gene and may be considered cis-acting. Trans-acting factors, which will be considered in a later section, act through an intermediary effector such as a protein or RNA.

From information regarding DNA sequences of cloned genes, it became apparent that there were highly conserved sequences (consensus sequences) located at the 5' region of many eukaryotic genes that are transcribed under the direction of RNA polymerase II. These include the sequence 5'-PyCATTCPu-3' which has been identified as the cap signal marking the site of addition of the 7-methylguanosine cap, an AT-rich sequence containing the consensus sequence TATA or some close variation, and further upstream, the CCAAT sequence (Fig. 3). In several cases there have been sequences located at large distances from the coding region that also influence transcription.

There have been two main approaches to mapping the function of these highly conserved sequences utilizing molecular techniques in conjunction with microinjection into *Xenopus* oocytes (outlined in Fig. 3). The first involves the use of deletion mutations in which a specific region of the 5' prelude sequence containing either one or more of the consensus sequences is removed using site-specific restriction endonucleases and subcloning of the fragments. Another utilizes a technique called linker scanning in which small clusters of nucleotides are substituted for existing DNA sequences (28). This

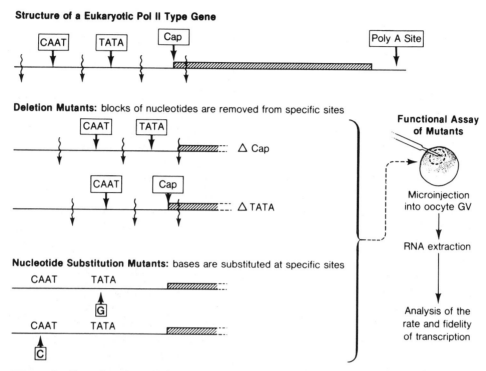

Figure 3 Functional analysis of regulatory sequences of a Pol II type gene. Mutant genes are prepared so that they will lack specific blocks of nucleotides (deletion mutants) or they will carry single nucleotide substitutions at positions corresponding to consensus sequences. Mutant genes are assayed for their fidelity and rate of transcription by microinjecting each of them into the GVs of oocytes.

effectively results in the production of a series of point mutations, some of which occur in the consensus sequences. The resulting "mutated" molecules are then microinjected into the GV of *Xenopus* oocytes. Transcriptional products may be analyzed with regard to the accuracy of initiation and termination and the rate of transcription using techniques such as Northern blot, S1 protection, and primer extension assays. It is also possible to analyze for the presence of functional messenger RNA by detection of a protein product. The remainder of this section will be devoted to discussion of specific examples of the use of these techniques to map regulatory regions. These include the genes coding for sea urchin histone proteins, thymidine kinase, and *Xenopus* ribosomal RNA genes.

In the sea urchin, the five histone genes that are expressed early in development are organized as a unit approximately 6.5 kb in length, which is repeated several 100-fold, while genes expressed later during development (late genes) are not tandemly clustered. The early genes are located on the same DNA strand and have the order H1, H4, H2B, H3 and H2A. The genes have been almost completely sequenced in two species, *Psammechinus miliaris* and *Strongylocentrotus purpuratus*. The availability of cloned DNA fragments containing portions of the complete repeat unit as well as complete nucleotide sequence information has made these genes useful in studies of transcriptional control.

The sea urchin histone genes are transcribed under the direction of RNA polymerase II and their RNA products are translated into proteins. Accurate and functional transcripts are synthesized in recipient *Xenopus* oocytes following the injection of cloned fragments of the sea urchin early histone genes. There are also a significant number of inaccurate transcripts, including transcripts produced by improper initiation and termination, improper strand selection and "read through" products of the eukaryotic portion of the injected plasmid. Inaccurate transcripts are limited to the nucleus and are not detectable in the cytoplasm. We have observed similar rates of transcription of both early and late sea urchin H2B histone genes following injection into oocytes.

Several highly conserved DNA sequences were found in the 5' (upstream) prelude or spacer sequences of the five early genes of *S. purpuratus* and *P. miliaris* (25). These include: (a) the sequence of 5'PYCATTCPu3', which has been identified as the putative cap signal; (b) an AT-rich sequence containing the consensus sequence TATA, which has been identified as a putative promoter known as the Hogness-Goldberg box (approximately 25 nucleotide pairs upstream from the CAP signal); (c) a GC-rich region containing the sequence 5'GATCC3', which is 10 to 20 nucleotide pairs upstream from TATA (this sequence located at this position may be unique to the histone genes), and (d) the consensus sequence CCAAT, which has been identified in the 5' spacer sequences of many eukaryotic genes. The H2A gene in *P. miliaris* also contains a 30 nucleotide long DNA block upstream from the GATCC and CCAAT sequence referred to as the H2A gene-specific conserved sequence.

Utilizing site-specific restriction enzymes and molecular cloning procedures, Grosschedl and Birnstiel (19, 20) constructed several deletion nutants of the H2A gene. They microinjected the modified DNA sequences into the nuclei of *Xenopus* oocytes. As an internal control the injected fragment of DNA contained an unmanipulated copy of the H2B gene. Deletion of the TATA sequence and the GATCC sequence resulted in the synthesis of a series of mRNAs containing several different initiation sites. Deletion of the cap sequence also resulted in the production of a transcript with a novel initiation site, but the site was the same distance downstream from the TATA as was the wild-type mRNA. This suggested that the TATA motif may specify the site of initiation of transcription at a fixed position downstream.

When a unique gene unit consisting of the sea urchin H2A and H2B protein coding regions and a mutated TATA (TAGA) sequence was microinjected into the nuclei of *Xenopus* oocytes, initiation of transcription occurred at a new site but always approximately 26 nucleotides downstream from the TATA or mutated TAGA sequences (21). The amount of specific transcription was decreased 5-fold with the mutant (TAGA) sequence. Deletion of a large segment of the 5′ upstream spacer sequence (from positions −185 to −524 leaving the GATCC and TATA sequence intact) reduced initiation of H2A transcription to about one-fifteenth to one-twentieth of the wild-type transcription. When the sequence was inverted, a 4- to 5-fold increase in H2A gene expression was observed reminiscent of enhancer affects seen in other genes. This segment of DNA contains several stretches of A and T as well as several inverted repeats in the region −345 to −184.

Using *in vitro* deletion mutations and site-specific mutagenesis, the regulatory signals of the Herpes simplex virus thymidine kinase CHSV TK gene have been analyzed (28). A cloned portion of DNA containing the TK gene and 680 nucleotides of the 5′ flanking region and 294 nucleotides of the 3′ flanking region produced transcripts with accurate 5′ and 3′ termini following injection into nuclei of *Xenopus* oocytes. The mRNA was also translated into protein. The site of transcriptional initiation of the Herpes simplex virus TK gene was specified by sequences residing within 60 to 100 bp upstream from the gene. DNA sequencing data showed the presence of a variation of the consensus sequence TATAA (TATTAA) located 24 nucleotides upstream from the coding region and a variation of the consensus sequence CCAAT (CGAAT) located 80 nucleotides from the coding region. The region containing the TATTAA sequence was not essential for accurate initiation, but did enhance the *amount* of accurate initiation of transcription of the TK gene. Mutants lacking the CGAAT sequence showed reduced amounts of transcription. Small clusters of base substitution mutations produced in the 5′ region of this gene have resulted in the identification of the mutation sensitive regions. They probably indicate the sites of transcriptional control signals. One of these is the TATA box and the other two are hexanucleotide sequences that are located 50 and 90 bp upstream from the gene. The hexanucleotide sequences are inverted complements of each other and function in affecting the efficiency of transcription, but not the site of initiation as does the TATA sequence (29).

We can categorize putative regulatory segments into several groups based on their possible function. Sequences that are located 85 to 90 bp upstream (the CAAT sequence) modulate the rate of transcription. Elements such as the TATA box select or specify a unique or predominant 5′ mRNA terminus. Initiator elements that include the cap sequence are located next to the transcribed region. Recently, regulatory sequences in pol II type genes have also been identified within the structural gene (9, 37).

Deletion mutagenesis also was used to analyze the regulatory sequences in the genes coding for *Xenopus* 18S, 5.8S, and 28S ribosomal RNA. These genes are transcribed under the direction of RNA polymerase I and do not produce any protein products. In *Xenopus* there are 450 copies per haploid genome. They are organized as tandomly repeated units. Each unit contains a spacer that is not transcribed and a region that gives rise to a 40S precursor transcript containing the 28S, 5.8S, and 18S ribosomal RNAs. These RNA species result from processing of the 40S precursor. The promoter lies within a 13 nucleotide sequence (−7 to +6 nucleotides relative to the site of initiation). Within the large nontranscribed spacer at the 5′ (upstream) region of each unit are duplications of this promoter sequence (Bam islands). Residing within the spacer between each of the duplicated Bam islands are 60 to 81 bp elements. In a series of elegant deletion experiments, Reeder et al., (27) demonstrated that transcription of the gene depends on the number of regulatory sequences present. Specially constructed plasmids which contain varying numbers of the 60 to 81 bp elements compete for the substances that are necessary for transcription. Thus DNA molecules containing large numbers of these elements will outcompete and be transcribed more efficiently than molecules containing fewer numbers of these elements. This competition or sink effect occurs irrespective of the orientation of the 60 to 81 sequences implying an enhancer-like activity.

Another group of genes that have been amenable to this type of analysis are those coding for small nuclear RNAs (snRNA). In *Xenopus* several of these genes have been cloned and characterized and have been found to be members of multigene families exhibiting differential expression during development. Deletion analysis and site-directed mutagenesis experiments revealed two promoter elements in an embryonic U1B gene. Removal of a region from −200 to −250 produced a 10-fold reduction in transcription, while removal of a region down to −34 nucleotides from the cap site abolished transcription. Another gene from the same clone (U1A) contained only the latter control element (9a). The "activator" or "enhancer" element was detected on the U1A clone at a position of further upstream (26a). Krol et al. (26a) demonstrated that the U1B gene out-competes the U1A gene for transcription factor(s) in injected *Xenopus* oocytes. This may due to the presence of an 18-bp palindromic sequence located near the upstream activator sequence. Mattaj et al. (27a) detected the presence of a similar sequence in the upstream activator of the *Xenopus* U2 snRNA gene which acts in an immunoglobulin heavy chain and SV40 enhancers.

Effect of Trans-Acting Substances or Effector Molecules on Transcription

In the previous section we discussed the role of DNA sequences on the regulation of transcription. Studies in bacterial and viral systems have shown that there are many gene products that act in trans (trans-activating) by producing effector molecules (protein or RNA) that regulate the expression of other genes. This section focuses on the use of the *Xenopus* oocyte system as a means to analyze the effects of trans-acting factors on both the fidelity and activation of transcription of eukaryotic genes.

As discussed previously, transcription of the H3 sea urchin histone gene is accurately initiated following injection into *Xenopus* oocytes. Termination however, is defective, resulting in the generation of readthrough transcripts. Stunnenberg and Birnstiel (36) found that coinjection of a chromosomal salt wash fraction from sea urchin embryos produced accurate 3' termini. This factor is a small poly(A)-RNA species about 60 nucleotides in length, and appears to work by affecting proper processing of the 3' end of the transcript.

Jones et al. (26) showed the effect of the adenovirus E1a gene product on the expression of a chimeric gene consisting of the adenovirus early promoter linked to the prokaryotic gene coding for chloramphenicol acetyltransferase (CAT). This plasmid injected into the oocyte produced CAT enzyme activity. When coinjected with either a cellular extract produced from adenovirus-infected HeLa cells (which contains the E1a protein), or with a second plasmid containing the E1a gene, they observed an increase in CAT activity up to 8.5-fold over basal levels. Thus, they were able to demonstrate in the *Xenopus* oocyte the trans effect of a regulatory protein (E1a) on the expression of the adenovirus early promoter. A similar effect was observed in developing embryos at the mid-blastula transition following coinjection of purified E1a protein and this plasmid into the fertilized egg (16) (discussed in later section).

Recently, Maxson et al. (27b) demonstrated the presence of a trans-acting regulatory molecule capable of enhancing the expression of a specific developmentally regulated gene (Fig. 4). As mentioned previously, sea urchin histone genes are organized into both early and late gene families. The early genes are expressed maximally during early cleavage stages, while the late genes are expressed maximally during gastrulation. Utilizing salt washes of chromatin from gastrula stage nuclei, they found that a 1M salt wash fraction preferentially affects expression of the late sea urchin H2B histone gene 5- to 20-fold above a coinjected early H2B gene. It was also found that the late H2B and early H2B genes were competing for a common component in this salt wash but the late genes had approximately the 4-fold higher affinity for the factor(s) than the early genes. This suggests that the expression of the late sea urchin H2B gene at the gastrula stage in sea urchins is, at least in part, regulated by one or more components found in this particular chromatin salt wash fraction. This reveals a fundamental mechanism by which regulatory factors may act in trans to control the expression of specific sets of genes at the transcriptional level during development.

Birnstiel's group (30a) has described a component in a salt wash fraction derived from sea urchin blastulae chromatin that stimulates transcription of the early H2B gene from the sea urchin *P. miliaris*. It appears that there may be two target sites for this factor, which are located downstream of the initiation site.

Partington et al. (33) observed that human adult β- and fetal γ-globin genes were transcribed in injected oocytes but transcription was not initiated at the proper site. Incubation of the oocytes in sodium butyrate (increases enhancer-dependent transcription in transfected tissue culture cells) prior to injection resulted in a significant amount of accurate initiation from the γ-globin but not the β-globin gene. This suggests that sodium butyrate in some way affects the proper interaction of the transcriptional machinery with γ-globin gene regulatory sequencers.

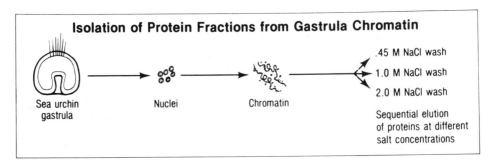

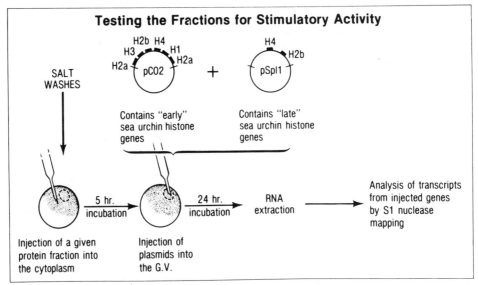

Figure 4 Effect of trans-acting factors on developmentally regulated genes. Sea urchin chromatin proteins from a specific stage of development (gastrula) were fractionated by sequentially washing chromatin at different salt concentrations. The fractions were then assayed for stimulatory activity of the transcription of sea urchin early and late histone genes; each fraction was microinjected into the cytoplasm of oocytes about 5 hours prior to the injection of an equimolar mix of early and late histone clones. After another incubation RNA was extracted and analyzed by S1 mapping (from Maxson et al., submitted for publication).

The 5S genes coding for ribosomal RNA consist of the oocyte type 5S genes (20,000 copies per haploid genome) and the somatic-type 5S genes (400 copies per haploid genome). The somatic type genes are transcribed during oogenesis, throughout development, and in adult tissues, while the oocyte type 5S genes are transcribed primarily in oocytes. Transcription of both types 5S genes depends on the presence of a trans-acting protein transcription factor TFIIIA. This factor binds to a 50-bp region within the gene. Transcription also depends on the presence of at least two other factors, whose nature has not been clearly ellucidated (8).

Transcription of the 5S genes occurs very efficiently on microinjected DNA templates, but only a portion of the injected DNA is actually transcriptionally active. Worcell's group refers to this active fraction as the "dynamic" chromatin (35). They observed that microinjected somatic type 5S genes inhibit transcription of oocyte genes when coinjected in equal or greater amounts, due to the preferential binding of TFIIIA to the somatic type gene. The oocyte type genes apparently are complexed first with histone proteins inhibiting binding of TFIIIA and rendering them inactive. The oocyte genes therefore, are in the so-called "static" conformation which is transcriptionally inactive. Both the active "dynamic" and inactive "static" conformations are stable complexes that once formed, cannot be reversed in injected oocytes. This phenomenon occurs in oocytes that were microinjected with *Xenopus* H4 histone genes, and may involve factors specific for histone gene transcription.

Injection of Genes into Fertilized Eggs

We have discussed the injection of macromolecules into the oocyte in order to assess the effect of both cis- and trans-acting factors on transcription. This section deals with the injection of purified genes into the fertilized egg of *Xenopus* to analyze the factors that are involved in developmental and tissue-specific regulation of gene expression. The general strategy of these experiments involves injection of a gene into the cytoplasm of fertilized eggs, and subsequent analysis of its fate in terms of its copy number, distribution, integration, and finally, its expression at different developmental stages and in various tissues (Fig. 5). The early transformed embryo can be utilized as a transient expression system to analyze the regulation of transcription during the period of cleavage, when little or no transcription is taking place, and at the mid-blastula transition (MBT) when transcriptional activity is first detectable from the embryonic nuclei. Transformed embryos, tadpoles, and frogs may be used to study the tissue-specific regulation of gene expression.

Replication and Fate of Injected DNA Molecules Following Injection into Eggs

Following fertilization or activation, the egg ceases detectable transcription and synthesizes DNA. Exogenous DNAs injected into this cell are replicated and transcribed at low levels or not at all. Injection into the unfertilized egg causes activation. The egg will survive in this condition for 5 to 22 hours. During this period, however, cytolysis is occurring. It is, therefore, imperative that injected unfertilized eggs be monitored closely while experiments are in progress. Several groups have studied DNA replication by microinjecting cloned DNAs into unfertilized eggs. There is good evidence that replication occurs in a semiconservative manner and probably follows the cell cycle kinetics at least for the first few replication cycles (4, 30).

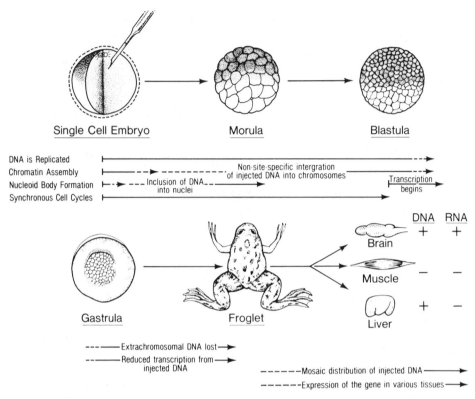

Figure 5 DNA replication and synchronous cell cycles. In the early embryo, DNA is actively replicated, and synchronous cell cycles occur every 20 minutes. When exogenous DNA is injected into the cytoplasm of the fertilized egg it will also replicate, become localized in early cleavage nuclei, and integrate at nonspecific sites of the chromosomes; all nonintegrated DNA will be lost following gastrulation. Transcription of endogenous DNA is first detected by mid-blastula, and so is transcription from exogenous injected DNA. At more advanced stages of development (beyond gastrula) transcription from exogenous DNA declines. In the froglet a mosaic distribution of the integrated DNA is characteristic. The DNA will be expressed variably in different tissues.

DNA injected into eggs appears to become sequestered into vesicles or nucleoid structures which possess many of the morphological and physiological properties of nuclei (17). Injected ³H-labeled recombinant molecules are scattered in the cytoplasm immediately after injection, but by stage 6 (midcleavage), the majority of the labeled molecules are located within the cleavage nuclei. All nuclei, however, are not labeled, implying a mosaic pattern of distribution of the molecules.

Injected supercoiled circular molecules are immediately relaxed. Within 60 minutes supercoiled molecules are detected again, probably as a result of chromatin formation. Linear molecules also replicate and tend to form concatameric oligomers. Circular molecules persist within the embryo as discrete entities, though there is a portion that forms concatamers and comigrates with the high molecular weight frog DNA. The majority of the circular molecules are unchanged as determined by restriction enzyme analysis and rescue experiments in which the plasmid DNA is extracted from blastula staged embryos and used to retransform *Escherichia coli*. Following the blastula stage, circular DNA molecules are degraded, leaving a residual of detectable sequences that comigrate with the high molecular weight frog DNA. We have observed transmission of exogenous DNA sequences through the germ line. This, in conjunction with restriction enzyme digestion analysis of the DNA, suggests that the exogenous genes are integrated into the frog genome. Exogenous DNA sequences are not methylated *de novo* following integration into the frog DNA.

The pattern of distribution of the DNA demonstrates that the initial transformants are mosaics in that some tissues contain the gene while in others there are no detectable copies. There is also mosaicism with regard to copy number in that some animals have tissues with 30 copies of the gene per cell while other tissues have 2 to 5 copies per cell. The appearance of mosaicism in the tissues of transformed frogs suggests that integration probably occurred during cleavage stages. We speculate that nucleoid-like structures that contain the injected DNA fuse with the cleavage nuclei at various times during early cleavage stages resulting in the production of mosaicism in the embryo.

The exogenous DNA sequences may be transmitted through the germ line or by means of nuclear transplantation. The latter procedure permits the production of clones of embryos from a single blastula, or if taken a step further, clones of individuals from a single nucleus by serial nuclear transplantation. This leads to the production of clones of embryos with the gene integrated into the same site which is useful for studies of the regulation of the exogenous sequences.

Expression of Exogenous Genes in Transformed Embryos

The Early Embryo

In the amphibian embryo, transcription is not detectable until the mid-blastula stage of development. At that stage, there is an increase in transcription from the embryonic genome, though it is not known whether only a few genes are expressed or if a large fraction of the genome is active.

Exogenous DNA injected into the fertilized egg also follows this same pattern of expression in that there is no detectable transcription until the blastula stage of development. The early embryo therefore, may be viewed as a transient expression system in which detectable transcription on injected templates peaks at the blastula-gastrula stage of development and then declines in parallel with the loss of the extrachromosomal plasmid molecules.

Several lines of experimentation have attempted to elucidate the mechanism that regulates the expression of genes during this early developmental period. Newport and Kirschner (31, 32) have tested the idea that nucleocytoplasmic ratios affect transcription. They injected massive amounts of DNA into cleavage-arrested fertilized eggs, and tested for the time of expression of injected yeast tRNA and endogenous small nuclear RNAs (snRNA). Inhibition of cleavage is necessary since large amounts of injected DNA interfere with this process and kill the embryo. Injection of amounts of DNA approaching the calculated DNA content of blastula-stage embryos resulted in early activation of the exogenous tRNA genes. Therefore, the amount of DNA present was important for regulating transcription at the MBT. They postulated the presence of an inhibitor in the egg which is titrated out by increasing amounts of DNA. Recently, however, Bendig and Williams (2) observed no increase in the amount of transcription following injection of large amounts of cloned *Xenopus* globin genes into unfertilized eggs.

In another attempt to analyze the nature of this regulatory mechanism, Etkin and Balcells (16) injected a recombinant molecule containing the CAT gene linked to the adenovirus E3 promoter into fertilized eggs. This promoter is regulated by the presence of a positive regulatory molecule, the E1a protein, which enhances expression of this gene. Injected E3CAT was expressed at the MBT. When mixed with the E1a protein prior to injection, the gene was also expressed at the MBT; however, expression of the CAT gene was enhanced 2 to 8-fold. This indicates that even in the presence of a positive regulatory molecule such as E1a, and the responding viral promoter, gene expression is inhibited until the MBT stage of development is reached. Thus, even though the protein is capable of functioning, it does not until the MBT is reached. We have also attempted to perturb the timing of transcription by utilizing plasmids containing the CAT gene linked to a relatively strong promoter, the SV40 early promoter, and by injecting linearized and circular plasmids. In both cases expression was detected precisely at the MBT, though quantitative differences in CAT enzyme activity and transcription were observed. The mechanism responsible for inhibiting transcription prior to the MBT is quite stable and could not be perturbed by the presence of a regulatory molecule, but is perturbable by injection of massive amounts of DNA.

We believe that transcription of injected DNA at the MBT is taking place on extrachromosomal plasmids which are exposed to transcription factors present in the embryo. Therefore, for the most part, any DNA containing the proper cis-acting regulatory sequences should be expressed at this time. The only exception noted to date is the gene for *Xenopus* larval β-globin which, when injected as a recombinant plasmid, is expressed at the tailbud stage. This is in contrast to the adult β-globin gene which is expressed at the MBT (3). The reason for this difference is not clear and will have to await further investigation.

Expression of Genes in Tissues of Young Frogs

Following the blastula stage of development, injected DNA sequences that are extrachromosomal are degraded. A residual fraction of injected DNA, however, persists and is believed to be integrated into the frog genome (15). Many of the initial transformants are found to be mosaic organisms that contain the gene integrated into different positions and/or have even different copy number in various tissues. We have analyzed expression of a plasmid, pSV2CAT, (contains the CAT gene with SV40 early promoter) which, in principle, should not contain any frog regulatory sequences on the DNA since it is a hybrid containing both viral and prokarytic DNA sequences. This gene, however, does exhibit a variety of patterns of expression in different transformed frogs. Expression does not depend on copy number since tissues containing 2 to 5 copies of the gene may express relatively large amounts of the CAT gene product while other tissues containing 10 or more copies do not express the gene. Restriction enzyme digestion analysis of the DNA from those tissues not expressing the gene indicate that large rearrangements have not occurred. This in turn suggests that the lack of expression of the gene may be due to the effects of the site of integration. In another study, injected vitellogenin genes were expressed during early development, but no expression was detected in tissues of young frogs, nor was expression induced by estrogen treatment (1).

Conclusions

The microinjection of purified genes, nuclei, and other macromolecules into *Xenopus* oocytes and eggs has been a useful tool in examining and analyzing the mechanisms involved in the regulation of gene expression during development and cell differentiation. In this chapter, we have attempted to explicate the role of several integral components, such as consensus sequences, trans-acting factors, and the nature of the genomic environment on the expression of genes both temporally and spatially during development.

Injection of DNA into oocytes has permitted analysis of the function of several conserved nucleotide sequences in DNA in terms of their ability to determine either the specificity or rate of initiation of transcription. These sequences may be envisioned as the minimum requirement for accurate transcription and include the TATA and CAAT sequences located at specific positions relative to the site of initiation. They may interact with the basic transcriptional machinery such as polymerases and other macromolecules necessary for transcription. These components may be responsible for the constitutive expression of many types of cloned DNA sequences in injected oocytes and at the mid-blastula transition of transformed embryos. When a gene is injected into the oocyte in the form of chromatin or in intact nuclei, it is recognized in the context of its location relative to other genes and not as an isolated DNA sequence. The regulatory sequences, therefore, may not be accessible to the transcriptional machinery.

Injection of cloned genes into fertilized eggs provides a system whereby one may assess the factors that are involved in the developmental and tissue-specific regulation of gene expression. Injected DNA is stable and is distributed in a mosaic pattern to different tissues of the frog. Expression of the injected sequences is probably influenced by several factors including the site of integration in the genome. Questions still remain regarding the role of DNA sequences in the tissue-specific regulation of gene expression. Evidence to date suggests that such sequences are present, but none has been identified. It is probable that sequence information alone will not be the sole factor in determining tissue-specific expression of genes. Other factors include the location in the genome, modifications suc as methylation of specific regions of the DNA, and the trans-acting factors that may interact with the gene.

The availability of transcription factors at precise times during development will also influence the expression of genes. This is suggested by recent results demonstrating stable stages of repression or activation of 5S genes that depend on the order of binding of histones or the transcription factor TFIIIa to the DNA. Another example is the putative trans-acting factor present at a specific developmental stage in the sea urchin which may differentially regulate the early and late sea urchin histone H2B genes.

The injection of purified genes and other macromolecules into the *Xenopus* oocyte and embryo has provided a model system permitting us to gain insights into the complex mechanism of transcriptional regulation during development and cell differentiation. It may be possible to use the egg or oocyte to reconstruct a regulatory system by the sequential addition of purified genes, proteins, or RNAs which act as regulatory factors. The injection of genes into unfertilized and fertilized eggs is useful for studies of DNA replicatiion, of the regulation of transcription, and of factors involved in tissue specific expression. The continued use of this system in combination with other *in vivo* and *in vitro* approaches should provide the means by which gene regulatory mechanisms will be unraveled.

General References

ETKIN, L. D., DiBERARDINO, M. A. 1983. Nuclear transplantation and gene injection systems in: *Eukaryotic Genes: Their Structure, Activity, and Regulation* N. MacLean, S. P. Gregory, and R. A. Flavell, eds. pp. 127–156, Butterworths, U.K.

GURDON, J. B. and MELTON, D. A. 1981. Gene transfer in amphibian eggs and oocytes, *Annu. Rev. Genet.* 15:189–218.

References

1. Andres, A., Muellener, D. B., and Ryffel, G. A. 1984. Persistence, methylation, and expression of vitellogenin gene derivatives after injection into fertilized eggs of *Xenopus laevis, Nucleic Acids Res.* 12:2283–2302.
2. Bendig, M., and Williams, J. 1984. Fidelity of transcription of *Xenopus laevis* globin genes injected into *Xenopus laevis* oocytes and unfertilized eggs, *Mol. Cell. Biol.* 4:2109–2119.

3. Bendig, M., and Williams, J. 1984. Differential expression of the *Xenopus laevis* tadpole and adult β-globin genes when injected into fertilized *Xenopus laevis* eggs, *Mol. Cell. Biol.* 4:567–570.

4. Benbow, R. M., Gaulette, M. F., Hines, P. J., and Shioda, M. 1985. Initiations of DNA replication in eucaryotes, in: *Cell Proliferation: Recent Advances*, (A. L. Boynton and H. L. Leffert, eds.), Academic Press, Orlando, Fla., 449–461.

5. Bienz, M. 1984. *Xenopus* Hsp70 genes are constitutively expressed in injected oocytes, *EMBO J.* 3:2477–2483.

6. Birchmeier, C., Schumperli, D., Sconzo, G., and Birnstiel, M. 1984. 3′ Editing of mRNAs: Sequence requirements and involvement of a 60-nucleotide RNA in maturation of histone mRNA precursors, *Proc. Nat. Acad. Sci. U.S.A.* 81:1057–1061.

7. Briggs, R. 1979. Genetics of cell type determination, *Int. Rev. Cytol. Suppl.* 9:107–127.

8. Brown, D. D. 1984. The role of stable complexes that repress and activate eukaryotic genes, *Cell* 37:359–365.

9. Charnay, P., Treisman, R., Mellon, P., Chao, M., Axel, R., and Maniatis, T. 1984. Differences in human α and β gobin gene expression in mouse erythroleukemia cells: The role of intragenic sequences, *Cell* 38:251–163.

9a. Cilibero, G., Buckland, R., Cortese, R., and Philipson, L. 1985. Transcription signals in embryonic *Xenopus laevis* U1RNA genes, *EMBO J.* 4:1537–1543.

10. Davidson, E. 1976. *Gene Activity in Early Development*, Academic Press, Orlando, Fl.

11. Etkin, L. D. 1982. Analysis of the mechanisms involved in gene regulation and cell differentiation by microinjection of purified genes and somatic cell nuclei into amphibian oocytes and eggs, *Differentiation* 21:149–159.

12. Etkin, L. D., and DiBerardino, M. A. 1983. Nuclear transplantation and gene injection systems, in: *Eukaryotic genes: their structure, activity, and regulation* (N. MacLean, S. P. Gregory, R. A. Flavell, eds), pp. 127–156, Butterworths, U.K.

13. Etkins, L. D., Maxson, R. E., Jr., 1980. The synthesis of authentic sea urchin transcriptional and translational products by sea urchin histone genes injected into *Xenopus laevis* oocytes. *Dev. Biol.* 75:13–25.

14. Etkin, L. D., Roberts, M. 1983. Transmission of integrated sea urchin histone genes by nuclear transplantation in *Xenopus laevis, Science* 221:67–69.

15. Etkin, L. D., Pearman, B., Roberts, M., and Bektesh, S. 1984. Replication, integration, and expression of exogenous DNA injected into fertilized eggs of *Xenopus laevis, Differentiation* 26:194–202.

16. Etkin, L. D., and Balcells, S. 1985. Transformed *Xenopus* embryos as a transient expression system to analyze gene expression at the mid-blastula transition, *Dev. Biol.* 108:173–178.

17. Forbes, D. J., Kirschner, M. W., Newport, J. W. 1983. Spontaneous formation of nucleus-like structures around bacteriophage DNA microinjection into *Xenopus* eggs, *Cell* 34:13–23.

18. Gargiulo, G., Wasserman, W., and Worcell, A. 1983. Properties of the chromatin assembled on DNA injected into *Xenopus* oocytes and eggs, *Cold Spring Harbor Symp. Quant. Biol.* 47:549–556.

19. Grosschedl, R., and Birnstiel, M. L. 1980. Identification of regulatory sequences in the prelude sequences of an H2A histone gene by the study of specific deletion mutants *in vivo, Proc. Nat. Acad. Sci. U.S.A.* 77:1432–1436.

20. Grosschedl, R., and Birnstiel, M. L. 1980. Spacer DNA sequences upstream of the TATAAATA sequence are essential for promotion of H2A histone gene transcription *in vivo*, *Proc. Nat. Acad. Sci. U.S.A.* 77:7102–7106.

21. Grosschedl, R., Wasylyk, B., Chambon, P., and Birnstiel, M. L. 1981. Point mutation in the TATA box curtails expression of sea urchin H2A histone gene *in vivo*, *Nature (London)* 294:178–180.

22. Gurdon, J. B. 1974. *The Control of Gene Expression in Animal Development*, Harvard Univ. Press, Cambridge, Mass.

23. Gurdon, J. B., Melton, D. A. 1981. Gene transfer in amphibian eggs and oocytes, *Annu. Rev. Genet.* 15:189–218.

24. Gurdon, J. B., and Wickens, M. P. 1983. The use of *Xenopus* oocytes for the expression of cloned genes, in: *Methods in Enzymology*, vol. 101, part C (R. Wu, L. Grossman, and K. Moldave, eds.), Academic Press, New York.

25. Hentschel, C. C., and Birnstiel, M. L. 1981. The organization and expression of histone gene families, *Cell* 25:301–313.

26. Jones, N. C., Richter, J. D., Weeks, D. L., and Smith, L. D. 1983. Regulation of adenovirus transcription by EIa in microinjected *Xenopus* oocytes, *Mol. Cell. Biol.* 3:2131–2142.

26a. Krol, A., Lund, E., and Dahlberg, J. 1985. The two embryonic U1RNA genes of *Xenopus laevis* have both common and gene specific transcription signals, *EMBO J.* 4:1529–1535.

27. Labhart, P., and Reeder, R. 1984. Enhancer-like properties of the 60/81 6p elements in the ribosomal gene spacer of *X. laevis*, *Cell* 37:285–289.

27a. Mattaj, I., Lienhard, S., Jiricny, J., and De Robertis, E. 1985. An enhancer-like sequence within the *Xenopus* U2 gene promoter facilitates the formation of stable transcription complexes, *Nature (London)* 316:163–167.

27b. Maxson, R., Ito, M., Balcells, S., Thayer, M., and Etkin, L. Differential effect of a sea urchin gastrula nuclear extract on expression of sea urchin early and late histone genes following injection into *Xenopus* oocytes. Submitted for publication.

28. Mcknight, S., and Kingsbury, R. 1982. Transcriptional control signals of a eukaryotic protein-coding gene, *Science* 217:316–324.

29. McKnight, S. L., Kingsbury, R., Spence, A., and Smith, M. 1984. The distal transcription signals of the herpes virus tk gene share a common hexanucleotide control signal, *Cell* 37:253–262.

30. Mechali, M., and Kearsey, S. 1984. Lack of specific sequence requirement for DNA replication of *Xenopus* eggs compared with high sequence specificity in yeast, *Cell* 38:55–64.

30a. Mous, J., Stunnenberg, H., Georgiev, O., and Birnstiel, M. 1985. Stimulation of sea urchin H2B histone gene transcription by a chromatin-associated protein fraction depends on gene sequences downstream of the transcription start site, *Mol. Cell. Biol.* 5:2764–2769.

31. Newport, J., Kirschner, M. 1982. A major developmental transition in early *Xenopus*. I. Characterization and timing of cellular changes at the mid-blastula stage, *Cell* 30:675–686.

32. Newport, J., Kirschner, M. 1982. A major developmental transition in early *Xenopus* embryos. II. Control of the onset of transcription, *Cell* 30:687–696.

33. Partington, G. A., Yarwood, N. J., and Rutherford, T. R. 1984. Human globin gene

transcription in injected *Xenopus* oocytes: Enhancement by sodium butyrate, *EMBO J.* 3:2787–2792.

34. Proudfoot, N. I., Rutherford, T. R., and Partington, G. A. 1984. Transcriptional analysis of human zeta globin genes, *EMBO J.* 3:1533–1540.

35. Ryoji, M., and Worcell, A. 1984. Chromatin assembly in *Xenopus* oocytes: *in vivo* studies, *Cell* 37:21–32.

36. Stunnenberg, H. G., and Birnstiel, M. 1982. Bioassay for components regulating eukaryotic gene expression: A chromosomal factor involved in the generation of histone mRNA 3′ termini, *Proc. Nat. Acad. Sci. U.S.A.* 79:6201–6204.

37. Wright, S., Rosenthal, A., Flavell, R., and Grosveld, F. 1984. DNA sequences required for regulated expression of globin genes in murine erythroleukemia cells, *Cell* 38:265–273.

Questions for Discussion with the Editors

1. *Clearly, injection of purified genes into amphibian oocytes and eggs is feasible and has provided useful information about gene regulation. What are the prospects, however, for introducing cloned genes selectively (vs. Fig. 5, "mosaic" transformation) into the nuclei of specific cell lineages (e.g., prospective dorsal lip cells)?*

A useful feature of amphibian oocytes and eggs is that macromolecules can be introduced into specific regions of the egg or into specific blastomeres during cleavage stages. Injection into a specific blastomere could be used to produce mosaic embryos that contain a gene or gene product localized in a certain cell lineage. This could be done with a gene that might produce a product useful in marking the fate of the cell and its progeny or with a gene that produces a morphogen in attempts to alter the developmental fate of the cell. This approach will probably require the construction of vectors that (a) permit expression at the desired time and with high efficiency and (b) remain extrachromosomal so as not to be influenced by position effects caused by integration. Inducible promoters such as the heat shock or the metallothionein promoters may be useful in producing a product at a desired time or developmental stage.

2. *To what extent do you expect that injection of "antisense" mRNAs (such as those homologous to maternal mRNA) into Xenopus eggs will be informative?*

The concept of injecting an antisense strand RNA to effectually eliminate the presence of a specific gene product is intriguing. Antisense RNA experiments have proved successful in perturbing the translatability of mRNAs in tissue culture cells, *Xenopus* oocytes, and *Drosphila* embryos. The use of this approach in *Xenopus* embryos will depend on overcoming several obstacles. The first is the stability of the injected antisense RNAs. Indications are that the antisense strand may be unstable following injection into the egg, as opposed to the oocyte in which the RNA is relatively stable. Another difficulty is whether the injected antisense strand RNA will distribute evenly throughout the egg prior to cleavage. This will be crucial in experiments in which the target gene product is not localized in one region of the egg. A third problem is the analysis of the perturbation. Unless a specific defect is observed, the result will be difficult to interpret. For example, if an antisense RNA is injected against a gene product which is expressed at the mid-blastula transition it is quite possible that the result will be exogastrulation. Exogastrulation in amphibians may be induced by a variety of treatments including high salt concentration, injection of proteins, and injection of both nonspecific DNA and RNA. Therefore, it is imperative that the analysis is of a specific defect such as the expression of specific molecular markers or the effect on morphogenesis and/or morphology of a specific tissue or cell type.

Assuming that most of these obstacles are overcome, this could be an extremely powerful approach in analyzing the function of specific gene products. Injection of an antisense RNA against a maternal RNA will create an artificial maternal effect mutant. Thus, one will be able to perform genetic analyses on artificially derived mutants, overcoming the relatively long generation time in amphibians.

Virus-Associated Systems

CHAPTER 11

Molecular Biology of Hepatitis B Virus

Roberto Cattaneo, Rolf Sprengel, Hans Will, and Heinz Schaller

THERE ARE SEVERAL REASONS why the molecular biology of hepatitis B virus (HBV) is currently the subject of intense research. First, HBV is a major world health problem because of the high prevalence of acute and chronic HBV infection. Second, HBV infection is closely linked to the appearance of liver cancer (primary hepatocellular carcinoma). Third, it has been recently demonstrated that HBV replication, which leads to the production of infectious particles containing DNA, involves reverse transcription of an RNA intermediate, and thus shares some steps with retroviral replication.

This chapter will review briefly some medical aspects of HBV infection, and more thoroughly the molecular biology of HBV. First we will focus on structural data regarding the virion, its proteins, its genome, and the transcripts produced during the viral cycle. Then we will introduce the animal HBV-like viruses, which have helped to elucidate some aspects of the HBV replication cycle. The evidence suggesting that replication of the HBV-like viruses goes through an RNA intermediate will also be reviewed in the context of mechanisms that regulate HBV gene expression. Finally, we will discuss the correlation between HBV infection and the development of liver carcinoma and compare the replicative strategies of HBV and retroviruses.

Medical Aspects

HBV infection was originally termed "serum hepatitis" because it is usually transmitted by blood contact. The manifestations of the disease are variable, ranging from undetectable to fulminant, fatal hepatitis.

The pattern of the most common form of HBV infection, a self-limited HBsAg-positive primary infection, is shown in Fig. 1 (38). The surface antigen HBsAg is usually the first viral marker to appear in the blood, normally 4 to 8 weeks after infection. It reaches a peak 4 to 6 weeks thereafter, at which time another antigen (HBeAg) and the HBV virions (Dane particles) also reach their maximum, and clinical hepatitis becomes evident. As symptoms and jaundice clear, the HBsAg titer falls, and anti-HBs and anti-HBe appear. Antibodies directed against the core antigen (anti-HBc) usually appear before anti-HBs and anti-HBe, in spite of the fact that the core antigen HBcAg can never be detected in serum. The appearance of an antibody to an antigen that cannot be detected in blood can be tentatively explained by the strong immunogenicity of HBcAg, and by a short half-life in serum.

Depending on the dose of virus received and on the reaction of the immune system, other forms of illness can develop [for reviews see (1, 38)]. They range from subclinical, self-limited primary infections without detectable HBsAg to persistent HBsAg-positive infections and to the rare lethal infections. Persistent infections are detected in 5 to 20 percent of the patients. These subjects have an increased risk of liver disease and hepatoma, and as "chronic carriers" promote further spread of the virus. The worldwide "reser-

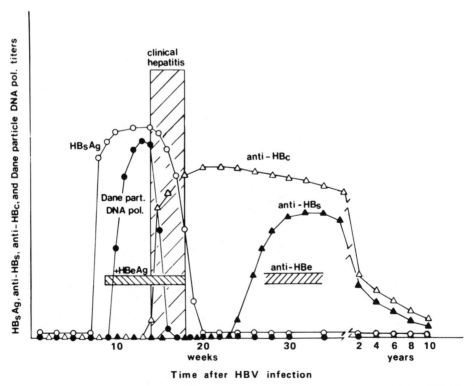

Figure 1 Representation of viral markers in the blood during a typical self-limited HBV infection [from Robinson (38)].

voir" of such chronic carriers has been estimated to include about 200 million individuals, and, as a result HBV infection, presently represents a major world health problem (23).

To control the incidence and the spread of HBV infection, viral antigens are needed as diagnostic tools and for vaccination. Since HBV cannot be propagated in tissue culture, the supply of HBV-specific antigens depended until recently exclusively on the material isolated from HBV-infected people. HBsAg from chronic carriers, although available only in limited amounts and difficult to obtain in a pure, virus-free form, has been used as a basis for an efficient and safe vaccine (54). The risk, the cost, and the limited supply of the natural antigens has led to substantial efforts to produce HBV-specific antigens uncoupled from the viral life cycle by recombinant DNA technology, and now HBsAg, HBcAg, and HBeAg produced in bacteria, yeast, or animal cells are progressively replacing the natural antigens for diagnostic and vaccination purposes (29, 50).

Structural Aspects

Structure of the Virion and of the Viral Proteins

The structure of the hepatitis B virion is depicted in Fig. 2. The virion or Dane particle, corresponds to the 42-nm spherical particle described by Dane et al. (6). The inner core, composed of the DNA genome and core antigen, is wrapped in an outer lipid-containing envelope bearing the various forms of the surface antigen.

The viral envelope consists of a lipid bilayer in which glycosylated and nonglycosylated HBsAg proteins are embedded. The two major HBsAg proteins are detected in roughly equal amounts and migrate on denaturing proteins gels with an apparent molecular weight of 27 (27) and 24 (24) kilodaltons (kd) for the glycosylated and nonglycosylated protein, respectively. From nonreduced material, HBsAg purifies as a 49-kd molecule, indicating the existence of an HBsAg dimer, whose subunits are probably linked by intermolecular disulfide bonds (Fig. 2). In addition to these major HBsAg proteins, the viral envelope contains minor amounts of 32- and 35-kd proteins (32 and 35) related to HBsAg which possess 55 additional amino acids at their N-terminus. This pair of proteins differs in that 32 is glycosylated only at one site, the 35-kd protein at two [Fig. 3 and (51)]. The 55 N-terminal amino acids have been shown to interact specifically with polymerized human albumin, whose receptor may mediate virus access to hepatocytes (20). Finally, a 39- and 43-kd pair of HBsAg-related proteins (39 and 43) have also been recently detected in the viral particle, and it has been estimated that a virion contains 300 to 400 27/24 molecules and 40 to 80 molecules of the larger HBsAg proteins (15).

The 27-nm nucleocapsid or viral core (Fig. 2) is an icosahedric structure containing 180 molecules of core antigen (33). HBcAg is a polypeptide of 183

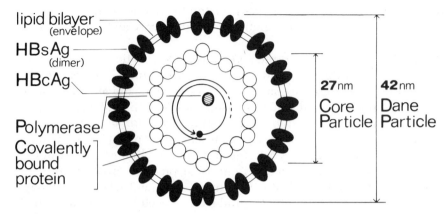

Figure 2 Structure of the HBV virion.

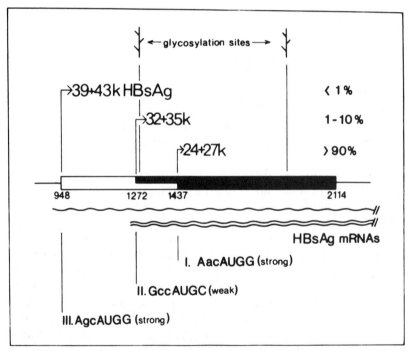

Figure 3 Map of part of the HBV genome, of the HBsAg-related proteins, of the HBsAg transcripts, and sequences surrounding the translation initiation sites. The pre-S HBsAg region is drawn with the same symbols as in Fig. 5. The mass of the HBsAg-related proteins and the position of the glycosylation sites are indicated in the upper part of the figure. The relative percentage of the HBsAg-related proteins refers to the very abundant 22-nm particles (9), and not to the rare 42-nm viral particles (15). The sequence context of the sites of initiation of translation are drawn in the lower part, and their strength is indicated according to the rules of Kozak (17). The major HBsAg mRNA is indicated as a double wavy line, and the minor HBsAg mRNA as a wavy line.

(or 185; see Fig. 4) amino acids and a calculated molecular weight of 20.5 kd. The C-terminal 40 residues of HbcAg are very basic, with repeated arginines separated by serines and prolines. It has been speculated that this portion of HBcAg is involved in binding to DNA. A similar cluster of basic amino acids is also present in the core antigens of all HBV-related viruses even in the case of duck hepatitis B virus (see below) where the rest of the core antigen sequence differs substantially.

HBcAg can be converted into another HBV-specific antigen, the HBeAg, by treatment with detergent or by limited proteolysis. However, the HBe antigen found freely circulating in the blood consists of at least two distinct proteins with an apparent molecular weight of 16 and 18 kd on SDS gel electrophoresis. The evidence presently available indicates that these forms represent HBcAg lacking a carboxy-terminal segment (55) and/or alternatively, different translation products of the HBcAg gene lacking amino-terminal amino acid sequences (40).

Three other proteins have been detected either directly or by their enzymatic activities in viral cores, but they have not yet been isolated in amounts allowing structural studies. They include a DNA-dependent DNA polymerase activity, a protein covalently bound to HBV DNA, and a protein kinase activity [for review see (56)].

Finally, it is important to note that the 42-nm particles represent only a small fraction of the HBV-specific protein in the serum from acutely infected patients and they are frequently absent in sera from chronic carriers: the vast majority is found as 22-nm spherical or filamentous aggregates of HBsAg, which are devoid of nucleic acids. These particles are currently used for the production of HBV vaccine (54).

Organization of the Genome

The genome of HBV is one of the shortest and most compactly organized genomes known, and it has an unusual physical structure (Fig. 5). It is composed of one 3.2-kb DNA strand (the "minus" strand) base-paired with a shorter "plus" strand of variable length. The 5' ends of these two strands overlap by 200 to 250 bases and, therefore, the genome is usually in a circular configuration, although neither strand is a closed circle.

Cloning in bacteria and DNA sequencing of the HBV genome has allowed prediction of the genetic map as drawn in Figs. 4 and 5. Four open reading frames (ORF), all oriented in the same direction, have been defined. Two of these have been assigned to the structural proteins HBsAg and HBcAg (and its derivative HBeAg). The HBsAg ORF is preceded by a "pre-S region" of 163 amino acids [this region is slightly larger in ad subtypes (Fig. 4)] which encodes the N-terminal extension of the larger HBsAg forms.

The HBcAg ORF is preceded in some HBV genomes by a short "pre-C" sequence of 29 amino acids (Fig. 4). Expression studies in prokaryotes (34) and eukaryotes (64) suggest that translation of the HBcAg gene initiates

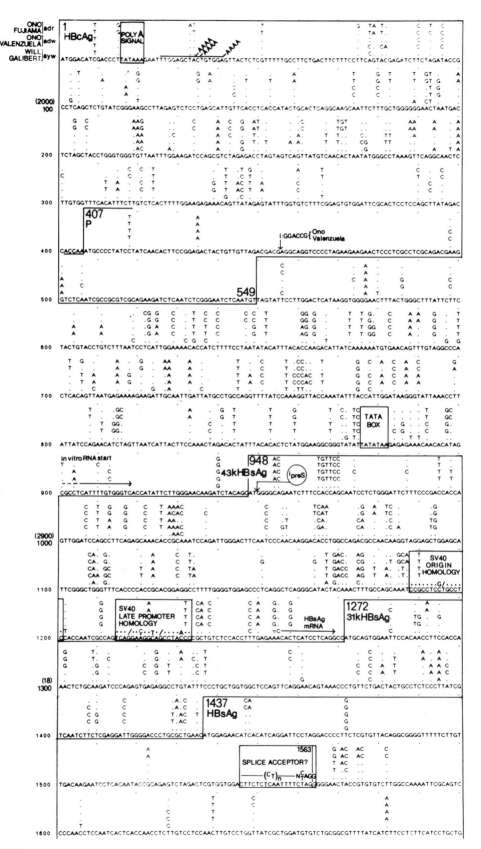

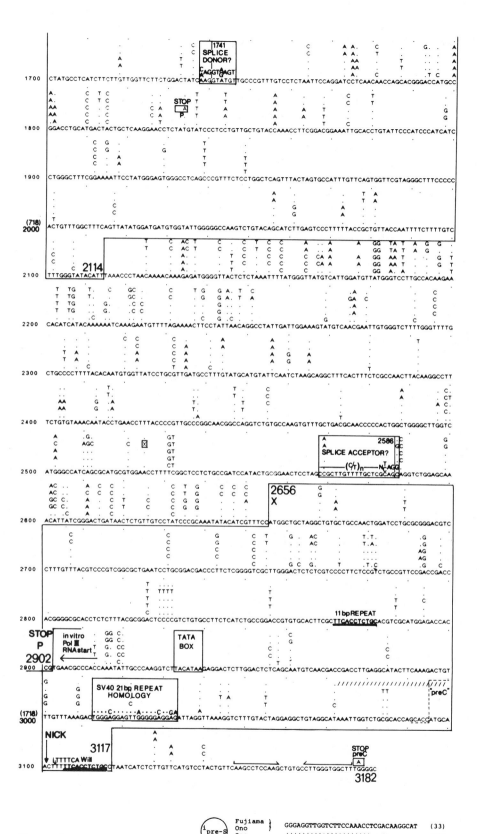

Figure 4

277

Figure 4 Alignment of six HBV DNA sequences and potential transcription and replication signals. Subtype and source of the six sequences are indicated on the top left [(32, 10, 32, 59, 63, and 11) from top to bottom; the upper ayw clone (59) was from a gene bank from which an incomplete HBV DNA sequence was determined earlier (34)]. The numbering system of Pasek et al. (34) has been used, but we have also indicated in parentheses on the left margin the numbering system of Galibert et al. (11). The nucleotide sequence determined by Galibert et al. (11) is shown completely; for the other five sequences only the differences with the Galibert sequence are indicated. Small insertions are indicated at their exact positions, the 12- to 33-nucleotide insertions present in four clones in the pre-S region are indicated at the right bottom of the figure. Deletions are indicated with slashes. In the sequence of Fujiama et al. (10), we have arbitrarily inserted a nucleotide (x) at position 2525 to allow alignment with the other five sequences. The lengths of the six nucleotide sequences presented here are (from top to bottom) 3188, 3215, 3200, 3221, 3188, and 3182 nucleotides.

The HBsAg and HBcAg proteins are boxed, as the N-termini of the 43-K and the 31-K HBsAg-related proteins. The X ORF is also boxed. The beginning and the end of the P ORF are indicated at nucleotides 407 and 2902, respectively. The beginning of the pre-C region, present in 4 of 6 sequences is indicated at position 3096; the G-to-A mutation introducing a stop in the pre-C region in the sequence of Will et al. (63) and in the sequences of Pasek et al. (34) is indicated (nucleotide 3178). The T-to-A mutation introducing a stop in the P ORF of the adw sequence of Ono et al. (32) is also indicated (nucleotide 1834). The biological significance of all these deviations from the consensus sequence is uncertain, since only the clone sequenced by Galibert et al. has been shown to be infectious in chimpanzees (65).

Sequences involved in replication, and *in vivo* or *in vitro* transcription: Position 16, the HBV polyadenylation signal (4, 14). Positions 34 to 41, the precise polyadenylation sites (45). Position 876, a TATA box strongly active in heterologous systems (18, 21), but not always active in the infected liver (4, 67); position 905, corresponding mRNA start site (37, 67). Positions 1188 and 1215, regions of homology with the SV40 origin and the SV40 late promoter situated shortly upstream of the major liver HBsAg mRNA start site (position 1255) (4, 49) (the SV40 homologous sequences are indicated in the boxes). Position 1563, potential splice acceptor sequence identified in the COS cells expression system (3) (the splice acceptor consensus sequence is indicated in the box). Positions 1741 and 2586, splice donor and acceptor sequences, which are spliced together in COS cells (45). Positions 2872 and 3108, 11-bp direct repeat (text). Position 2913, start site of an *in vitro* polymerase III "minus sense" transcript (48). Position 2944, TATA box active *in vitro* (36) but not *in vivo* (67), where an ATA box (position 3071 and Fig. 8) promotes initiation of transcription of the pregenome-HBcAg mRNA (67). Position 3013, region of homology (unpublished) with an element of the SV40 early promoter, the 21-bp repeat encoding two binding sites for the transcription factor SP1 (7a). Position 3102, most frequent location of the 3'-end of the DNA minus strand as deduced from the distribution of these ends in cloned HBV DNA (34). Recently it was demonstrated that this site does not correspond to a nick as indicated, because the HBV DNA minus strand has a 6- to 9-nucleotide terminal repeat (67). Position 3147, the most striking of the several hairpin structures that can be drawn in this region of the genome [(12, 45) and (our unpublished observations)]. (Figure appears on pages 276 and 277.)

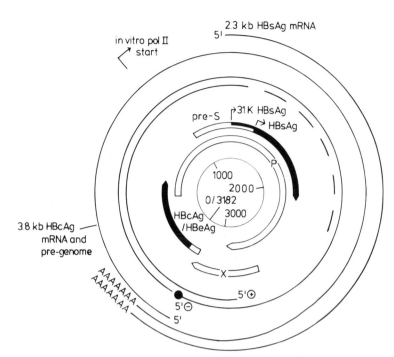

Figure 5 Genetic, physical and transcription map of the HBV genome. The genetic map is drawn in the center. The numbering system of Pasek et al. (34) is used. Solid arrows indicate known genes, open arrows open reading frames specifying proteins not identified yet. The "minus" strand DNA and the "plus" strand DNA are represented as solid lines; the part of the "plus" strand which is not present in all molecules is drawn as an interrupted line. The protein bound to the 5'-end of the minus strand is indicated by a black circle. The transcription map is drawn on the outside. The "2.3-kb" HBsAg mRNA and the "3.8-kb" HBcAg mRNA and pre-genome are represented as solid lines. A minor site of initiation of transcription upstream of the pre-S region also is indicated.

usually, as indicated in Fig. 5, at the ATG in position 1. However, it is conceivable that minor, larger forms of HBcAg are translated from the "pre-C" ATG.

The longest ORF (P), covering almost 80 percent of the viral genome and overlapping the three other ORFs, is thought to code for the HBV polymerase because it could encode a large protein. Furthermore, it shows significant sequence homology in its C-terminal third with reverse transcriptases (58). However, other parts of the P-ORF sequence are highly variable among the six HBV isolates sequenced so far (Fig. 6), and thus it is conceivable that only part of the P-ORF gene product has reverse transcriptase activity.

The function of the fourth ORF, the X frame, has not been identified. An analogous ORF is present in the genome of the woodchuck and ground squirrel HBV-like viruses but not in the genome of duck hepatitis B virus. Thus,

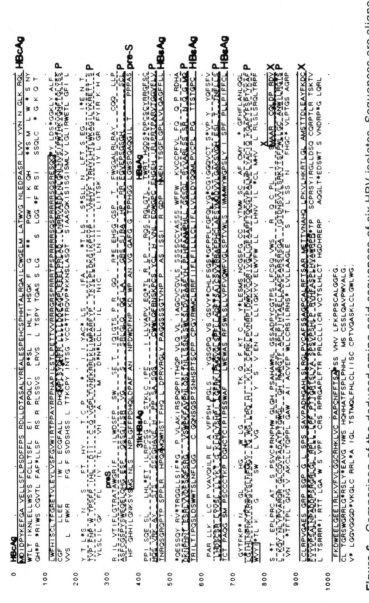

Figure 6 Comparison of the deduced amino acid sequences of six HBV isolates. Sequences are aligned as shown in Fig. 4, and the deduced amino acids are indicated in all three possible reading frames if at least five out of six subtypes share the same amino acid. The main open reading frames are boxed in solid or dashed lines. The region showing homology to reverse transcriptases (58) is shown between slashes (positions 634 and 724) and the most conserved amino acids are marked by black dots. Amino acids are given in one-letter codes: A, Ala; C, Cys; D, Asp; E, Glu; F, Phe; G, Gly; H, His; I, Ile; K, Lys; L, Leu; M, Met; N, Asn; P, Pro; Q, Gln; R, Arg; S, Ser; T, Thr; V, Val; W, Trp; Y, Tyr;* Och, Opa, Amb.

the potential function of the X product would be unique to the mammalian hepatitis B–like viruses.

No gene has been related so far to the other two proteins common to the avian and mammalian viruses, a protein kinase and a DNA-bound protein, which was shown to be involved in the priming of reverse transcription (26). Thus, the viral origin of these two proteins is not certain.

Transcripts Produced in the Liver

Recently, a map of the HBV transcripts expressed in the liver has been obtained (5, 67). The two major HBV mRNAs are unspliced and share the same polyadenylation signal (Fig. 5, outer circles).

The shorter mRNA initiates within the pre-S region and is terminated within the HBcAg gene. This transcript is 2.1 to 2.2 kb long but migrates on RNA gels with an apparent length of 2.3 kb. It was detected not only in the infected liver but also in several heterologous systems, and correlated with the expression of HBsAg and some of the minor HBsAg-related proteins (4).

The other major liver transcript is somewhat larger than the 3.2-kb HBV genome. It was originally detected as a minor transcript in rodent cell lines containing HBV DNA where it was correlated with HBcAg and HBeAg expression (14). An analogous transcript is present in the infected liver at levels approaching those of HBsAg mRNA (5). Recently the structure of this transcript was precisely defined (67), and it was suggested that it could function both as HBcAg mRNA and as template for reverse transcription. Slightly larger than genome transcripts have also been identified in the HBV-like viruses of ducks, ground squirrels, and woodchucks (DHBV, GSHV, and WHV) (2, 8, 28).

It is noteworthy that all the major transcripts of the four HBV-like viruses are unspliced and use a polyadenylation site situated within the core antigen gene. Splice sites and other transcription signals defined on HBV DNA using heterologous expression systems (3, 18, 21, 37, 45, 48) were found to be used poorly, if at all, in the infected liver (5). This does not exclude the possibility that some of them may be used at very low level, and could be crucial for the expression of minor HBV proteins or for other aspects of the viral replication cycle. Recently a minor transcript starting upstream of the pre-S region has been detected in the HBV-infected liver (67), but other transcripts accounting for the expression of the P or X ORF have not yet been detected. Instead, fusion proteins comprising part of the HBcAg and P ORFs have been characterized in liver tumors (66). Thus it is conceivable that these proteins result from a frameshift event at translation, as they do for the reverse transcriptase of Rous sarcoma virus and other retrotransposons (15a). A similar mechanism could also underly the expression of the X protein (27). Alternatively, the X and P products could be translated from internal initiation sites on the major transcripts as with other viral bicistronic mRNAs (13, 44).

Viral Replication Cycle

Reverse Transcription

Biochemical analysis of the events underlying HBV replication has been hampered for years by the failure of HBV to grow in cultured cells; only the discovery of animal viruses related to HBV [see (52) for a review] has allowed a molecular analysis of the replication cycle of these viruses, the results of which have been extrapolated to HBV biology. HBV-like viruses have been detected in woodchucks (*Marmota monax*), ground squirrels (*Spermophilus beecheyi*), and Pekin ducks (*Anas domesticus*). Woodchuck hepatitis virus infection is associated with hepatic injury and hepatocellular carcinoma in these animals, and thus is clearly a useful model for the pathogenesis (and oncogenesis) of HBV infection in humans.

Similarly drastic pathogenic or oncogenic processes do not seem to occur in GSHV-infected ground squirrels or in DHBV-infected ducks, which may be related to the inability of the DNA of these viruses to integrate.

Nevertheless, the DHBV system has been used for the study of the molecular biology of HBV-like viruses because, while woodchucks and ground squirrels are difficult to obtain and house, embryonated duck eggs are available from commercial suppliers and can be easily hatched and maintained in the laboratory. This feature facilitates experimental DHBV infection of eggs and ducklings, and thus the study of the molecular events underlying infection. This study provided strong evidence for a peculiar mechanism of DHBV replication involving reverse transcription of an RNA intermediate.

On first sight there is no obvious reason why a virus containing a DNA genome should replicate through an RNA intermediate, but Mason et al. (22) observed that free "minus" stranded DNA was produced in the liver of DHBV-infected ducklings. Summers and Mason (53) investigated the mode of synthesis of HBV DNA using actinomycin D, an agent that specifically blocks the use of DNA as a template and found that it did not influence the synthesis of the DHBV minus strand. From this and other experiments they postulated that a genome-sized RNA is first encapsulated in the "immature" viral cores and then reverse-transcribed into full-length DNA-minus strand (Fig. 7). The DNA-plus strand is then synthesized on this template to give the circular, partially double-stranded DNA genome found in mature virus (Fig. 7).

The above model also accounted for both the supercoiled DNA molecules and the "genomic length" transcripts detected in the DHBV infected liver since it postulated that supercoiled HBV DNA is produced immediately after entry of the virus into the cell and is used as template for transcription. Finally, the model postulated that the DHBV cores were coated with the surface antigen envelope and exported from the cell only when a packaging signal had been reconstructed as double-stranded DNA (Fig. 7).

Some steps of this model required further investigation and others were highly speculative. However, the Summers and Mason's model of HBV replication via reverse transcription integrated and explained for the first time several observations that had not been correlated before, and posed some questions. In the meantime Mason, Summers, and their colleagues (23) have established experimentally that supercoiled DHBV DNA is produced in one

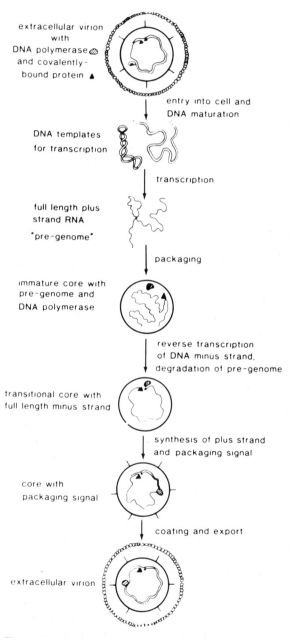

Figure 7 Proposed pathway for the replication of the genome of the HBV-like viruses [From Summers and Mason (53). Copyright 1982 MIT.]

of the earliest stages of DHBV infection and have identified the site of initiation of reverse transcription (26). Other groups have confirmed the existence of DNA intermediates resulting from asymmetric replication in HBV and GSHV (25, 62), and have identified the structure of the potential pregenomic RNA of HBV-like viruses as polyadenylated RNA molecules slightly larger than the DNA genome.

Another technical improvement has rendered the DHBV model still more attractive. Will and colleagues have adapted to the duck system a transfection technique which had been worked out with the HBV system and allows the rescue of infectious virus from cloned DNA (46, 65). This technique will allow *in vivo* analysis of the effects of mutations introduced *in vitro* into the DHBV genome, and should in the long term lead to the precise definition of the signals that control DHBV replication and gene expression, even in the absence of an *in vitro* system for virus propagation.

Regulation of Gene Expression

HBV has a very small genome and thus should be an ideal tool to study regulation of gene expression. However, the lack of an *in vitro* system for HBV propagation has greatly hampered these studies, and today there is mostly indirect evidence for sometimes peculiar mechanisms regulating HBV gene expression.

One puzzling observation in HBV biology is the overproduction of the 22-nm HBsAg aggregates, which occur in vast excess (100- to 1000-fold) over the 42-nm viral particles during infection. In analogy to this, the HBsAg gene is readily expressed under the control of its own promoter in heterologous systems, whereas the HBcAg gene is expressed at only very low levels in similar systems (14). A study of the HBV RNAs expressed in the infected liver has indicated that, at least in the single case analyzed, similar amounts of HBcAg mRNA and HBsAg mRNA were produced (5). Thus, the overproduction of the 22-nm HBsAg aggregates cannot be explained by a constitutively higher rate of HBsAg transcription. On the other hand it has been proposed that HBcAg and HBeAg expression is linked to viral replication because both rely on similar RNA templates. If so, inhibition of viral replication by cellular defense mechanisms should result in concomitant loss of HBcAg/HBeAg expression. In contrast HBsAg expression could continue, probably from HBV genomes that have been stably integrated in the DNA of subpopulations of liver cells, which would explain the overproduction of HBsAg relative to HBcAg and the resulting accumulation of 22-nm particles in the blood.

Another interesting problem is the expression of the HBsAg-related proteins. As shown in Fig. 3 the pre-S-HBsAg region encodes three AUGs marked III, II, and I. Initiation of translation on these three AUGs results in expression of three pairs of differentially glycosylated proteins called 39/45, 32/35, and 24/27. The relative proportion of these three pairs of proteins in 42-nm viral particles is about 1:1:10 (15, 51) and, in the more abundant 22-nm particles, about 1:10:100 (9). Why is the downstream AUG used 10 to 100 times more efficiently than the two upstream ones?

The reason for the poor utilization of AUG III is that it is not contained in the major HBsAg mRNA. This mRNA covers AUG II and I, but AUG I is used preferentially, possibly because AUG II is situated at an unusually short distance from the 5' end (15 nucleotides) and has an unfavorable sequence context for ribosome recognition (17). Moreover, some minor forms of the major HBsAg mRNA do not cover the second AUG (5, 49). Thus, it seems that expression of 32/35 kd HBsAg is downregulated mostly at the level of initiation of translation, whereas expression of 39/43 kd HBsAg is downregulated mostly at the transcription level. It is important to notice that the 32/35 HBsAg bears a receptor for human polymerized albumin and thus has a function different from the 24/27 HBsAg (20, 35), and that, contrary to most eukaryotic membrane proteins, all forms of HBsAg do not possess an N-terminal signal peptide, thus leading to the assumption that the signal for membrane insertion is internal.

Little is known about the sequences regulating HBV transcription *in vivo;* most of the signals defined on the HBV genome using heterologous expression systems (Fig. 4) turned out to be poorly if at all used in the liver (5). Conversely, the transcription signals active in the infected liver turned out to be in some aspects peculiar: the strong HBsAg promoter, situated in the pre-S region (Figs. 4 and 5) does not contain a TATA box but shares some homology with a promoter element of the SV40 late transcription unit (4). In contrast, the weak promoter situated upstream of the pre-S region (Figs. 3 and 4) contains a conventional TATA box. The TATA-like box identified as a potential promoter element for HBcAg transcription *in vitro* (Fig. 4) (37) seems not to be used *in vivo*, where an ATA box (Fig. 8) (67) is used. The unique HBV polyadenylation site (Fig. 4) is a variant of the AAUAAA consensus sequence (36), and the recently defined, liver cell-specific enhancer [(43, 57); position 2400 to 2600 on HBV DNA] does not have perfect matches with consensus sequences from other enhancers. It might not be surprising that HBV regulatory sequences often do not conform to rules applying to cellular (and other viral) transcription signals, because these HBV sequences are often simultaneously used as protein coding regions, and must have evolved under peculiar constraints.

A striking example of the interconnection of sequences involved in replication, transcription, and translation is illustrated in Fig. 8. This figure shows the region preceding the core antigen of HBV and of the related ground squirrel, woodchuck, and duck hepatitis viruses. From this comparison it becomes clear that several structural features are strictly conserved among these four viruses.

Two perfect, direct repeats, 11 to 12 nucleotides long, exist in the four viruses, but are separated by a different (46 to 223) number of nucleotides. These two direct repeats are involved in the priming of the DNA minus and plus strand synthesis (19, 26, 67) and possibly also in the integration of HBV DNA (7). They bracket the HBcAg mRNA ATA-box and the major transcription initiation site, which is situated in the same relative position in at least

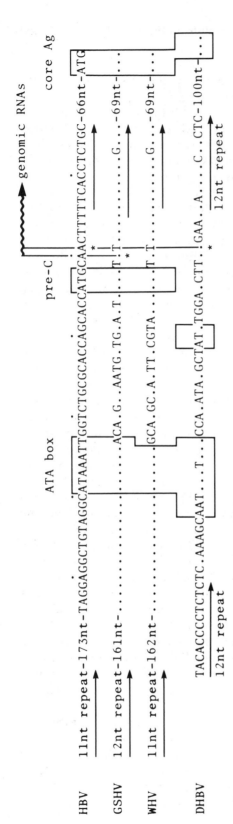

Figure 8 Sequences surrounding the genomic/core antigen RNAs major initiation site in the DNA of hepatitis B-like viruses. The 11- to 12-bp perfect direct repeats involved in the priming of the DNA minus and plus strand synthesis are underlined by an arrow; the ATA box, the pre-C and core antigen ATGs are boxed; the major initiation site of the genomic/core antigen RNAs is indicated with an asterisk, and the corresponding RNA by a wavy line. The HBV, GSHV, WHV, and DHBV sequences are from (11, 41, 12, and 47), respectively. The first HBV nucleotide shown corresponds to position 3056. The precise position of the major site of initiation of transcription was determined with a precision of ±1 nucleotide for HBV (67), ±3 nucleotide for GSHV (8) and ±1 nucleotide for DHBV (q), and was not determined precisely for WHV (28). Interestingly, the three mammalian hepatitis B-like viruses share a stretch of 21 identical nucleotides (indicated by dots) at the beginning of the sequence shown here, and a stretch of 37 identical nucleotides between the pre-C and core antigen ATGs.

286

three of the four viruses. The major site of initiation of HBcAg transcription is situated invariably downstream of the pre-C AUG, but minor mRNA 5′ ends (not indicated) are located upstream of the AUG. As for the surface antigen, the choice of the site of initiation of transcription could determine the expression of different core antigen forms.

HBV Infection and Liver Carcinoma

Three lines of evidence link HBV infection with one of the most common fatal human cancer, primary hepatocellular carcinoma (PHC). First, there is compelling epidemiological evidence that chronic HBV infection constitutes a high risk factor in the development of PHC [reviewed in (1)]. Second, hepatomas are strongly associated with WHV infection in woodchucks. Third, integrated HBV (or WHV) DNA has been found in most human and woodchuck PHC samples tested to date. These observations do not establish the HBV-like viruses as etiological agents of PHC, but, in a multistep model of carcinogenesis, HBV infection appears to be one of the most important events favoring the development of PHC.

What is the relationship between HBV infection and neoplasia? One obvious possibility would be that HBV carries an oncogene responsible for induction of tumors in animals and for neoplastic transformation of cultured cells. This appears unlikely, since PHC arises usually many (10 to 50) years after infection with HBV, and since there is no reproducible evidence for transformation of cultured cells by HBV DNA.

Some retroviruses lack oncogenes, but are able to induce tumors by inserting their DNA and thus their transcriptional enhancer in the proximity of cellular protooncogenes (30, 61a). HBV also contains a tissue-specific enhancer and thus it was of obvious interest to clone the HBV integrates in hepatomas and hepatoma cell lines and screen their neighboring sequences for the presence of known cellular oncogenes. In spite of extensive screening, no cellular oncogene has yet been mapped in the vicinity of an HBV insert.

To gain some clues on the mode of induction of carcinogenesis by the HBV-like viruses, the integration of HBV and WHV DNA in the cellular chromosome has been studied in some detail, but this led to the somewhat surprising observation that no rule can yet be drawn for the mode of integration of HBV and WHV DNA in tumorous tissues (31, 39, 42). In general, in independent hepatomas one or several copies of viral DNA are integrated at different sites in both the viral and cellular sequences. The amount of viral DNA in individual integrates varies from less than to more than one genome, and extensive rearrangements are found in the viral and cellular sequences around the integration site.

In contrast to this work with extensively propagated tumorous tissue or cell lines, a recent determination of HBV integration sites in two "early" human tumors indicates that a specific integration event may exist (7). However, such a primary integration event cannot solely account for the development of PHC because the long latency of PHC (10 to 50 years after HBV infection) points to an indirect and ineffective mechanism connecting the two events. It is of interest in this context to note that in the WHV system complex rearrangements of both viral and cellular sequences have been observed at the site of viral DNA integration in hepatomas, but not in nontumorous tissue (39). Thus, HBV-like viruses could perhaps contribute to carcinogenesis by favoring secondary rearrangements of flanking sequences (16).

Finally, it is important to note that, in contrast to retroviruses, there is at present no evidence for a function of integrated DNA molecules in the HBV life cycle.

Replication Cycle of HBV-like Viruses and Retroviruses

Since HBV-like viruses replicate through an RNA intermediate, it is interesting to compare their replication cycle with that of retroviruses. An earlier theoretical study (61) has underlined the similarities of the two replication cycles, but the data recently obtained about the structure of the HBV transcripts and the mode of integration of HBV DNA in the cellular genome have now revealed some marked differences (2, 8, 67).

When the general aspects of the two viral cycles are considered, some broad similarities are evident. Both classes of viruses share a common reverse transcription step, and the RNA used as a template for DNA synthesis is of plus-strand polarity (i.e., can be used as mRNA for expression of some viral proteins). The distribution of the ORFs on the genomic transcripts of the prototype retrovirus Rous sarcoma virus (60) and of the HBV-like viruses is also strikingly similar (Table 1): genomic transcripts from both groups of viruses contain the core (gag), pol, surface (env) and X (onc) genes in this order.

When the life cycles of HBV-like viruses and retroviruses are drawn in parallel (Fig. 9), it becomes clear that for some aspects the two life cycles represent mirror images of each other: retroviruses contain RNA in their virions and HBV-like viruses DNA; retroviruses reverse-transcribe their RNA genomes immediately after infection and HBV-like viruses reverse transcribe their RNA pre-genome at the end of the replication cycle.

The most important difference between retroviruses and HBV-like viruses resides in the mode and function of DNA integration: retroviruses insert their DNA in the host genome in a well-determined way which follows a well-defined pattern, and the integrated retroviral DNA is then used as template for transcription. The DNA integrates of HBV-like viruses are of very heterogeneous length and structure (Table 1), and at present there is no evidence for a function of integrated DNA molecules in the HBV life cycle.

Several other peculiarities distinguish HBV-like viruses from the retroviruses: the transcription template is usually integrated viral DNA for retroviruses and extrachromosomal, probably supercoiled DNA for HBV-like viruses. Retroviruses express their env gene via a spliced RNA, whereas HBV-like viruses use a second promoter to express their surface antigen

Table 1 Similarities and differences between HBV-like viruses and retroviruses

Parameter	HBV-like viruses	Retroviruses
Gene organization:	core X / pol / surface	pol onc / gag env
Overlap of genes	Extensive	Limited
Redundancies (RNA)	Yes	Yes
Redundancies (DNA)	Short	Long
Transcription:	Unidirectional	Unidirectional
DNA template	Extrachromosomal	Integrated
Promoters	Two or three	One
Splicing in major RNAs	No	Yes
Reverse transcription:	Yes	Yes
Primer for (−) strand	Protein, 2 potential primer sites	tRNA, 1 primer site
Primer for (+) strand	"Displaced" 5′ end of genomic RNA	Polypurine
Template switch in (−)(+) strand synthesis	Unlikely	Essential
Integration and carcinogenesis:		
Integrated viral DNA, length and structure	Varying	Constant
Function of integrated DNA	Unknown	Replication intermediate
Mode of carcinogenesis	Unknown	Promoter insertion, oncogenes

Note: For references on the retroviral life cycle see (60). The gene organization of Rous sarcoma virus is indicated as a retroviral prototype.

(Table 1). The mode of reverse transcription also differs, in particular in the primer for initiation of minus strand DNA synthesis, which is a protein for the HBV-like viruses and tRNA for the retroviruses. The primer for initiation of plus strand DNA synthesis is in both cases derived from genomic RNA, but is displaced from its original site in HBV-like viruses and not in retroviruses (Table 1) (19, 60, 67).

Conclusions

In spite of the lack of a tissue culture system for HBV propagation, the study of the molecular biology of HBV-like viruses is gaining impetus; the fact that HBV-like viruses reverse-transcribe their RNA within an easy-to-isolate core particle is catalyzing quick advances toward the understanding of the precise mechanisms of reverse transcription. The definition of tissue-specific promoter and enhancer regions on the HBV genome provides clues for the analysis of liver-specific promoter elements. The next years will show if the study of the

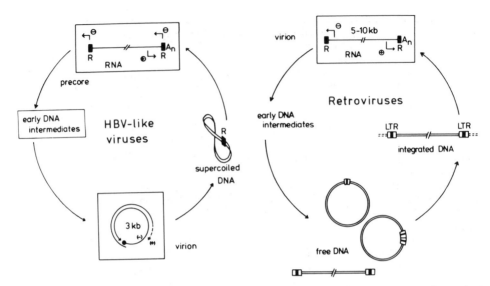

Figure 9 Replicative intermediates of retroviruses and HBV-like viruses [modified from Varmus (61)]. R, Sequence at both ends of viral RNA; LTR, long terminal repeat present in retroviral DNA intermediates; (+) / (−), sites of initiation of plus and minus strand synthesis; A_n, polyadenylate tract at the 3′ end of the genomic RNA; (●), protein linked to the 5′ end of the minus DNA strand of the HBV-like viruses. Boxs indicate mature virions or pre-core structures; single lines denote RNA, double lines DNA (except in HBV virion, where the partially single-stranded DNAregion is indicated as a single line).

HBV-like viruses, as was the case for retroviruses, will shed some light on the mechanism of malignant transformation of the mammalian cell.

Acknowledgments

We thank our colleagues in Heidelberg for their contribution of unpublished data and Oliver Steinau for the sequence alignments of Figs. 4 and 6. We also are indebted to Ken Murray, Edinburgh, and Martin Billeter and Jean-François Conscience, Zurich, for a careful reading and constructive criticism of the manuscript and to Karin Coutinho for excellent secretarial work.

General References

MELNICK, J. L., DREESMAN, G. R., and HOLLINGER, F. B. 1977. Viral hepatitis, *Sci. Amer.* 237: (July) 44–52.

TIOLLAIS, P., CHARNAY, P., and VYAS G. N. 1981. Biology of hepatitis B virus, *Science* 213:406–411.

References

1. Beasley, R. P. 1982. Hepatitis B virus as the etiologic agent in hepatocellular carcinoma—Epidemiologic considerations, *Hepatology* 2:21s–26s.

2. Buescher, M., Reiser, W., Will, H., and Schaller, H. 1985. Characterization of transcripts and the potential RNA pregenome of duck hepatitis B virus: Implications for replication by reverse transcription, *Cell* 40:717–724.

3. Cattaneo, R., Will, H., Darai, G., Pfaff, E., and Schaller, H. 1983. Detection of an element of the SV40 late promoter in vectors used for expression studies in COS cells, *EMBO. J.* 2:511–514.

4. Cattaneo, R., Will, H., Hernandez, N., and Schaller, H. 1983. Signals regulating hepatitis B virus surface antigen transcription, *Nature (London)* 305:336–338.

5. Cattaneo, R., Will, H., and Schaller, H. 1984. Hepatitis B virus transcription in the infected liver, *EMBO. J.* 3:2191–2196.

6. Dane, D. S., Cameron, C. H., and Briggs, M. 1970. Virus-like particles in serum of patients with Australia antigen associated hepatitis, *Lancet* 1:695–698.

7. Dejean, A., Sonigo, P., Wain-Hobson, S., and Tiollais, P. 1984. Specific hepatitis B virus integration in hepatocellular carcinoma DNA through a viral 11-base-pair direct repeat, *Proc. Nat. Acad. Sci. U.S.A.* 81:5330–5354.

7a. Dynan, W. S., and Tijan, R. 1985. Control of eucaryotic messenger RNA synthesis by sequence-specific DNA-binding proteins, *Nature (London)* 316:774–778.

8. Enders, G. H., Ganem, D., and Varmus, H. 1985. Mapping of the major transcripts of ground squirrel hepatitis virus: The presumptive template for reverse transcriptase is terminally redundant, *Cell* 42:297–308.

9. Feitelson, H. A., Marion, P., and Robinson, W. S. 1983. The nature of polypeptides larger in size than the major surface antigen components of hepatitis B and like viruses in ground squirrels, woodchucks and ducks, *Virology* 30:76–90.

10. Fujiyama, A., Miyanohara, A., Nozaki, C., Yoneyama, T., Ohtomo, N., and Matsubara, K. 1983. Cloning and structural analysis of hepatitis B virus DNAs, subtype ADR, *Nucleic Acids Res.* 11:4601–4610.

11. Galibert, F., Mandart, E., Fitoussi, F., Tiollais, P., and Charnay, P. 1979. Nucleotide sequence of the hepatitis B virus genome (subtype AYW) cloned in *E. coli,* *Nature (London)* 281:646–650.

12. Galibert, F., Chen, T. N., and Mandart, E. 1982. Nucleotide sequence of cloned woodchuck hepatitis virus genome: Comparison with the hepatitis B virus sequence, *J. Virol.* 41:51–65.

13. Giorgi, C., Blumberg, B. M., and Kolakofsky, D. 1983. Sendai virus contains overlapping genes expressed from a single mRNA, *Cell* 35:829–836.

14. Gough, N. M. 1983. Core and E antigen synthesis in rodent cells transformed with hepatitis B virus DNA is associated with greater than genome length viral messenger RNAs, *J. Mol. Biol.* 165:683–699.

15. Heermann, K. H., Goldmann, V., Schwartz, W., Seyffarth, T., Baumgarten, H., and Gerlich, W. H. 1984. Large surface proteins of hepatitis B virus containing the pre-S sequence, *J. Virol.* 52:396–402.

15a. Jacks, T., and Varmus, H. E. 1985. Expression of the Rous sarcoma virus POL gene by ribosomal frameshifting, *Science,* 230:1237–1242.

16. Koch, S., Freytag von Loringhoven, A., Hofschneider, P. H., and Koshy, R. 1984.

Amplification and rearrangement in hepatoma cell DNA associated with integrated hepatitis B virus DNA, *EMBO J.* 3:2185–2189.

17. Kozak, M. 1983. Comparison of initiation of protein synthesis in procaryotes, eukaryotes and organelles, *Microbiol. Rev.* 47:1–45.

18. Laub, O., Rall, L. B., Truett, M., Shaul, V., Standring, D. N., Valenzuela, P., and Rutter, W. J. 1983. Synthesis of hepatitis B surface antigen in mammalian cells: Expression of the entire gene and the coding region. *J. Virol.* 48:271–280.

19. Lien, J. M., Aldrich, C. E., and Mason, W. S. 1986. Evidence that a capped oligoribonucleotide is the primer for duck hepatitis B virus plus strand DNA synthesis, *J. Virol.*, 57:229–236.

20. Machida, A., Kishimoto, S., Ohnuma, H., Baba, K., Ito, Y., Miyamoto, H., Funatsu, G., Oda, K., Usuda, S., Togami, S., Nakamura, T., Miyakawa, Y., and Mayumi, M. 1984. A polypeptide containing 55 amino acid residues coded by the pre-S region of hepatitis B virus deoxyribonucleic acid bears the receptor for human as well as chimpanzee albumin, *Gastroenterology* 86:910–918.

21. Malpiece, Y., Michel, M. L., Carloni, G., Revel, M., Tiollais, P., and Weissenbach, J. 1983. The gene S promoter of hepatitis B virus confers constitutive gene expression, *Nucleic Acids Res.* 11:4645–4654.

22. Mason, W. S., Aldrich, C., Summers, J., and Taylor, J. M. 1982. Asymmetric replication of duck hepatitis B virus DNA in liver cells: Free minus-strand DNA, *Proc. Nat. Acad. Sci. U.S.A.* 79:3997–4001.

23. Mason, W. S., Halpern, M. S., England, J. M., Seal, G., Egan, J., Bates, L., Aldrich, C., and Summers, J. 1983. Experimental transmission of duck hepatitis virus, *Virology* 131:375–384.

24. McCollum, R. W., and Zuckerman, A. J. 1981. Viral hepatitis: Report on a WHO informal consultation, *J. Med. Virol.* 8:1–29.

25. Miller, R. H., Marion, P. L., and Robinson, W. S. 1984. Hepatitis B viral DNA-RNA hybrid molecules in particles from infected liver are converted to viral DNA molecules during an endogenous DNA polymerase reaction, *Virology* 139:64–72.

26. Molnar-Kimber, K., Summers, J., and Mason, W. S. 1984. Mapping of the cohesive overlap of duck hepatitis B virus DNA and of the site of initiation of reverse transcription, *J. Virol.* 51:181–191.

27. Moriarity, A. M., Alexander, H., Lerner, R. A., and Thornton, G. B. 1985. Antibodies to peptides detect new hepatitis B antigen: Serological correlation with hepatocellular carcinoma, *Science* 227:429–433.

28. Mŏrŏy, T., Etiemble, J., Trepo, C., Tiollais, P., and Buendia, M.-A. 1985. Transcription of woodchuck hepatitis virus in the chronically infected liver, *EMBO J.* 4:1507–1514.

29. Murray, K., Bruce, S. A., Hinnen, A., Wingfield, P., van Erd, P.M.C.A., de Reus, A., and Schellekens, H. 1984. Hepatitis B virus antigens made in microbial cells immunize against viral infection, *EMBO J.* 3:645–650.

30. Neel, B. G., Hayward, W. S., Robinson, H. L., Fang, J., and Astrin, S. M. 1981. Avian leukosis virus-induced tumors have common proviral integration sites and synthesize discrete new RNAs: Oncogenesis by promoter insertion, *Cell* 23:323–334.

31. Ogston, C. W., Jonak, G. J., Rogler, C. E., Astrin, S. M., and Summers, J. 1982. Cloning and structural analysis of integrated woodchuck hepatitis virus sequences from hepatocellular carcinomas of woodchucks, *Cell* 29:385–394.

32. Ono, Y., Onda, H., Sasada, R., Igarashi, K., Sugino, Y., and Nishioka, K. 1983. The complete nucleotide sequences of the cloned hepatitis B virus DNA; subtype ADR and ADW, *Nucleic Acids Res.* 11:1747–1757.

33. Onodera, S., Ohori, H., Yamaki, M., and Ishida, N. 1982. Electron microscopy of human hepatitis B virus cores by negative staining-carbon film technique, *J. Med. Virol.* 10:147–155.

34. Pasek, M., Goto, T., Gilbert, W., Zink, B., Schaller, H., MacKay, P., Leadbetter, G., and Murray, K. 1979. Hepatitis B virus genes and their expression in *E. coli Nature (London)* 282:575–579.

35. Persing, D. H., Varmus, H. E., and Ganen, D. 1985. A frameshift mutation in the pre-S region of the human hepatitis B virus genome allows production of surface antigen particles but eliminates binding to polymerized albumin, *Proc. Nat. Acad. Sci. U.S.A.* 82:3440–3444.

36. Proudfoot, N. J., and Brownlee, J. J. 1976. 3' non-coding region sequences in eucaryotic messenger RNA, *Nature (London)* 263:211–214.

37. Rall, L. B., Standring, D. L., Laub, O., Rutter, W. J. 1983. Transcription of hepatitis B virus by RNA polymerase II. *Mol. Cell. Biol.* 3:1766–1773.

38. Robinson, W. S. 1983. Hepatitis B virus, in: *Viral Hepatitis; Laboratory and Clinical Science* (F. Deinhardt and J. Deinhardt, eds.), pp. 57–116. Marcel Dekker, New York.

39. Rogler, C. E., and Summers, J. 1984. Cloning and structural analysis of integrated woodchuck hepatitis virus sequences from a chronically infected liver, *J. Virol.* 50:832–837.

40. Salfield, J. 1985. Genprodukte des hepatitis B virus: Struktur, zuornung zum, genom und biosynthese, *Ph.D. Thesis,* Univ. Heidelberg, Heidelberg, West Germany.

41. Seeger, C., Ganem, D., and Varmus, H. E. 1984. Nucleotide sequence of an infectious molecularly cloned genome of ground squirrel hepatitis virus, *J. Virol.* 51:367–375.

42. Shaul, Y., Ziemer, M., Garcia, P. D., Crawford, R., Hsu, H., Valenzuela, P., and Rutter, W. J. 1984. Cloning and analysis of integrated hepatitis virus sequences from a human hepatoma cell line, *J. Virol* 51:776–787.

43. Shaul, Y., Rutter, W. J., and Laub, O. 1985. A human hepatitis B viral enhancer element, *EMBO J.* 4:427–430.

44. Shaw, M. W., Choppin, P. W., and Lamb, R. A. 1983. A previously unrecognized influenza B virus glycoprotein from a bicistronic RNA that also encodes the viral neuraminidase, *Proc. Nat. Acad. Sci. U.S.A.* 80:4879–4883.

45. Simonsen, C. C., and Levinson, A. D. 1983. Analysis of polyadenylation signals of the hepatitis B virus surface antigen gene by using simian virus 40-Hepatitis B virus chimeric plasmids, *Mol. Cell. Biol.* 3:2250–2258.

46. Sprengel, R., Kuhn, C., Manso, C., and Will, H. 1984. Cloned duck hepatitis B virus DNA is infectious in Pekin ducks, *J. Virol.* 52:932–937.

47. Sprengel, R., Kuhn, C., Will, H., and Schaller, H. 1985. Comparative sequence analysis of duck and human hepatitis B virus genomes. *J. Med. Virol.* 15:323–333.

48. Standring, D. N., Rall, L. B., Laub, O., and Rutter, W. J. 1983. Hepatitis B virus encodes an RNA polymerase III transcript, *Mol. Cell. Biol.* 3:1774–1782.

49. Standring, D. N., Rutter, W. J., Varmus, H. E., and Ganem, D. 1984. Transcription

of the hepatitis B surface antigen gene in cultured murine cells initiates within the presurface region, *J. Virol* 50:563–571.

50. Stevens, C. E., Taylor, P. E., Tong, M. J., Toy, P. T., and Vyas, G. N. 1984. Hepatitis B vaccine: An overview, in: *Viral Hepatitis and Liver Diseases* (G. N. Vyas, J. L. Dienstag, and J. H. Hoofnagle, eds.), pp. 275–291, Grune and Stratton, Orlando, Fl.

51. Stibbe, W., and Gerlich, W. H. 1983. Structural relationships between minor and major proteins of hepatitis B surface antigen, *J. Virol.* 49:626–628.

52. Summers, J. 1981. The recently described animal virus models for human hepatitis B virus, *Hepatology* 1:179–183.

53. Summers, J., and Mason, W. S. 1982. Replication of the genome of a hepatitis B-like virus by reverse transcription of an RNA intermediate, *Cell* 29:403–415.

54. Szmuness, W., Stevens, C. E., Harley, E. J., Zang, E. A., Oleszko, W. R., William, D. C., Sadovsky, R., Morrison, J. M., and Kellner, A. 1980. Hepatitis B vaccine, *New Engl. J. Med.* 303:833–841.

55. Takahashi, K., Machida, A., Funatsu, G., Nomura, M., Usada, S., Aoyagi, S., Tachibana, K., Miyamoto, H., Imai, M., Nakamura, T., Miyakawa, Y., and Mayumi, M. 1983. Immunochemical structure of hepatitis B e antigen in the serum, *J. Immunol.* 130:2903–2907.

56. Tiollais, P., Charnay, P., and Vyas, G. N. 1981. Biology of hepatitis B virus, *Science* 213:406–411.

57. Tognoni, A., Cattaneo, R., Serfling, E., and Schaffner, W. 1985. A novel expression selection approach allows precise mapping of the hepatitis B virus enhancer, *Nucleic Acids Res.* 13:7457–7472.

58. Toh, H., Hayashida, H., and Miyata, T. 1983. Sequence homology between retroviral reverse transcriptase and putative polymerase of hepatitis B virus and cauliflower mosaic virus, *Nature (London)* 305:827–829.

59. Valenzuela, P., Quiroga, M., Zaldivar, J., Gray, P. and Rutter, W. J. 1980. The nucleotide sequence of the hepatitis B viral genome and the identification of the major viral genes, in: *Animal Virus Genetics* (B. Fields, R. Jaenisch, and C. W. Fox, eds.), pp. 57–70, Academic Press, New York.

60. Varmus, H. E. 1982. Form and function of retroviral proviruses, *Science* 216:812–820.

61. Varmus. H. E. 1983. Reverse transcription in plants? *Nature (London)* 304:116–117.

61a. Weber, H., and Schaffner, W. 1985. Enhancer activity correlates with the oncogenic potential of avian retroviruses, *EMBO J.* 4:949–956.

62. Weiser, B., Ganem, D., Seeger, C., and Varmus, H. E. 1983. Closed circular viral DNA and asymmetrical heterogeneous forms in livers from animals infected with ground squirrel hepatitis virus, *J. Virol.* 48:1–9.

63. Will, H., Kuhn, C., Cattaneo, R., and Schaller, H. 1982. Structure and function of the hepatitis B virus genome, in: *Primary and Tertiary Structure of Nucleic Acids and Cancer Research* (M. Miwa, S. Nishimura, A. Rich, D. G. Söll, and T. Sugimura, eds.), pp. 237–247, Japan Science Society Press, Tokyo.

64. Will, H., Cattaneo, R., Pfaff, E., Kuhn, C., Roggendorf, M., and Schaller, H. 1984. Expression of hepatitis B antigens with a simian virus 40 vector, *J. Virol.* 335–342.

65. Will, H., Cattaneo, R., Darai, G., Deinhardt, F., Schellekens, H., and Schaller, H. 1985. Infectious hepatitis B virus from cloned DNA of known nucleotide sequence, *Proc. Nat. Acad. Sci. U.S.A.* 82:891–895.

66. Will, H., Salfeld, J., Pfaff, E., Manso, C., Theilmann, L., and Schaller, H. 1986. Putative reverse transcriptase intermediates of human hepatitis B virus in primary liver carcinomas, *Science,* in press.
67. Will, H., Reiser, W., Weimer, T., Pfaff, E., Buscher, M., Sprengel, R., Cattaneo, R., and Schaller, H. Replication strategy of human hepatitus B virus, *J. Virol.,* submitted for publication.

Questions for Discussion with the Editors

1. *How complete is the inventory of HBV proteins? Please elaborate on the possibility that proteins of nonviral origin might be included in the mature virion. Could a requirement for trace quantities of host (e.g., humoral) proteins account for the inability of cultured cells to support HBV replication?*

The inventory of HBV proteins is by far incomplete, and several groups are now investigating HBV proteins other than HBsAg, HBcAg, and HBeAg, mostly using antibodies against proteins expressed in *E. coli,* and corresponding to the HBV unassigned reading frames. Such antibodies have detected fusion proteins of the X reading frame, and a serological correlation was observed between the detection of anti-X antibodies and the diagnosis of hepatocellular carcinoma.[1] Moreover, fusion proteins comprising part of the HBcAg and of the P reading frames have been recently described, and it was observed that they accumulate only in tumorous tissues.[2] We anticipate that in the next few months, or years other HBV proteins will be disclosed and that fusion of different HBV reading frames with other HBV or cellular reading frames will be an important expression mechanism.

Concerning the possibility that proteins of nonviral origin might be included in the mature virion, it is possible that the protein attached to the DNA minus strand is of cellular origin, and it is likely that some cellular membrane proteins could be trapped in the viral envelope during the maturation process. It is also conceivable that humoral proteins could in some way stick to the virions and influence the first steps of infection, but this is only one of several hypotheses that could explain the failure of HBV to grow in cultured cells.

2. *Is it possible that the overproduction of HBsAg aggregates in some way facilitates suppression of host defense mechanisms?*

Patients who do not produce antibodies against HBsAg represent an important fraction of the people with HBV infection, and are designated as chronic HBsAg carriers. Most of these chronic carriers show suppression of some defense mechanisms because they continue to produce low or higher titers of virus. In spite of this, chronic carriers usually have high titers of anti-HBcAg antibodies. Thus the carrier state is strictly related to the lack of an immune response to HBsAg. On the other hand, relatively small doses of HBsAg used as vaccine induce protection against HBV infection, thus indicating that overproduction of surface antigen could be responsible for the deregulation of the immune response.

[1]Moriarty, A. M., Alexander, H., Lerner, R. A., and Thornton, G. B. 1985. Antibodies to peptides detect new hepatitis B antigen: Serological correlation with hepatocellular carcinoma, *Science* 227:429–433.
[2]Will, H., Salfelo, J., Pfaff, E., Manso, C., Theilmann, L., and Schaller, H. 1986. Putative reverse transcriptase intermediates of human hepatitis B virus in primary liver carcinomas, *Science,* in press.

CHAPTER **12**

Molecular Analyses of Tumor Metastasis

James E. Talmadge, Berton Zbar, and Lance A. Liotta

CLINICALLY, METASTASES are the major cause of treatment failure in cancer and therefore represent one of the major areas of cancer research. Metastasis is a complex, multifaceted process which is dependent on both tumor cell and host properties (14, 52). The process of metastasis selects for the rare tumor cells that express the phenotypes required to complete the metastatic cascade including the ability to invade host-tissue barriers, enter and exit the vasculature, evade host defense mechanisms and proliferate to form a discrete focus at a discontiguous site by inducing vascularization (51). The progression of a tumor cell population and the resultant development of these malignant properties has been suggested by Foulds (15) to occur by the "acquisition of permanent, irreversible quantative changes of one or more characteristics in a neoplasm." Therefore, one of the major challenges to cancer researchers is to identify those discrete factors (including specific gene products) that are augmented in metastatic tumor cells and functionally associated with their malignant properties. This chapter reviews the molecular approaches that have been developed to study the genetic lesions that cause normal cells or benign cells to become metastatic.

Because metastasis is the end result of a complex series of steps, it is apparent that this process must involve multiple gene products. Metastatic propensity is assumed to be distinctly separate from tumorigenicity, and transfection studies have demonstrated that DNA from a variety of human tumors can induce neoplastic transformation in NIH3T3 cells (56). The transfected DNA has been shown to contain transforming genes (oncogenes) that usually represent the activated cellular homologs of retrovirus oncogenes. To date, most of the oncogenes identified that have been associated with human malignancy belong to the ras gene family or the myc gene family. Recent results from several laboratories have indicated that activation of the N-ras on-cogenes may play a role in tumor progression (56) and that specific point mutations may be associated with the neoplastic event but differ predicated on the transforming agent (62).

However, the role of oncogenes in the metastatic process has to date been unclear. A number of laboratories are now attempting to identify "metastasis genes" that result in or augment the metastatic phenotype. The following approaches are being utilized to identify the "metastatic" genes:

1. *Traditional molecular biology:* Cloning of genes that code for proteins known to be involved in the metastatic process.
2. *Somatic cell hybridization:* Identification of augmented or suppressed gene products from hybrid tumor cells formed from normal cells fused with metastatic tumor cells or the fusion of metastatic and nonmetastatic tumor cells.
3. *DNA transfection:* Transfection of high molecular weight DNA fragments or isolated genes (including known oncogenes) into normal or nonmetastatic tumor cells to induce the metastatic phenotype. A corollary of this approach is to transfect isolated genes which could inhibit the metastatic phenotype and observe the effect of the expression of this gene on the metastatic phenotype.
4. *Differential cDNA screening:* Selection of specific genes that are differentially expressed in metastatic vs. nonmetastatic tumor cells by the use of subtractive elimination of gene libraries.
5. *Dose-response:* The correlation of specific gene dosages (oncogenes have been studied exclusively to date) with host survival, tumor grade, or tumor stage.

Biochemistry of Metastasis

A variety of biochemical factors have been associated with the metastatic phenotype. These include cell-surface receptors for collagen (32), glycoproteins such as laminin (35), or proteoglycans, all of which have been suggested to facilitate tumor cell attachment to the basement membrane, which is one of the first steps involved in the process of invasion and metastasis. In addition, during the process of metastasis, the presence of several enzymes has been shown to correlate with the ability of tumor cells to invade through tissue barriers. Enzymes that have been suggested to be important in this process include collagenases (33), serum proteases (5, 40), and cathepsin B (48). Once the tumor cell has entered into the host organ, specific growth factors may then have a role in the outgrowth of individual metastatic lesions. Furthermore, throughout the steps of metastasis, the evasion of host immune defenses is important and may also involve a variety of tumor cell gene products.

The receptor for laminin is a 67,000 dalton cell-surface protein that has been shown to play a role in tumor cell attachment to basement membranes

(43). Furthermore, monoclonal antibodies to the laminin receptor inhibit the formation of tumor nodules following intravenous injections as does the preincubation of tumor cells with a small molecular weight fragment from laminin known to bind to the tumor cell receptor for laminin (36).

One example of a protein that may be involved in metastasis, whose gene has been cloned, is plasminogen activator (4). Plasminogen activator is believed to augment tumor cell invasion by the activation of plasmin, an enzyme that can degrade extracellular matrix proteins and that also activates other latent enzymes such as collagenase. The transfection of such genes into cell lines using genes cloned into an appropriate plasmid vector will allow the determination of the role of these individual genes in the metastatic process. Thus, the role of a specific gene product in the metastatic process can be studied by transfecting it into either benign tumor cell lines or normal cell lines and then investigating their metastatic and invasive properties.

Cellular Hybridization

Somatic cell hybridization is an approach that has been utilized to study genes involved in metastasis. The results from this approach must, however, be carefully analyzed because hybrid cells are unstable, and the exact karyotypic feature of each hybrid clone will be different. In one of the first studies Sidebottom and Clark (47) fused highly metastatic mouse melanoma cells with diploid mouse lymphocytes. The hybrid clones were isolated and tested for their tumorigenicity and metastatic potential by injection into newborn syngenic, sublethally irradiated mice. Most of the clones were tumorigenic but had reduced metastatic potential. They concluded that it was possible to make hybrids between malignant and normal cells and that the resulting hybrid clones varied considerably in their ability to produce metastases. Furthermore, in these segregated clones, tumorigenicity and metastatic potential were dissociated. They suggested that both tumorigenicity and metastatic potential are genetically determined, although the genes determining each phenotype were not closely linked. Hart (21) fused metastatic tumor cells with benign or poorly metastatic cells or with normal diploid fibroblasts. He reported that the fusion of metastatic cells with benign cells resulted in a metastatic phenotype, whereas the fusion of metastatic cells with normal diploid cells resulted in a tumorigenic but nonmetastatic phenotype. Hart's results (21) as well as the results from the laboratory of Liotta (18) have shown that the fusion of two tumor cell lines with different metastatic capacities resulted in the maintenance and occasionally the enhancement of metastatic potential. However, when metastatic cells are fused with normal cells, the metastatic phenotype, but not tumorigenicity, is suppressed. This, however, has not been consistently observed by all investigators (9, 19, 24, 30, 31, 42, 61). In several studies fusion products of two established tumor cells were found to be tumorigenic, but with proliferation rates lower than that of the original tumors. This may be due to the relative inefficiency in the mitotic partition of the chromosomes and to the high rate of segregation events, resulting in genotypically imbalanced progeny.

In recent years, there have been a number of reports of spontaneous *in vivo* hybridization of tumor and host cells (30). Many of these *in vivo* fusions of tumor cells and normal host cells have resulted in the emergence of highly metastatic tumor variants (23–26, 29). The development of a high degree of malignancy in tumor host hybrids, as compared to the parental tumor, appears to be due to chromosomal segregation. The invasive and metastatic human-hamster tumors established by Goldenberg et al. (18, 19) following the transplantation of human cancer cells into hamster cheek pouches showed extensive, parental segregation of the human chromosomes, as did *in vitro* human-rodent hybrids (10). Goldenberg has suggested (18) that human chromosome loss accounts for the hamster-like character and behavior of these hybrids. The occasional tumors that develop following the injection of karyotypically complete hybrids tend toward a reduction in chromosomal numbers in agreement with the observation of suppressed hybrids made *in vitro* (25, 55). A modest chromosomal segregation has been observed in hybrids formed *in vivo* that have a tumorigenic potential comparable to that of the injected tumor (2, 13). However, there have been two cases where no chromosomal loss could be detected, and this also occurred with hybrids formed *in vivo* with a tumorigenic potential comparable to that of the injected tumor (1, 6, 12, 25, 49, 55).

The few instances of isogenic hybrid systems observed have shown increased malignant potential compared to the injected tumor, including the MDW4 tumor-host hybrids analyzed by Kerbel et al. (24, 29, 42), which revealed significant chromosomal segregation. These studies from the laboratory of Kerbel (24, 29, 42) focused on the relevance of the *in vivo* tumor-host hybridization process to tumor progression. Tumor-host hybrids were selected *in vivo* that had progressed toward higher malignancy as indicated by the appearance of metastatic nodules in the liver. In this way, these authors examined the evolution of hybrids and their relationship to the host selective pressures as well as the detection of malignant tumor variants generated by segregational mechanisms. Similar studies by De Baetselier et al. (8, 9) found that the injection of the nonmetastatic BW415 T lymphoma resulted in metastatic variants following *in vivo* fusion and chromosomal segregation.

Several kinds of normal cells may be implicated as preferential partners in fusion events depending on the phenotype of the parental tumor, which may restrict the fusion to similarly committed cells. However, lymphocytes, or macrophages that can host tumors due to their "circulating" capability, appear to have a high probability of undergoing spontaneous fusion with tumor cells (23). Indeed, the majority of studies on tumor-host hybrids formed *in vivo* suggest that a cell of bone marrow origin, possibly a macrophage, is one of the most common host cells involved in the fusion process. Support of the monocyte origin of the normal host cell has been provided by Kerbel et al. (24, 29, 42) and from the laboratory of Schirrmacher (30, 31). Lymphocyte-lymphoma fusions have also been observed. In one study a plasmacytoma was fused with normal splenic B lymphocytes and was shown to result in hybridomas which metastasized to the spleen and liver (7). In addition, the development of metastatic variants was observed following the fusion of a T lymphoma with normal T cells resulting in hepatic metastasizing fusion variants (9). De Baetselier et al. (7, 8, 9) have concluded from these studies that the acquisition of metastatic properties by nonmetastatic tumors can occur after fusion with normal lymphoreticular cells and that the differentiated types of the normal fusion partner may influence the target organ specificity for metastatic growth.

It is apparent from this discussion that the studies to date on the role of cellular fusion in metastasis are contradictory. In general, it appears that the *in vivo* or *in vitro* fusion of tumor cells with lymphocytes or macrophages results in an increased metastatic phenotype. In contrast, the fusion of "normal" nonlymphoid cells with metastatic cells results in a loss of the metastatic phenotype but retention of tumorigenicity, whereas the fusion of benign and metastatic cells results in the expression of the metastatic phenotype. We suggest that it should be possible to examine segregant variants following cellular fusion and associate specific chromosomes with the metastatic process.

It is possible to interpret the data from an individual hybridization study by correlating specific gene products with the resultant biological phenotype. A report by Ramshaw et al. (42) demonstrated that nonmetastatic hybrid cells expressed low levels of plasminogen activator (PA) whereas the parental, highly metastatic rat mammary adenocarcinoma cells expressed high levels of PA. Although the hybrid cells produced little PA activity, this could not be directly correlated with the decreased metastatic potential since one clone, following extensive *in vitro* culture, reverted to a more metastatic phenotype without a concomitant increase in PA activity. The authors concluded that the suppressed PA activity may be due to the presence of an inhibitor that is spontaneously produced by the hybrid cells. Similar studies by Turpeenniemi-Hujanen et al. (58) examined six parental and eight hybrid lines and found that the type IV collagenase activity was altered in a manner that was parallel with the metastatic behavior of the cell lines and hybrid lines in nude mice. In no case did hybrid cells exhibit the metastatic phenotype without the expression of type IV collagenase. These data support the concept that type IV collagenase expression may be one cell property that is linked to the metastatic phenotype (32, 33, 34).

Transfection Studies

A series of recent reports have demonstrated that the transfection of human tumor DNA into suitable nontumorigenic and nonmetastatic recipient cells can cause these cells to become metastatic (Fig. 1). Studies from several laboratories (3, 38, 39, 56) have shown that the transfection of human tumor DNA from a variety of tumor histiotypes into NIH3T3 cells results in the de-

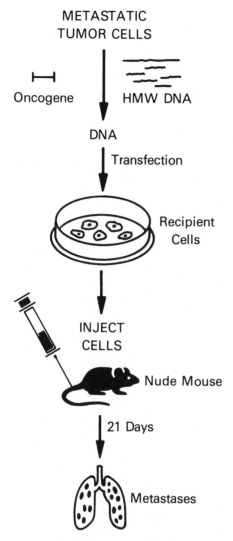

Figure 1 Oncogenes or high molecular weight DNA (HMW DNA) from a metastatic tumor or a human tumor is transfected into "normal" or benign recipient cell lines. Transfectants are then studied to determine metastatic properties by injection into nude mice.

velopment of not only tumorigenic but also metastatic properties in the transfected NIH3T3 cells following either intravenous or subcutaneous injection into nude mice. In these studies, it was found that the ras^N oncogene was one of the genes transfected from the human tumors into the NIH3T3 cells that subsequently became metastatic. Recently, Liotta's laboratory (38, 39, 56) transfected members of the ras gene family into NIH3T3 cells and observed the effects of the transfected gene on the tumor phenotype. In these studies the NIH3T3 cells transformed by the ras^H gene formed local tumors and pulmonary metastases in nude mice after intravenous injection, although spontaneous metastases were not formed.

However, NIH3T3 cells are not an adequate representative of normal fibroblasts because some of the changes involved in malignant transformation have occurred. Therefore, Muschel et al. (38, 39) transfected ras^H into rat diploid embryo fibroblasts as well as into the diploid adult murine C127 fibroblast cell line. The rat diploid embryo fibroblasts transfected with the activated form of the ras^H gene formed metastases in nude mice following intravenous injection; NIH3T3 transfectants with the proto-oncogene (normal cellular component of the ras oncogene) did not form metastases following intravenous injection. The cells transfected with the proto-oncogene expressed large amounts of the P21 protein and were fully tumorigenic but not metastatic. The oncogene-transformed C127 cells expressed less of the viral gene product P21 than NIH3T3 ras^H transformants. To ensure that the decreased levels of P21 were not responsible for the lack of metastatic potential, C127 cells were transformed with an altered construct of the plasmid containing the Harvey sarcoma virus, which resulted in greater levels of P21. These transfectants had augmented expression of P21 levels that were equivalent to or greater than that expressed by the NIH3T3 transformants but did not confer metastatic potential on the C127 transformants. These studies support the concept that tumor growth potential alone is not sufficient to induce metastases in nude mice. Interestingly, embryonic diploid fibroblasts transfected with the ras^H gene produced extensive spontaneous metastases. Thus the induction of the complete metastatic phenotype by ras^H does not require the use of aneuploid or otherwise unusual recipients such as NIH3T3 cells. However, embryonic cells may express fetal genes that could have a role in the metastatic process. These results suggest that at least two complementation groups are required for the induction of metastatic potential in this system, one involving the activated oncogene form of the ras^H gene and another as yet undefined factor found in the background of NIH3T3 cells and embryonic cells but not adult "normal fibroblasts." This conclusion was also reached in the preliminary studies from Thorgeirsson et al. (56), where they found that the transfection of NIH-3T3 cells with human DNA required additional sequences associated with human alu sequences to render these cells highly metastatic.

The studies by Bernstein and Weinberg (3) have identified a gene in human DNA that confers the metastatic phenotype to NIH3T3 cells following transfection. These investigators transfected the ras^H oncogene isolated from the EJ human bladder carcinoma cell line into NIH3T3 cells. These transfectants were fully metastatic in nude mice but did not produce spontaneous metastases when injected into "syngenic" mice. DNA from a human metastatic tumor was then transfected into these ras^H transformants and one of the resulting colonies formed a metastasis following subcutaneous inoculation into "syngenic" mice. The distinct segment of human DNA was linked to highly repetitive human alu sequences and remained associated with the metastatic phenotype through two cycles of DNA transfection. We suggest that this transfected human DNA segment had a role in the host immune response to the tumor that facilated metastasis in syngenic mice but was not required for metastasis in nude mice.

Greig et al. (17) investigated the role of oncogene activation in the expression of the metastatic phenotype by transfecting NIH3T3 cells with the activated ras[H] gene derived from the human T24 bladder carcinoma cell line and compared them with the untransfected NIH3T3 cells. The ras[H] transfected cell lines formed rapidly growing primary tumors that produced pulmonary metastases. In an assay for experimental pulmonary metastases in nude mice, intravenously injected transfected cell lines formed multiple lung tumor colonies detectable 14 days after injection. However, the subcutaneous injection of "normal" NIH3T3 cells into a posterior footpad gave rise to malignant tumors that metastasized to the lungs although this required many months. In addition, the intravenous injection of "normal" NIH3T3 cells gave rise to pulmonary nodules. However, this occurred only at high cell inocula and in long-term survivors (90 days following injection). The authors suggest that the "normal" untransfected NIH3T3 cells contained subpopulations of cells that express malignant properties, and that transfection of NIH3T3 cells with activated ras[H] accelerates the expression of a metastatic phenotype.

Studies by Pozzatti et al. (41) have shown that the cotransfection of rat embryo fibroblasts with ras[H] and the adenovirus type 5 E1a gene results in a much higher frequency of transformation (at least 10-fold higher) than when the cells were transfected with the ras[H] gene alone. Interestingly, however, only one of six cell lines cotransformed by E1a and ras[H] was able to form lung metastases when injected into the tail vein of nude mice, while all the cell lines transformed by ras alone were capable of forming large numbers of metastatic nodules in the lungs of intravenous injected mice, and several of these cell lines were able to form lung metastases following subcutaneous injection into nude mice. In contrast, none of the cotransfected cell lines formed lung metastases following subcutaneous injection. These authors suggest that these results were due to two "cooperating" oncogenes which are capable of transforming a relatively broad spectrum of the cell types present in embryos. In contrast, when ras[H] alone is transfected, a smaller pool of cells are transformed that are predisposed to the metastatic phenotype or can be induced to the metastatic phenotype by the action of the activated ras[H] gene. In other studies by Eccles et al. (11), it was found that not only could plasmid vectors containing EJ-ras genes cause a murine mammary tumor to become metastatic but that the plasmid vector PSV2 neo also was able to transform the mouse mammary tumor (MT1 clone 5/7), resulting in metastatic variants. The authors suggested that transfection with the vector alone may influence metastasis in the absence of any detectable effect on tumorigenicity and that the integration of transforming genes may further potentiate tumor progression.

Several hypotheses can be developed to explain how a single transformed gene can induce the complex metastatic process in the appropriate cell type. The oncogene can confer growth potential on cells that already express all the necessary gene products for metastasis, such as may occur with certain variant cell lines of NIH3T3 cells or embryonic fibroblasts. A second explanation is that the integrated oncogene induces a genetic instability resulting in the rapid production of metastatic variants which are then selected *in vivo*. The third possibility is that transformation by the ras oncogene induces a cascade of cellular gene products normally expressed by migrating cells during embryogenesis, by macrophages, or by cells involved in tissue remodeling.

The molecular genetics approach may be valuable in studying progression of the tumor cell and in the study of tumor cell interactions. Bernstein and Weinberg (3) found that ras[H] transformed NIH3T3 cells, which are fully metastatic in nude mice, are only tumorigenic in immunocompetent syngenic mice. This is not unexpected since host immune factors are well known to modulate the metastatic process. Bernstein and Weinberg went further and transfected a human tumor DNA segment that could restore the metastatic capacity in the immunocompetent mice. This gene may play an important role not only in the intrinsic progression or invasiveness of tumor cells but also in their interaction with the host. Genes that regulate the major histocompatibility antigens on the tumor cell surface have previously been shown to effect the metastatic process. Whether the Bernstein and Weinberg (3) gene is related to the MHC antigen expression remains to be determined.

In the report by Hui and coworkers (22), an AKR leukemia subline that lacked detectable H-2K^k antigen was studied. The H-2K^k gene was introduced into this cell line by transfection and the effects of the transfected gene on growth in normal syngenic mice was evaluated. Selected clones showed a 20-fold increase in H-2K^k mRNA, and the clones varied in antigen expression as measured by radioimmunoassay. In these studies the tumorigenicity of transfected cell lines that expressed H-2K^k antigen was reduced.

Recent studies by Wallich et al. (60) examined the influence of transfection of class 1 genes on an F1 hybrid tumor, T10. This tumor would be expected to express K^b D^b K^k and D^k molecules. However, the clones lack K^b and K^k antigen expression. Therefore, the effect of transfection of K^b and K^k antigens on the properties of the non–H-2K-expressing T10 tumor clones were evaluated. A nonmetastatic clone (IC9) and a metastatic clone (IE7) were transfected with H-2K^k or H-2K^b. Transfection of H-2K^b or H-2^k into IE7 reduced metastatic properties. In addition, the transfection of H-2K^b but not H-2K^k into the IE7 clone led to decreased tumorigenicity. Furthermore, the transfection of K^b, but not K^k, reduced the tumorigenicity of IC9. The transfected cell lines IE7 K^k or K^b, and IC9 K^b had increased immunogenicity as defined by transplantation tests.

The implication of these studies is that decreased expression of cell surface class 1 MHC antigens is one factor involved in the selection of tumor variants that grow progressively. The host immune response appears to have an impaired ability to recognize tumor cells that have decreased expression of specific class 1 antigens. The consequence of this difficulty in tumor cell recognition is increased local tumor growth and, under certain circumstances, metastasis. Wallich et al. have concluded that the expression of H-2K may lead to the recognition of the tumor cells by the host immune system. The suppression of the major histocompatibility antigens is not a universal characteristic of malignant tumors, and further specific immune mechanisms are not the only host defense capable of modulating metastatic spread. Indeed, in other reports, increased levels of MHC class 1 gene expression has been shown to be associated with metastatic capacity. In our own studies (53), we found that the transfection of allogenic class 1 major histocompatibility antigens resulted in a temporary slowing of primary tumor growth and a prolongation of survival. However, ultimately the tumors did grow and metastases did develop such that the host had selected for those rare tumor variants that had lost the gene for the transfected allogenic major histocompatibility antigen.

Talmadge et al. (53) transfected the allo-antigen H-2D^d into the mouse B16 melanoma cell line (H-2^b), and clones were selected for the expression of the D^d antigen. Clones that expressed the D^d antigen were injected into nonimmune C57B16 mice (H-2^b) and into C57B16 mice immunized to lymphocytes expressing D^d antigens. The primary footpad tumors were excised when they reached a size of 1 cm in diameter. The presence of the H-2^d gene in B16 melanoma cells led to the prolongation of life after excision of local tumors. In immune mice the growth of the primary tumor was retarded. Of particular interest was the observation that genomic DNA extracted from lung metastases of both normal and immune animals had complete or partial deletion of the transfected plasmid. The process of metastasis per se was not associated with plasmid gene deletion. This conclusion was reached on the basis of studies that showed that pulmonary metastases from B16 cell lines containing the parent plasmid without the allo-antigen retained the restriction pattern characteristic of that plasmid-containing cell line. The significance of these studies is that there appear to be reciprocal effects of a transfected plasmid encoding a foreign class 1 antigen in tumor cells. In the B16 system the presence of the D^d gene in B16 cells (H-2^b) led to prolongation of survival after tumor excision. However, the influence of host immunity was to select for those rare cells that deleted the plasmid and permitted such cells to grow progressively.

Tanaka et al. (54) transfected the L^d gene into an adenovirus transformed C57B16 cell line (H-2^b). Clones with high levels of mRNA for the L^d gene were injected into BALB/c mice (H-2^d). The experiment was designed to study the influence of the transfected L^d gene in a mouse (BALB/c) of the H-2^d haplotype. Retardation of growth of the C57B16 line expressing high levels of mRNA for L^d was observed in BALB/c mice, whereas no retardation of growth occurred in C57BL6 tumor clones transfected with L^d with low levels of mRNA for L^d.

In studies from our laboratories (56), we found that the expression of the metastatic phenotype conferred on NIH3T3 cells by the transfection of human DNA which included the activated N-ras gene as well as additional sequences was not correlated with the tumor cell sensitivity to either natural killer cells (NK) or macrophages. This suggested that if the cell is highly metastatic, this phenotype can be expressed in the presence of host immune response including not only T cells but also the host NK cells and macrophage responses.

In other studies from the laboratory of Zbar (50), the process of tumor progression was studied with tumor cells containing exogenous genes that encode cell-surface retroviral antigens. One property of retrovirus-infected cells is that the populations are heterogeneous in many aspects of retrovirus expression. Cells isolated from retrovirus-infected cultures differ in the number of copies of the provirus per cell, in whether infectious virus is released, and in viral antigen expression (63). The retrovirus-infected guinea pig cell lines were injected subcutaneously and thereby subjected to immunological selection. The primary tumor grew, regressed, and in approximately 70 percent of the hosts subsequently recurred. These recurrent tumor variants lacked the provirus. In addition, a clone was isolated from the retrovirus-infected guinea pig cell line that contained a single abbreviated copy of the provirus, which was also deleted from this clone under conditions of *in vivo* immunological selection.

These results suggest that tumor recurrence at the site of injection of retrovirus-infected guinea pig cells was a consequence of proviral gene loss. Cells with single copies of this provirus are thought to be at particular risk for proviral gene loss. Superinfection with an independent retrovirus introduced additional copies of the provirus into cells making the cells effectively diploid, and tumor recurrence was decreased in doubly infected cells. This observation raises the possibility that inactivation of genes that exist as single copies (haploid) can produce adverse consequences for the host.

Differential cDNA Screening

Several laboratories are studying the genes involved in metastasis by forming DNA libraries of cosmid DNA from metastatic cells which, following subtractive elimination from the libraries of benign tumors, allows the identification of specific clones that differ between the metastatic and nonmetastatic cell lines (Fig. 2). Studies have been reported to date from the laboratories of Sanchez et al. (45), LaBiche et al. (28), and Liotta. These studies have found that less than 0.1 percent of the gene products examined differ between the metastatic and benign tumor cell lines that were compared. Obviously, this approach is very powerful, as well as labor intensive, and requires the subsequent identification of the protein products from the gene products and the establishment of the roles of these different gene products. These gene products can differ between the benign cells and the metastatic cells but may have either a positive or negative influence on the process of metastasis.

Gene Dosage

The last approach to the molecular study of metastasis is a correlative study regarding the role of the ras or myc oncogene and the malignant expression of tumors. To date the data are conflicting. Vousden and Marshall (59) have found that there is an activation of ras^k in invasive metastatic tumors compared to benign tumors, at least in the progression of a mouse lymphoma.

ISOLATION OF MET⁺ cDNA

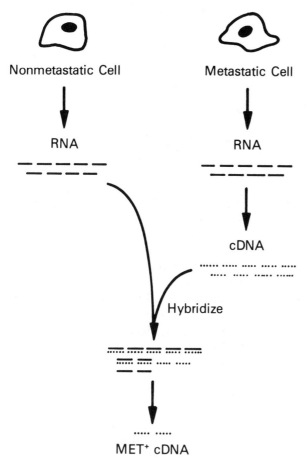

Figure 2 Gene libraries are made from metastatic and nonmetastatic tumors, and unique gene sequences are identified by differential hybridization.

Studies from the laboratory of Saksela (44) have shown that there is an amplification of the myc oncogene in a subpopulation of human small-cell lung cancer. Minna et al. (16, 37) have suggested that there are increased myc gene dosages in metastases of patients compared to their primary tumors. Evidence from the laboratory of Trent et al. (57) has suggested that myc is found in double minute chromosomes in patients with retinoblastoma and may be associated with tumor progression, whereas studies from the laboratory of Seeger et al. (46) suggest that there is an amplification of myc in patients with neuroblastoma that correlates with host survival and staging. In contrast, Gutterman (27) did not find a change in the expression of ras p21 in fresh primary and metastatic human carcinomas, and similarly Kris et al. (26) did not find a correlation in the expression of ras^k oncogene in tumor cell variants (murine fibrosarcomas) exhibiting different metastatic capabilities. Unfortunately, many of these studies did not distinguish activated ras or myc oncogene from the proto-oncogene form.

Conclusions

At present, it seems unlikely that a single gene will be found to be responsible for the expression of the complete metastatic phenotype. Indeed, the evidence to date suggests that multiple genes are required for the expression of metastasis. It would seem that a few key genetic changes (which may include the activation of an oncogene) can induce a cascade of genetic expression for multiple gene products which are necessary for all the steps in the metastatic process. This concept is supported by the studies of differential mRNA levels in metastatic versus nonmetastatic tumor cells derived from the same patient or variants from a single murine lymphoma. These studies reveal that some genes were expressed in greater amounts in metastatic tumor variants compared to the nonmetastatic variants or primary tumors, whereas other genes were preferentially expressed in the nonmetastatic or parental tumor cells. We hope that the further analysis of these gene products, the cloning of the products, and subsequent transfection, as well as an examination of their regulation will provide new strategies for the study of metastasis and the development of novel approaches for tumor diagnosis and metastatic therapy.

Acknowledgments

Research for this chapter was sponsored by the Biological Resources Branch, Biological Response Modifiers Program, Division of Cancer Treatment of the National Cancer Institute, DHHS, under contract no. N01-23910 with Program Resources, Inc. The contents of this chapter do not necessarily reflect the views or policies of the Department of Health and Human Services, nor does mention of any organization imply endorsement by the U.S. government.

General References

THORGEIRSSON, U. P., TURPEENNIEMI-HUJANAN, T., WILLIAMS, J. E., WESTIN, E. H., HEILMAN, C. A., TALMADGE, J. E., and LIOTTA, L. A. 1985. NIH-3T3 cells transfected with human tumor DNA containing activated ras oncogenes express the metastatic phenotype in nude mice, Mol. Cell. Biol. 5:259–262.

GREIG, R. G., KOESTLER, T. P., TRAINER, D. L., CORWIN, S. P., MILES, L., KLINE, T., SWEET, R., YOKOYAMA, S., and POSTE, G. 1985. Tumorigenic and metastatic properties of "normal" and ras-transfected NIH-3T3 cells, Proc. Nat. Acad. Sci. U.S.A. 82:3698–3701.

References

1. Aviles, D., Jami, J., Rousset, P., and Ritz, E. 1977. Tumor x host cell hybrids in the mouse: Chromosomes from the normal cell parent maintained in malignant hybrid tumors, *J. Nat. Cancer Inst.* 58:1391–1397.

2. Ber, R., Wiener, F., and Fenyo, E. M. 1978. Proof of in vivo fusion of murine tumor cells with host cells by universal fusers: Brief communication, *J. Nat. Cancer Inst.* 60:931–933.

3. Bernstein, S. C., and Weinberg, R. A. 1985. Expression of the metastatic phenotype in cells transfected with human metastatic tumor DNA, *Proc. Nat. Acad. Sci. U.S.A.* 82:1726–1730.

4. Blasi, F. 1985. Plasminogen activator genes: Structure and regulation, *Fogarty International Center Conference*, National Institutes Of Health, Bethesda, Md.

5. Carlsen, S. A., Ramshaw, I. A., and Warrington, R. C. 1984. Involvement of plasminogen activator production with tumor metastasis in a rat model, *Cancer Res.* 44:3012–3016.

6. Croce, C. M. 1980. Cancer genes in cell hybrids, *Biochem. Biophys. Acta* 605:411–430.

7. De Baetselier, P., Gorelik, E., Eshhar, Z., Ron, Y., Katzav, S., Feldman, M., and Segal, S. 1981. Metastatic properties conferred on non-metastatic tumors by hybridization of spleen-B-lymphocytes with plasmacytoma cells, *J. Nat. Cancer Inst.* 67:1079–1087.

8. De Baetselier, P., Roos, E., Brys, L., Remels, L., Gobert, M., Dekegel, D., Segal, S., and Feldman, M. 1984. Nonmetastatic tumor cells acquire metastatic properties following somatic hybridization with normal cells, *Cancer Metast. Rev.* 1:6–24.

9. De Baetselier, P., Roos, E., Brys, L., Remels, L., and Feldman, M. 1984. Generation of invasive and metastatic variants of a nonmetastatic T-cell lymphoma by in vivo fusion with normal host cells, *Int. J. Cancer* 34:731–738.

10. De Carli, L., and Larizza, L. 1978. Induced chromosome variation in cultured cell populations, in: *Origin and National History of Cell Lines*, A. R. Liss, ed. *Progress Clin. Biol. Res.* 26:93–124.

11. Eccles, S. A., Marshall, C., Vousden, K., and Purvies, H. 1985. Enhanced metastatic capacity of mouse mammary carcinoma cells transfected with H-*ras* in: *Proceedings of Treatment of Metastasis: Problems and Prospects*, paper 6.8, Sutton and London, U.K.

12. Evans, E. P., Burtenshaw, M. D., Brown, B. B., Hennion, R., and Harris, H. 1982. The analysis of malignancy by cell fusion. IX. Re-examination and clarification of the cytogenetic problem, *J. Cell Sci.* 56:113–130.

13. Fenyo, E. M., Wiener, F., Klein, G., and Harris, H. 1973. Selection of tumor-host cell hybrids from polyoma virus- and methylcholanthrene-induced sarcomas, *J. Nat. Cancer Inst.* 51:1865–1872.

14. Fidler, I. J., Gersten, D. M., and Hart, I. R. 1978. The biology of cancer invasion and metastasis, *Adv. Cancer Res.* 28:149.

15. Foulds, L. 1975. *Neoplastic Development*, vol. 2, Academic Press, London.

16. Gazdar, A. F., Carney, D. N., Nau, M. M., and Minna, J. D. 1985. Characterization of variant subclasses of cell lines derived from small cell lung cancer having distinctive biochemical, morphological and growth properties, *Cancer Res.* 45:2924–2929.

17. Greig, R. G., Koestler, T. P., Trainer, D. L., Corwin, S. P., Miles, L., Kline, T., Sweet, R., Yokoyama, S., and Poste, G. 1985. Tumorigenic and metastatic properties of "normal" and *ras*-transfected NIH/3T3 cells, *Proc. Nat. Acad. Sci. U.S.A.* 82:3698–3701.

18. Goldenberg, D. M., Bhan, R. D., and Pavia, R. A. 1971. *In vivo* human-hamster somatic cell fusion indicated by glucose 6-phosphate dehydrogenase and lactate dehydrogenase profiles, *Cancer Res.* 31:1148–1152.

19. Goldenberg, D. M., Pavia, R. A., and Tsao, M. C. 1974. *In vivo* hybridization of human tumor and normal hamster cells, *Nature (London)* 250:649–651.

20. Halaban, R., Nordlund, J., Franke, U., Moellmann, G., and Eisenstadt, J. M. 1980. Supermelanotic hybrids derived from mouse melanomas and normal mouse cells, *Somatic Cell Genet.* 6:29–44.

21. Hart, I. R. 1984. Tumor cell hybridization and neoplastic progression, in: *Cancer Invasion and Metastasis: Biologic and Therapeutic Aspects*, G. L., Nicolson and L. Milas, eds., pp. 133–143, Raven Press, New York.

22. Hui, D., Grosveld, F., and Festenstein, H. 1984. Rejection of transplantable AKR leukaemia cells following MHC DNA-mediated cell transformation, *Nature (London)* 311:750–752.

23. Kerbel, R. S., Lagarde, A. E., Dennis, J. W., and Donaghue, T. P. 1983. Spontaneous fusion *in vivo* between normal host and tumor cells: Possible contribution to tumor progression and metastasis studied with a lectin-resistant mutant tumor, *Mol. Cell. Biol.* 3:523–538.

24. Kerbel, R. S., Twiddy, R. R., and Robertson, D. M. 1978. Induction of a tumor with greatly increased metastatic growth potential by injection of cells from a low metastatic H-2 heterozygous tumor cell line into an H-2 incompatible parental strain, *Int. J. Cancer* 22:583–594.

25. Klinger, H. P., and Shows, T. B. 1983. Suppression of tumorigenicity in somatic cell hybrids. II. Human chromosomes implicated as suppressors of tumorigenicity in hybrids with Chinese hamster ovary cells, *J. Nat. Cancer Inst.* 71:559–569.

26. Kris, R. M., Avivi, A., Bar-Eli, M., Alon, Y., Carmi, P., Schlessinger, J., and Raz, A. 1985. Expression of Ki-*ras* oncogene in tumor cell variants exhibiting different metastatic capabilities, *Int. J. Cancer* 35:227–230.

27. Kurzrock, R., Gallick, G., and Gutterman, J. 1985. Expression of p21 *ras* in fresh primary human lung tumors, *Proc. Amer. Assoc. Cancer Res.* 26:65.

28. LaBiche, R. A., Frazier, M. L., Brock, W. A., and Nicolson, G. L. 1985. Differential gene expression between high and low metastatic sublines of murine RAW117 large cell lymphoma, *Proc. Amer. Assoc. Cancer Res.* 26:48.

29. Lagarde, A. E., Donaghue, T. P., Dennis, J. W., and Kerbel, R. S. 1983. Genotypic and phenotypic evolution of a murine tumor during its progression *in vivo* toward metastasis, *J. Nat. Cancer Inst.* 71:183–191.

30. Larizza, L., and Schirrmacher, V. 1984. Somatic cell fusion as a source of genetic rearrangement leading to metastatic variants, *Cancer Met. Rev.* 3:193–222.

31. Larizza, L., Schirrmacher, V., Graf, L., Pfluger, E., Peres-Martinez, M., and Stohr, M. 1984. Suggestive evidence that the highly metastatic variant ESb of the T-cell lymphoma Eb is derived from spontaneous fusion with a host macrophage, *Int. J. Cancer* 34:299–307.

32. Liotta, L. A. 1985. Mechanisms of cancer invasion and metastases, in: *Important*

Advances in Oncology V. Devita, K. Hellman, and S. Rosenberg, eds. pp. 28–41. J. B. Lippincott, New York.

33. Liotta, L. A., Kleinerman, J., Catanzaro, P., and Rynbrandt, D. 1977. Degradation of basement membrane by murine tumor cells, *J. Nat. Cancer Inst.* 58:1427.

34. Liotta, L. A., Tryggvason, K., Garvisa, S., Hart, I., Foltz, C. M., and Shafie, S. 1980. Metastatic potential correlates with enzymatic degradation of basement membrane collagen, *Nature (London)* 284:67–68.

35. Liotta, L. A., Rao, C. N., and Barsky, S. H. 1983. Tumor invasion and the extracellular matrix, *Lab. Invest.* 49:636.

36. Liotta, L. A., Hand, P., Rao, C., Bryant, G., Barsky, S., and Schlom, J. 1985. Monoclonal antibodies to the human laminin receptor recognize structurally distinct sites, *Exp. Cell Res.* 156:117–126.

37. Minna, J. D. 1984. Recent advances of potential clinical importance in the biology of lung cancer, *Proc. Amer. Assoc. Cancer Res.* 25:393–394.

38. Muschel, R. J., Pozzatti, R., Khoury, G., Nakahara, K., Chu, E., Lowy, D. A., Williams, J., and Liotta, L. A. The effect of ras[H] genes on metastatic potential in nude mice, in: *First Annual Meeting on Oncogenes*, NCI, Frederick, Md., July 10–13, 1985.

39. Muschel, R. J., Williams, J. E., Lowy, D. R., and Liotta, L. A. 1985. Harvey *ras* induction of metastatic potential depends upon oncogene activation and the type of recipient cell. *Amer. J. Pathol.* 121:1–8.

40. Ossowski, L., and Reich, E. 1983. Antibodies to plasminogen activator inhibit human tumor metastasis, *Cell* 35:611–619.

41. Pozzatti, R., Muschel, R., Williams, J., Padmanabhan, R., Howard, B., Liotta, L., and Khoury, G. Early passage rat embryo cells transformed by viral and cellular oncogenes exhibit different metastatic potentials (submitted manuscript).

42. Ramshaw, I. A., Carlsen, S., Wang, H. C., and Badenoch-Jones, P. 1983. The use of cell fusion to analyse factors involved in tumor cell metastasis, *Int. J. Cancer* 32:471–478.

43. Rao, C. N., Margulies, I. M. K., Tralka, T. S., Terranova, V. P., Madri, J. A., and Liotta, L. A. 1982. Isolation of a subunit of laminin and its role in molecular structure and tumor cell attachment, *J. Biol. Chem.* 257:9740.

44. Saksela, K., Bergh, J., Lehto, V.-P., Nilsson, K., and Alitalo, K. 1985. Amplification of the *c-myc* oncogene in a subpopulation of human small cell lung cancer, *Cancer Res.* 45:1823–1827.

45. Sanchez, J., Varani, J., Wicha, M., and Miller, D. 1985. Relationship of altered gene expression to metastatic potential, *Proc. Amer. Assoc. Cancer Res.* 26:59.

46. Seeger, R., Brodeur, G., Sather, H., Dalton, A., Siegel, S., Wong, K., and Hammond, D. 1985. Genomic amplification of N-*myc* in untreated primary neuroblastomas correlates with advanced disease stage at diagnosis and rapid tumor progression, *Proc. Amer. Assoc. Cancer Res.* 26:64.

47. Sidebottom, E., and Clark, S. R. 1983. Cell fusion segregates progressive growth from metastasis, *Brit. J. Cancer* 47:399–406.

48. Sloane, B. F., Honn, K. V., Sadler, J. G., Turner, W. A., Kimpson, J. J., and Taylor, J. D. 1982. Cathepsin B activity in B16 melanoma cells: A possible marker for metastatic potential, *Cancer Res.* 42:980–986.

49. Stanbridge, E. J., Flandermeyer, R. R., Daniels, D. W., and Nelson-Rees, W. A.

1981. Specific chromosomes loss associated with the expression of tumorigenicity in human cell hybrids, *Somat. Cell Genet.* 7:699–712.

50. Talmadge, C., Tanio, Y., Meeker, A., Talmadge, J. E., and Zbar, B. 1985. Use of cells transfected with the neomycin resistance gene (neo) to study tumor recurrence and metastasis in: *Symposium on Biochemistry and Molecular Genetics of Cancer Metastasis*, Bethesda, Md, March 18–20, National Institutes of Health, Bethesda, Md.

51. Talmadege, J. E., and Fidler, I. J. 1982. Cancer metastasis is selective or random depending on the parent tumour population, *Nature (London)* 297:593

52. Talmadge, J. E., Liotta, L., and Kohn, R. R. 1984. The biology of biochemistry of metastatic cells, in: *Head and Neck Management of the Cancer Patient*, (D. E. Peterson, E. G. Elias, and S. T. Sonis, eds.) Martinus Nijhoff Publishers, The Hague, Netherlands.

53. Talmadge, J. E., Talmadge, C. B., McEwan, R. N., and Meeker, A. 1985. Tumor growth and metastasis of B16-BL6 tumors following transfection of an allogenic major histocompatibility complex (D^d) antigen, *Proc. Amer. Assoc. Cancer Res.* 26:59.

54. Tanaka, K., Isselbacher, K. J., Khoury, G., and Gilbert, J. 1985. Reversal of oncogenesis by the expression of a major histocompatibility compex Class I gene, *Science* 228:26–30.

55. Tenchini, M. L., Larizza, L., Mottura, A., Colombi, M., Barlati, S., and De Carli, L. 1983. Studies on transformation markers and tumorigenicity in segregant clones from a human hybrid line, *Eur. J. Cancer Clin. Oncol.* 19:1143 –1149.

56. Thorgeirsson, U. P., Turpeenniemi-Hujanen, T., Williams, J. E., Westin, E. H., Heilman, C. A., Talmadge, J. E., and Liotta, L. A. 1985. NIH-3T3 cells transfected with human tumor DNA containing activated *ras* oncogenes express the metastatic phenotype in nude mice, *Mol. Cell. Biol.* 5:259–262.

57. Trent, J., Meltzer, P., Rosenblum, M., Harsh, G., Kinzler, K., Feinberg, A., and Vogelstein, B. 1985. Evidence for rearrangement and amplification of the c-*myc* cellular oncogene in a human glioblastoma, *Proc. Cancer Res.* 26:64.

58. Turpeenniemi-Hujanen, T., Thorgeirsson, U. P., Hart, I. R., Grant, S. S., and Liotta, L. A. 1985. Expression of collagenase IV (basement membrane collagenase) activity in murine tumor cell hybrids that differ in metastatic potential, *J. Nat. Cancer Inst.* 75:99.

59. Vousden, K. H., and Marshall, C. J. 1984. Three different activated ras genes in mouse tumors: Evidence for oncogene activation during progression of a mouse lymphoma, *EMBO J.* 3:913– 917.

60. Wallich, R., Bulbuc, N., Hammerling, G. J., Katzav, S., Segal, S. and Feldman, M. 1985. Abrogation of metastatic properties of tumor cells by *de novo* expression of H-2K antigens following H-2 gene transfection, *Nature (London)* 315:301–305.

61. Warner, T. F. C. S. 1975. Cell hybridization: An explanation for the phenotypic diversity of certain tumors, *Med. Hypoth.* 1:51–56.

62. Zarbl, H., Sukumar, S., Arthur, A. V., Martin-Zanca, D., and Barbacid, M. 1985. Direct mutagenesis of Ha-*ras*-1 oncogenes by N-nitroso-N-methylurea during initiation of mammary carcinogenesis in rats, *Nature (London)* 315:282.

63. Zbar, B., Terata, N., Nagai, A., Tanio, Y., and Hovis, J. 1984. Selection and rejection of retrovirus-expressing tumor cells from a heterogeneous murine leukemia virus-infected cell population, *Cancer Res.* 44:4622–4629.

Questions for Discussion with the Editors

1. *Do you imagine that the metastatic cascade develops under the influence of a few "primary" regulatory genes (e. g., oncogenes?) which directly control the expression of "secondary" genes? Or, do you visualize the cascade—although perhaps initially trigered by a single gene—as being the product of a series of stochastic gene expression events which, through in vivo selection, generate the rare successful metastatis?*

This is one of the most intriguing questions facing scientists investigating the metastatic cascade. The preliminary data suggests that the expression of multiple genes are necessary to result in a metastatic phenotype. Furthermore, it appears likely, due to the many and diverse phenotypes involved in the metastatic process, that the genotypes required for the metastatic phenotype may vary among tumors. Obviously, a primary regulatory gene is an attractive hypothesis and cannot be eliminated from consideration at present. However, the fusion studies and transfection studies undertaken to date suggest that a single oncogene, in the absence of other genes, may not be sufficient to induce the metastatic cascade.

2. *Do you expect the so-called stable metastatic state to be the product of extensive gene rearrangements?*

A few malignant histiotypes are associated with specific chromosomal rearrangements. In addition, there may be an association between chromosomal recombinants and the metastatic phenotype. However, to date, there has been no evidence to suggest that a stable metastatic state is associated with extensive gene or chromosomal rearrangements.

CHAPTER **13**

Transport and Processing of Viral Membrane Glycoproteins

Lewis J. Markoff and Ching-Juh Lai

SECRETORY AND INTEGRAL MEMBRANE PROTEINS are sorted with complete specificity. Sorting begins with translocation of a polypeptide across an intracellular membrane. As a consequence of translocation, the polypeptide passes from the cytoplasm into the membrane compartment of the cell. The translocation of membrane proteins is usually incomplete and results in the assymetrical integration of the polypeptide into translocation-competent membranes. Therefore, membrane proteins have a cytoplasmic domain. Sorting also includes intracellular transport that may follow translocation. Ultimately, sorting may result in intracellular association of a protein with organelles, such as mitochondria, lysosomes, endoplasmic reticulum, or the Golgi apparatus, insertion into the plasma membrane as a primary structural or functional component, or secretion. Membrane and secretory proteins are biochemically modified during sorting. Such processing may include glycosylation, sulfation, of added carbohydrate, acylation, and site-specific cleavage of polypeptides in association with the formation of a polymeric functional protein.

Viral Glycoprotein Maturation as a Model for Studying Intracellular Protein Topogenesis

According to the currently favored hypothesis (8), secretory and membrane proteins contain within their primary sequence the information that is required to target the nascent polypeptide to a specific translocation-competent membrane and to mediate the translocation event itself. This is the broad definition of a "signal peptide". After translocation, proteins that remain integrated in membranes not only span the lipid bilayer but must have a

314

hydrophilic domain on one or both sides. Two additional types of "topogenic sequences" (8), besides the signal sequence, must exist to account for the possible orientations of integral membrane proteins. These are a "stop-transfer" sequence, which would interrupt translocation to allow for membrane integration, and an "insertion" sequence, which would effect such integration of the polypeptide chain by spanning the lipid bilayer (75).

This chapter summarizes available information concerning the sorting and processing of viral membrane glycoproteins. This has been a subject of interest not only as an important aspect of virus maturation but also as a model for defining the topogenic sequences. The study of viral, rather than cellular, membrane glycoprotein maturation has been a preferred approach to the problem of protein topogenesis for many reasons. Viral genomes code for one or only a few glycoproteins. In virus-infected cells, host functions operate almost exclusively on viral gene products, because the synthesis of new cell-specific proteins is attenuated or completely shut off. This effect of viral infection in reducing the background noise of cellular protein synthesis facilitates the detection of virus-specific proteins in infected cells. In addition, virus mutants bearing altered membrane glycoproteins are more easily selected and classified than would be mutant eukaryotic cell lines with such generalized and potentially lethal defects. Finally, viral genes or their transcripts are more readily obtainable for cell-free transcription and/or translation experiments in which protein synthesis may be coupled to translocation and glycosylation selectively.

The influenza A virus hemagglutinin (HA) is taken as an example of a major class of membrane proteins with respect to the location within the linear polypeptide sequence of the topogenic sequences predicted by Blobel (8) (Fig. 1). In this regard, the major feature of the influenza HA sequence is the presence of uninterrupted hydrophobic regions at amino- (NH_2-) and carboxy- (C-) termini. These hydrophobic regions are each flanked by short terminal segments with net positive charge. [The NH_2-terminal hydrophilic region is not a prominent feature of the influenza HA (Fig. 2) and is more regularly found in prokaryotic membrane proteins (56).] Evidence will be presented that the hydrophobic NH_2-terminus of membrane proteins like the HA is the signal peptide. It will also be shown that the C-terminal hydrophobic region and the hydrophilic "tail" represent the additional topogenic sequences required for insertion and stop-transfer functions (15, 18, 21, 68, 82, 86, 87). As a second example, frequently alluded to in the text, the "map" of the vesicular stomatitis virus (VSV) glycoprotein (G) is completely analogous to that of the influenza HA (18). The functions of analogous topogenic sequences in the two proteins have been separately corroborated.

The influenza neuraminidase (NA) represents a less common class of membrane glycoprotein (Fig. 1). Sequence analysis of NA genes of different antigenic subtypes indicates that the NH_2-terminus includes a hydrophobic region 29 amino acids in length, from residue 7 through residue 35 (17, 31, 51). Residues 1 through 12 are highly conserved and consist of six (NH_2-terminal) predominantly hydrophilic residues followed by the first six residues

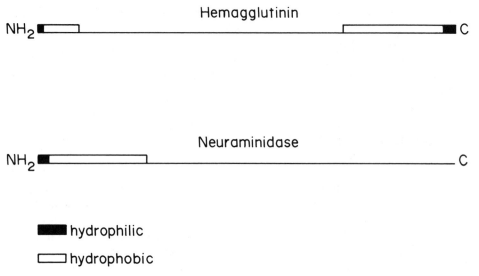

Figure 1 Topographical map of the influenza hemagglutinin (HA) as a representative of the major class of integral membrane proteins. These proteins share in common the presence of NH_2- and C-terminal hydrophobic regions and of a C-terminal hydrophilic "tail." The NH_2-terminal hydrophilic region depicted is variably present in eukaryotic membrane proteins and is a regular feature of prokaryotic ones. Map of the influenza neuraminidase (NA) as the representative of a less common class of membrane proteins that contain a hydrophilic NH_2-terminus upstream from a hydrophobic region longer in total extent than that found at either end of proteins like the HA. Diagrams are not drawn to scale.

of the hydrophobic region. (The functional significance of the conserved NH_2-terminus of the NA has not been determined.) In contrast to HA, an uninterrupted C-terminal hydrophobic sequence is not present in the NA. In addition, biochemical studies confirm that the NH_2-terminus of the influenza NA does not undergo cleavage during processing and that it serves to anchor the

(FPV) H7: Met·Asn·Thr·Gln·Ile·Leu·Val·Phe·Ala·Leu·Val·Ala·Val·Ile·Pro·Thr·Asn·Ala·(Asp) ------

H1: Met·Lys·Ala·Asn·Leu·Leu·Val·Leu·Leu·Cys·Ala·Leu·Ala·Ala·Ala·Asp·Ala·(Asp) ------

H2: Met·Ala·Ile·Ile·Tyr·Leu·Ile·Leu·Leu·Phe·Thr·Ala·Val·Arg·Gly·Asp·(Gln) ------

H3/72: Met·Lys·Thr·Ile·Ile·Ala·Leu·Ser·Tyr·Ile·Phe·Cys·Leu·Val·Leu·Gly·(Gln) ------

H3/75: Met·Lys·Thr·Ile·Ile·Ala·Leu·Ser·Tyr·Ile·Phe·Cys·Leu·Val·Phe·Ala·(Gln) ------

Figure 2 The amino acid sequences of the signal peptides of representative influenza A hemagglutinin molecules. Amino acids in parenthesis lie on the carboxy-terminal side of the cleavage site. FPV hemagglutinin of fowl plague virus (60); H1, hemagglutinin of strain A/PR/8/34 (H1N1) (17); H2-hemagglutinin of strain A/Japan/305/57 (20); H3/72, hemagglutinin of strain A/Udorn/307/72 (82); H3/75, hemagglutinin of strain A/Victoria/3/75. [From W., Min Jou et al. *Cell* 19:683–696, 1980, cited in (20)].

protein in the plasma membrane (10). Evidence will be presented that the NH_2-terminus also includes sequences that direct translocation (52). Therefore, available data suggest that all necessary topogenic sequences reside at the NH_2-terminus of the influenza NA. Other glycoproteins known to have a similar topography include the enzymes sucrase-isomaltase and leucine aminopeptidase (17), the hemagglutinin-neuraminidase (HN) glycoprotein of paramyxoviruses (32) and the G protein of respiratory syncytial virus (27).

Influenza A Glycoproteins and Their *In Vivo* Expression from Cloned DNA

The influenza A glycoproteins have been the subject of intensive study. The extensive biochemical data available concerning the processing of the HA and NA during sorting provided a background for our investigations. Both the HA and the NA are sorted to the plasma membrane in the infected cell where they are acquired by nascent virus particles during budding (65). The influenza A HA is the major surface antigen of the virus and is not only a primary stimulus to the humoral immune system but is also recognized on virus-infected cells by cytotoxic T cells with HA subtype and H-2 haplotype specificity (90). Naturally occurring point mutations in the HA accumulate with time and this results in antigenic "drift" of influenza A virus. Periodic replacement of the HA of a prevailing human influenza A subtype by an antigenically unrelated one from the animal or avian reservoir of influenza A viruses results in antigenic "shift." This occurs by reassortment of HA genes between human and donor influenza A viruses that coinfect an as yet unidentified host. Such an event must be rare, but when it occurs the emergence of a human influenza A strain with a novel hemagglutinin is associated with an influenza pandemic (20, 94). The three-dimensional structure of crystallized HA has been determined to 3Å resolution, and its epitopes have been mapped extensively using monoclonal antibodies and sequence analysis of natural variants (94, 96). The mature HA at the surface of infected cells or on viral particles is a trimer composed of identical subunits (HA_0) held together by ionic forces, disulfide bonding, and interaction between carbohydrate side chains. Cleavage of HA_0 into (amino-terminal) HA_1 and (carboxy-terminal) HA_2 segments that remain linked by disulfide is necessary for infectivity (11). Mature HA on infectious virus particles mediates attachment to cells, penetration, and ultimately fusion of engulfed particles with membranes of transporting vacuoles (50).

Both the influenza A glycoproteins and most of the viral glycoproteins mentioned in this discussion bear carbohydrate in N linkage to asparagine in the recognition sequence Asn-X-Ser or Asn-X-Thr (Fig. 3) (35, 42, 59). Only one or the other of two discrete species of carbohydrate is attached at a given site. Among different HA antigenic subtypes, there is slight variation in length of the polypeptide, number of potential glycosylation sites, and number of sites actually glycosylated (20, 42), but as an example, the H2 HA_0 is 563 amino acids in length and is glycosylated at five of six potential sites. Most HA antigenic subtypes contain both "complex" and "high mannose" carbohydrate (Fig. 3). The structure of complex carbohydrate on viruses such as influenza which bear a neuraminidase activity differs from that of VSV by absence of terminal sialic acid (NANA) residues. The carbohydrate content of HA varies according to the host cell in which the virus is grown, and even for a given viral strain grown in a given cell line, there is demonstrable heterogeneity of glycosyl residues but not of attachment sites (42). Pollack and Atkinson (59) have surveyed the available data on the glycosylation of 50 viral and cellular membrane glycoproteins and shown that "complex" carbohydrate attachment sites tend to be nearest the NH_2-terminus of glycoproteins, like the HA, which have a C-terminal orientation of insertion into the membrane.

The influenza NA undergoes antigenic "shift" and "drift" similar to that of the HA. It promotes release of budded virus from infected cells by cleaving nascent particles from cell receptors containing sialic acid. The NA in some cases activates the HA for cell receptor attachment presumably by trimming sialic acid from carbohydrate on HA. The three-dimensional structure of an N2 antigenic subtype NA has been determined to 2.9 Å resolution (91). The functional NA is a tetramer with a box-shaped head attached to a stalk. Viewed from above, the whole of the tetramer has circular 4-fold symmetry. The N2

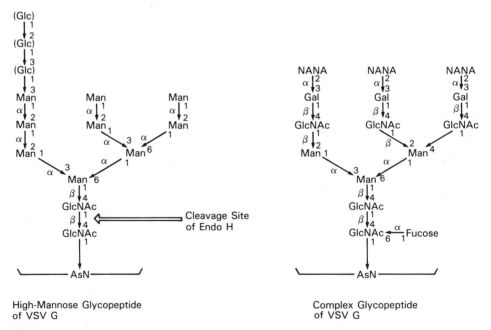

High-Mannose Glycopeptide
of VSV G

Complex Glycopeptide
of VSV G

Figure 3 The structure of N-linked "simple" or "mannose-rich" carbohydrate and "complex" carbohydrate as determined from analysis of the vesicular stomatitis virus G protein. The outlined arrow indicates the site of β-endo-N-acetylglucosaminidase H (endo H) cleavage of mannose-rich carbohydrate residues. [From Zilberstein et al. (100). Copyright 1980 MIT.]

monomer studied by Ward et al. (93) is glycosylated at four of five potential sites, on asparagine residues 86, 146, 200, and 234. Carbohydrate at positions 86 and 200 is of the high mannose type. Carbohydrate at position 146 is an unusual complex type in that it contains N-acetylgalactosamine.

We have studied the *in vivo* expression of the influenza HA and NA from full-length DNA copies of the genes cloned into the late region of an SV40 virus vector which was used to produce a lytic infection of monkey cells. Vector-specific gene products were not expected to perturb the maturation of the influenza A, HA, and NA, since SV40 does not encode a glycoprotein, nor does it replicate by budding. Influenza A membrane glycoprotein expression from the recombinant vectors was achieved in an influenza A virus permissive host cell in the absence of other influenza viral functions. We and others initially demonstrated that the vector-coded HA and NA display the properties of HA and NA produced during influenza A virus infection. For the HA, this includes glycosylation, hemagglutination, hemadsorption (Fig. 4), the ability

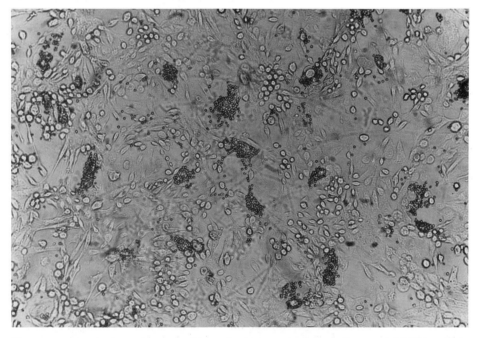

Figure 4 Adsorption of guinea pig reticulocytes to a monolayer of AGMK cells infected with HA SV40. AGMK cells were infected with HA SV40 and after 72 hours the medium was removed and the monolayer was washed with phosphate-buffered saline (PBS). The washed monolayer was then incubated at 4°C for 15 minutes in a 0.5 percent suspension of guinea pig reticulocytes in PBS. Reticulocytes were removed and the monolayer was washed again with PBS, then photographed under a microscope. Hemadsorption shown was comparable to that demonstrated for influenza A virus infected AGMK cells. Wild-type SV40 infected AGMK cells did not display hemadsorption.

to promote fusion, apical plasma membrane insertion in polarized cells, and recognition by cytotoxic T cells (69, 85, 90, 95). For the NA, this includes glycosylation, apical plasma membrane insertion, and enzyme activity (13, 39, 52). These data suggested not only that SV40-specific gene products do not perturb maturation of the HA and NA but also that information for the correct sorting and processing of these membrane glycoproteins resides within their primary amino acid sequences. Translocation, glycosylation, and transport of the vector-encoded influenza HA and NA were evidently analogous to that which occurs during influenza A virus infection. We therefore employed the HA and NA SV40 recombinant vectors to identify and characterize domains in the HA and NA polypeptides required for translocation and membrane insertion. A general description of this work is given in this review. The data support the following conclusions:

1. The signal peptide of the influenza HA is required for translocation and glycosylation.
2. Conservative amino acid changes in the HA signal peptide do not affect translocation.
3. Cleavage of the HA signal peptide is not necessary for translocation to occur.
4. The C-terminal hydrophobic region in HA is necessary for membrane insertion, and this region or the hydrophilic C-terminal "tail" may affect sorting to the plasma membrane.
5. The NH_2-terminal hydrophobic region in the influenza NA mediates translocation, as well as membrane insertion.

The Signal Hypothesis

The signal hypothesis initially predicted that secretory proteins would have an NH_2-terminal sequence recognized by the cytoplasmic membrane of the endoplasmic reticulum (ER) perhaps because of hydrophobicity. Such a sequence would function as a signal peptide. Recognition would result in binding followed by formation of a pore around the signal and of noncovalent bonds between the ribosome, from which the nascent NH_2-terminus of the secretory protein was emerging, and the ER. Polypeptide chain elongation would provide the energy to drive the protein through the pore, and an intraluminal protease would cleave the signal peptide. Translocation is co-translational in this model. Supportive data has been derived from in vitro experimentation. The cell-free translation of mRNA from mouse myeloma produced immunoglobulin G (IgG) light chain, a secretory protein. The product molecule was 1500 daltons larger than the native secreted product. When mouse myeloma rough ER was added to the reaction, authentic sized light chain was made, from which the NH_2-terminus had been cleaved (53). IgG light chain made in the presence of ER was protected from protease degradation and was thought to be sequestered within the lumen of added ER. Protease resistance and cleavage of the NH_2-terminus occurred only if microsomes were added during the protein synthetic reaction, i.e., translocation was cotranslational (9). NH_2-terminal cleavage and membrane insertion of a transmembrane protein, the VSV G protein, was observed in a similar system, and NH_2-terminal amino acid sequencing of product VSV G molecules made in the presence and absence of added microsomes disclosed the hydrophobic nature of the signal peptide (9, 49, 74).

Recently, two important components of the translocation mechanism, which are involved in the targeting of ribosomes bearing nascent proteins with an NH_2-terminal signal sequence to the ER membrane, were defined (92). The signal recognition particle (SRP) is a ribonucleoprotein made up of "7SL" RNA and six distinct polypeptides ranging in size from 9 kd to 72 kd. The RNA moiety provides an essential structural backbone for the proper assembly of the polypeptides to form an active particle. The SRP receptor is an integral 72-kd ER membrane protein which has a 60-kd cytoplasmic domain. SRP receptors can be removed from microsomal extracts active for translocation by digestion with protease, resulting in loss of activity. Addition of SRP receptor to protease treated extracts reconstitutes the microsomes for translocation. Free SRP recognizes and binds the signal sequences of nascent proteins that are addressed to the ER, and this results in arrest of translation. The basis for this recognition is unclear, since signal sequences completely lack primary sequence homology (29). SRP-arrested ribosomes then interact with SRP receptor on the ER, which reverses translation-arrest and liberates free SRP and SRP receptor. These components are presumed to recycle. In theory, the dissociation of SRP and receptor mediates a new association between the ribosome and membrane. Translation and translocation with cleavage of the signal peptide then proceed.

Processing and Transport of Viral Glycoproteins

Once translocation has occurred, N-linked glycosylation proceeds immediately, probably cotranslationally, by *en bloc* transfer of the single species of "simple" or mannose-rich carbohydrate side chain (Fig. 3) from lipid linkage to target asparagine residues on the nascent glycoprotein within the ER lumen. Some of the evidence for this sequence of events came from the study of the maturation of the Sindbis virus membrane glycoproteins, E1 and E2. The smaller glycoprotein E2 is derived by posttranslational processing of a precursor, PE2. Sefton (81) showed that [^{3}H]amino acids and [^{3}H]mannose were simultaneously incorporated into PE2 produced in virus infected cells after a 1-minute pulse and that both labels appeared simultaneously in E2 20 minutes after a 10-minute pulse. E2 labeled with [^{35}S]methionine but not [^{3}H]mannose was not detected, suggesting that glycosyl residues were added to PE2 during translation. The (mannose-rich) structure of oligosaccharide added to nascent glycoproteins was determined by chromatographic analysis of labeled carbohydrate appearing simultaneously on the lipid donor molecule [isoprenoid dolichol pyrophosphate (89)] and on VSV G protein after a short pulse of [^{3}H]mannose (48).

These and other data establishing the close temporal relationship of translation to glycosylation suggested that addition of mannose-rich carbohydrate takes place in the lumen of the rough ER. (14, 62, 81, 88). Furthermore, *in vitro* studies have demonstrated that glycosylation mediated in the ER is *limited to* this step (40, 41). Yet the primary attachment of mannose-rich carbohydrate is a determinant of the final carbohydrate architecture of the glycoprotein, since N-linked complex carbohydrate is derived exclusively by modification of mannose-rich residues that have been added to nascent glycoproteins in the ER. This mechanism is supported by kinetic studies which reveal that the appearance of complex carbohydrate on glycoproteins is temporally associated with the depletion of mannose-rich carbohydrate (62). It was possible to establish this association using the enzyme, endo-β-N-acetylglucosaminidase (endo H) which cleaves at the penultimate N-acetylglucosamine (GlcNAc) of N-linked oligosaccharides containing four or more mannose residues in linkage to GlcNAc (Fig. 3), thus distinguishing mannose-rich carbohydrate from complex carbohydrate. Pulse-chase of label in cells infected with VSV or Sindbis virus revealed that endo H sensitive carbohydrate appeared at early times in the lipid donor and was depleted from protein in parallel with the appearance of endo H resistant carbohydrate attached to protein. (The Sindbis virus E1 and E2 glycoproteins each bear one simple and one complex carbohydrate residue. Both the carbohydrate residues of the mature VSV G protein are of the complex type.)

Additional evidence in support of this hypothesis comes from experiments using tunicamycin. This antibiotic acts in the ER lumen to inhibit glycosylation by suppressing the formation of the mannose-rich oligosaccharide-lipid donor molecule (47, 89). Because of depletion of this substrate, the *en bloc* addition of mannose-rich carbohydrate in the ER does not occur (19, 47, 98). Following treatment with tunicamycin, viral glycoproteins are not uniformly stable in the ER without normally acquired mannose-rich carbohydrate. For example, the glycoproteins of avian sarcoma virus (ASV), Semliki forest virus (SFV), and VSV are stable under these conditions, while the HA of the avian influenza A strain, fowl plague virus (FPV), is not (19, 80). Stable VSV G is apparently transported to the plasma membrane normally in the presence of tunicamycin, because VSV particles bearing unglycosylated G are formed and are infectious (22). Ordinarily the glycosylated mature VSV G protein contains only complex carbohydrate. Hence, the observation that VSV G present on virus made in the presence of tunicamycin lacks glycosyl residues, suggests that mannose-rich carbohydrate is a necessary substrate for the formation of complex carbohydrate ligands.

In vivo experiments reveal a delay of 30 to 40 minutes between the appearance of VSV G bearing mannose-rich carbohydrate and the processing of those residues to complex forms (43, 73). The timing is consistent with information to be presented that the pathway to the plasma membrane followed by membrane and secretory proteins proceeds through the Golgi apparatus. The formation of complex carbohydrate and of mannose-rich carbohydrate on mature glycoproteins is the result of a series of processing events which occur in the Golgi. These include (a) trimming of mannose residues that were added *en bloc* in the ER and their replacement by sequentially added mannose residues, recreating simple carbohydrate on mature glycoproteins, and (b) trimming of mannose residues to form complex oligosaccharides by the sequential addition of fucose, galactose, *N*-acetylneuraminic acid (NANA), or GlcNAc. In each case, glycosyl residues are added to the trimmed core structure: $(Man)_3(GlcNAc)_2$-Asn (88).

Viral membrane proteins appear to become membrane-anchored in the ER in conjunction with termination of the translocation event, placing constraints on the mechanism of transport to succeeding membrane organelles. Knipe et al. (43) proposed that the C-terminal insertion orientation of the VSV G protein, which these investigators demonstrated was achieved in the ER, is maintained throughout protein maturation, that is, during the transport event to the Golgi, through the Golgi, and during transport to the plasma membrane. Rose et al. (68) confirmed this hypothesis by amino acid sequence analysis of a fragment of G protein that was viral membrane-associated and nucleotide sequence analysis of a partial cDNA clone of the VSV G gene encoding the C-terminus. The C-terminus of VSV G cDNA was shown to include: (a) an uninterrupted sequence of 20 hydrophobic amino acids, sufficient in length to span a biological membrane as an α helix, and (b) the putative cytoplasmic hydrophilic tail. Most of the hydrophobic C-terminal region predicted from the nucleotide sequence of cloned G DNA was included in the amino acid sequence of the membrane-associated fragment of G protein. Therefore, transport of membrane anchored glycoproteins from ER through the Golgi to the plasma membrane seems likely to occur by transfer of the membrane itself along with the associated nascent glycoprotein.

According to this model, transport begins with a process of vesicle formation from the membrane bilayer of one organelle and ends with fusion of the formed vesicle with the membrane of the succeeding cellular organelle. Vesicles involved in this transport are coated with proteins acquired during vesiculation (30). The major such protein components of coated vesicles are a 180-kd polypeptide and two distinct, associated 35-kd polypeptides referred to as clathrin and clathrin light chains. These associate to form trimeric subunits of three heavy (180-kd) chains and three light chains. Subunits assemble in a cage-like structure around transporting vesicles. During the fusion process, the clathrin coat is shed and the clathrin subunits are recycled, a process that occurs every few minutes. Rothman and Fine (73) followed the transport of [^{35}S]-methionine-labeled VSV G protein in infected Chinese hamster ovary (CHO) cells by isolating clathrin coated vesicles containing G protein at various time intervals after a chase. Two waves of such vesicles containing labeled G were distinguished: an early wave terminating at 20 minutes of chase and consisting of vesicles containing only endo H sensitive G protein and a later wave terminating at 60 minutes, consisting of vesicles bearing endo H resistant G protein. At a given time, 5 to 15 percent of pulse label could be detected in vesicles, suggesting a turnover rate consistent with clathrin recycling. These data suggest that clathrin-coated vesicles transport membrane glycoproteins bearing mannose-rich carbohydrate from ER to Golgi (early) and fully processed glycoproteins from Golgi to plasma membrane (late).

Circumstantial evidence for the processing of high mannose oligosaccharides in the Golgi apparatus came from the demonstration that enzymes such as α-mannosidase and various glycosyltransferases necessary for such processing are localized in that organelle (12, 61, 76). Bergmann et al. (7) used direct immunofluorescence and immune electron microscopy (EM) of VSV infected cells to demonstrate VSV G protein as fluorescent antigen within the Golgi apparatus. Evidence for passage of the G protein through the Golgi was provided by Bergeron et al. (6) who counted silver grains appearing in electron micrographs of cellular organelles at 10-minute intervals after a pulse of [^{3}H]mannose to VSV infected baby hamster kidney (BHK) cells. Grain density was greatest in the rough ER (RER) after 10 minutes and peaked in the Golgi apparatus after 30 minutes. The appearance of maximum grain density in the Golgi was correlated with the loss of grains from RER. This immediately preceded the appearance of endo H resistance in [^{3}H]-mannose-labeled carbohydrate extracted from infected cells, which occurred after 40 minutes. Green et al. (23) provided similar information on the processing of the E1 and E2 glycoproteins of SFV.

Processing in the Golgi complex is not limited to modification of carbohydrate but also includes acylation of the polypeptide and sulfation of preexisting carbohydrate. For example, some of the complex oligosaccharide side chains of the influenza HA are sulfated in the Golgi (55). Schmidt and Schlesinger (79) demonstrated the incorporation of [^{3}H] palmitic acid into the VSV G and Sindbis virus E1 and E2 glycoproteins. They showed that acylation was a "late" event that preceded the processing of carbohydrate side chains to endo H resistance. Schmidt (78) also subjected the glycoproteins of influenza A, Newcastle disease (NDV) SFV, and mouse hepatitis A viruses to chemical analysis and found that the HA$_2$ subunit of the influenza HA, the F (fusion) glycoprotein of NDV, the E1 and E2 glycoproteins of SFV, and the E2 glycoprotein of mouse hepatitis A virus were each acylated. Controlled proteolysis and cyanogen bromide fragmentation of the influenza HA suggested that fatty acids are linked to serine residues located in regions of the polypeptide in the immediate vicinity of or possibly inside the viral lipid bilayer. Using a different approach, the VSV G protein was shown to bear palmitate in linkage to a cysteine residue within its (cytoplasmic) hydrophilic C-terminal (65). Schmidt speculated that acylation may modulate the membrane affinity of the C-terminal anchorage region possibly by "masking" hydrophilic hydroxyl groups on serine, making these regions more hydrophobic. [The glycoproteins that were not acylated, the influenza NA and the NDV HN, are more difficult to incorporate into liposomes *in vitro* than are the HA and F glycoproteins, respectively (34).] Not consistent with this hypothesis is the observation that VSV G, which lacks the cysteine attachment site for palmitate, nonetheless undergoes completely normal maturation.

As mentioned, Schmidt and Schlesinger (77, 79) had demonstrated that acylation and formation of complex carbohydrate were temporally dissociable Golgi functions. These observations and the finding that the stacks of Golgi cisternae exhibit at least two distinctive sets of properties as determined by a combination of specific cytochemical and biochemical staining procedures as well as EM studies suggest that the morphologically homogeneous stacks of Golgi cisternae represent functionally distinct "cis" and "trans" organelles (72). The direction of transport of plasma membrane glycoproteins through the Golgi complex is consistent with the order of processing events and with cellular anatomical relationships of the cis and trans Golgi. The cis Golgi cisternae are "nearest the nucleus and adjacent to a specialized portion of the rough ER that lacks bound ribosomes and is called 'transitional ER'" (72). The trans Golgi is nearest the plasma membrane and is associated with secretion granules.

Griffiths et al. (25) used lectins to demonstrate the direction of transport through the Golgi and define the distinctive function of the trans region. Frozen sections of cycloheximide-treated SFV-infected BHK cells were incubated either with concanayalin A (Con A), which binds mannose-rich carbohydrate, or *Ricinus communis* agglutinin 1 (RCA), which binds carbohydrate bearing galactose. Lectin binding was established by locating lectin-antilectin antibody complexes with protein A–linked colloidal gold in electron micrographs. Con A was consistently bound throughout all Golgi cisternae, while RCA was heavily and uniformly bound exclusively by one-half to two-thirds of the stacks of Golgi cisternae away from the nucleus, defining the trans region as the site of galactose addition to oligosaccharide. The fact that RCA-binding carbohydrate was never seen in the cis region of the Golgi established the unidirectional movement of glycoprotein from cis to trans regions. Since mannose-rich carbohydrate is present in fully processed SFV glycoproteins, Con A binding did not specifically identify the cis Golgi cisternae.

Processing in the Golgi complex was further elucidated using monensin, a sodium ionophore (26). SFV-infected BHK cells were treated with monensin, and trapping of labeled E1 and E2 glycoproteins in "medial" Golgi cisternae that were negative for cytochemical and biochemical markers of either the cis or the trans regions was documented by electron microscopy. Medial cisternae were anatomically located between identifiable cis and trans regions, and kinetic evidence of glycoprotein transport from the cis into the medial region was obtained. Acylation of E1 and E2 was not affected, a finding consistent with the probability that acylation occurs in the cis organelle. However, terminal glycosylation, a trans function (25), did not occur, demonstrating that monensin was specifically toxic to trans Golgi functions in BHK cells. In summary, the Golgi apparatus acts on viral membrane glycoproteins bearing N-linked carbohydrate in the cis region by acylating them and possibly by removing mannose from mannose-rich carbohydrate. Additional mannose trimming, terminal sugar addition, and sulfation occur in the trans region (72).

Some viral glycoproteins bear carbohydrate in 0-linkage to serine or threonine residues. The addition of 0-linked carbohydrate to glycoprotein also takes place in the Golgi complex. Examples of such glycoproteins are the vaccinia virus hemagglutinin (83), Herpes simplex virus glycoproteins gB, gC, gD, and gE (38), the fiber protein of adenoviruses (37), glycoprotein E1 of coronaviruses (3, 57), and the G protein of respiratory syncytial virus (27). Each of these glycoproteins also contains N-linked carbohydrate. O-Linked oligosaccharides are synthesized by the sequential addition of monosaccharides directly to the polypeptide chain by glycosyltransferases located within the Golgi apparatus (12). Monensin completely blocks O-linked glycosylation of Herpes simplex glycoprotein gD, and predictably also prevents the Golgi-directed processing of N-linked "high mannose" carbohydrate to the complex type, yet acylation of gD does occur, suggesting that attachment of O-linked carbohydrate is specifically a trans Golgi function (38). The structures of O-linked oligosaccharide chains vary from disaccharides composed of GalNAc and sialic acid to large oligosaccharides of 20 sugar residues containing GalNAc, galactose, GlucNAc, fucose, and sialic acid (44). The functional significance of O-linked carbohydrate, if it is unique from that of N-linked sugars, is open to speculation. It is of possible importance that viruses so far known to bear O-linked carbohydrate bud internally to the plasma membrane of infected cells (with the probable exception of respiratory syncytial virus), yet O-linked carbohydrate is not universally required for normal maturation of such viruses. For example, Herpes simplex virus particles formed in the presence of monensin are fully infectious (38), whereas coronavirus particles are not (3).

An important but incompletely understood aspect of transport of viral and cellular glycoproteins between the Golgi and the plasma membrane is the fact that this process is directed in polarized epithelial cells. Rodriguez-Boulan and Sabatini (64) showed that maturation of VSV and of several type C retroviruses occurs almost exclusively at the basolateral surface of Maden-Darby canine kidney (MDCK) cells, while that of influenza A virus and of paramyxoviruses occurs at the apical surface. The apical and basolateral plasma membranes of such polarized cells are morphologically isolated from each other by tight junctional complexes, and the inference was made and subsequently confirmed that the restriction of the maturation of viruses to apical or basolateral plasma membrane domains is a concomitant of restricted sorting of their glycoproteins (63, 70). Rodriguez-Boulan and Sabatini (64) originally suggested that sialation of carbohydrate residues might determine the polarity of plasma membrane insertion of viral glycoproteins, since viruses that did not contain a neuraminidase activity were found to mature at the basolateral surface, and those that did matured at the apical surface. However, neither the complete inhibition nor the modification of glycosylation altered the polarity of the maturation of influenza and VSV in MDCK cells (24, 71).

Two findings suggest that the directed transport of the VSV G protein and of the influenza A HA is instead a consequence of their primary amino acid sequence which contains the information for their insertion into specific membrane domain-directed Golgi complexes or for conferring transport specificity on unprogrammed sets of Golgi vesicles. First, the expression of HA from DNA cloned into an SV40 vector in African green monkey kidney (AGMK) cells is polarized (69), implying that other influenza viral proteins or a set of cellular functions modified or turned on by influenza viral infection are not necessary. Second, monensin affects the transport to the membrane of the VSV G protein and of the influenza HA differently in infected MDCK cells. EM studies show that monensin causes the formation of identical vacuoles in the Golgi region during infection by either virus, but the VSV G protein is localized to the vacuoles, and aberrant budding into vacuoles is observed only in VSV infected cells. Infectious VSV is not produced. In contrast, monensin does not alter transport of the influenza HA to the apical plasma membrane, nor does it affect the yield of infectious influenza virus (2).

Influenza HA Signal Peptide Mutants

To study expression of the influenza virus hemagglutinin and neuraminidase in eukaryotic cells in the absence of influenza virus infection, we first cloned complete DNA copies of the HA or NA virion RNA (vRNA) of the influenza A virus strain A/Udorn/72 (H3N2) in the plasmid vector pBR322 (45, 52, 85). The identity of clones to be used for expression was confirmed by sequencing the 5' and 3' ends of the HA DNA or of the NA DNA and comparing the results to available complete nucleotide sequences. For expression of the HA, HA DNA was used to replace the late region coding sequences of a viable SV40 deletion mutant, dl 2330, which lacks SV40 late 16S and 19S mRNA splice donor sequences. A deletion was created in the late region of circular dl 2330 DNA by digestion with the restriction endonucleases HaeII and BamHI, cleaving at map positions 0.82 and 0.14, respectively. This removed sequences coding for late SV40 gene products but left intact late promoters upstream from the HaeII site and mRNA transcription termination and polyadenylation sites downstream from the BamHI site in linear DNA. This moiety was blunt-end ligated to hemagglutinin DNA containing G-C tails that had been added during the plasmid cloning procedure. Ligated HA SV40 molecules were used to transfect AGMK cells that were coinfected with the helper SV40 early region temperature-sensitive mutant, ts-A28 which provided late SV40 gene functions. SV40 virus plaques were picked, and a clone producing a high level of HA was identified by immune precipitation of a [^{35}S]-methionine-labeled protein from lysed HA SV40 infected AGMK cells that comigrated with authentic HA produced during influenza viral infection. Vector encoded HA incorporated mannose, fucose, and glucosamine, hemagglutinated and hemadsorbed erythrocytes (Fig. 4) and was present at the surface of HA SV40 infected AGMK cells as indicated by indirect immunofluorescence (85). These observations and the previously mentioned findings of other investigators (46, 69, 90, 95) provide a basis for examining the effect of strategic deletions or point mutations in HA DNA on sorting and processing of resulting mutant HA polypeptides (Table 1).

TABLE 1 Properties of Mutant HA Proteins Produced from HA-SV40 Vectors Bearing Deletion or Point Mutations in HA DNA

Site of mutation	Type of mutation	Cleavage of signal	Translocation	Glycosylation[a]	Cellular localization
Wild type	—	+	+	+	Plasma membrane
Signal peptide (dl-HA)	Deletion (11 of 13 aa's)	(Signal absent)	—	—	Cytoplasm
Signal peptide (8 mutants)	Point (conservative aa changes)	+	+	+	Plasma membrane
Signal peptide	Point (Gly→Ser at cleavage site)	—	+	±	?Golgi
C-terminus (3 mutants)	Deletion → loss of hydrophobic C-terminus + hydrophilic "tail"	+	+	± and ++	?ER or Golgi and secreted } Two species
C-terminus (dl −12)	Deletion → mutant hydrophobic C-terminus + hydrophilic "tail"	+	+	+	?Golgi

[a] + = Wild-type glycosylations (both mannose-rich and complex carbohydrate); ± = mannose-rich carbohydrate, only; ++ = complex carbohydrate, only.

Source: Adapted from 82, 86, and 87.

The following observations led to identification of the "signal peptide" in HA. First, Elder et al. (15) showed that, like the VSV G protein, the glycosylation of HA in a cell-free system depended on the presence of membranes in the reaction, that glycosylated HA was protected from protease, and that the membranes had to be added within a few minutes after the initiation of the reaction for glycosylation and protection to be achieved, demonstrating the requirement for nascent HA NH_2-termini for translocation. *In vivo* HA_1 lacked NH_2-terminal sequences present in the *in vitro* unglycosylated molecule. Second, Air (1) compared the 5' (+) strand nucleotide sequence of the H2 hemagglutinin gene to that of the NH_2-terminus of H2 HA glycoprotein. She verified the absence of the signal peptide in mature HA and located the cleavage site of the signal peptide in the H2 antigenic subtype HA molecule (-NH-Gly-Asp-COO-). Third, the 5'-end sequences of all hemagglutinins subsequently examined encode a short hydrophilic segment followed by an uninterrupted hydrophobic region 13 to 15 amino acids in length that is not conserved (Fig. 2).

These observations indicate that the hydrophobic signal in HA is cleaved at the time of translocation, but it is not clear that signal peptide is required. A conformational requirement for signal peptide was inferred by Wirth et al. (98) who studied a ts mutant of Sindbis virus which failed to cleave its structural polyprotein: NH_2-core-PE2-E1-COOH. As a result, the NH_2-termini of the viral glycoproteins PE2 and E1 are internalized with respect to core protein sequences in the polyprotein and presumably are not sterically available to associate with SRP. As a consequence, signal peptide is not cleaved, and PE2 and E1 remain in the cytoplasm.

To establish rigorously that the HA signal peptide is required for translocation to occur, Sekikawa and Lai (82) deleted sequences in HA DNA encoding 11 of 13 hydrophobic amino acids in the putative signal peptide of the A/Udorn/72 (H3) HA (Fig. 5). This was achieved in a manner that resulted in the conservation of the wild-type HA translation initiation codon in mutant "dl-HA" DNA. The phenotype of the mutant polypeptide produced from dl-HA DNA during SV40 vector infection was compared to that of wild-type initially by immune precipitation of [^{35}S]-methionine-labeled HA polypeptides from lysates of vector-infected cells in the presence or absence of tunicamycin. Immune precipitated HA proteins were analyzed in a 15 percent sodium dodecyl sulfate polyacrylamide gel (SDS/PAGE)(Fig. 6). As we had previously shown (85), HA SV40 produced a glycoprotein that comigrated as a broad band with [^{35}S]-methionine-labeled HA from influenza virus infected cells. In the presence of tunicamycin, immune-precipitated wild-type HA is greatly reduced in amount with respect to control (Fig. 6) and migrates as a homogeneous narrow band with the expected molecular weight of unglycosylated HA (53,000 daltons). In contrast, the major immune precipitated product of dl-HA comigrated with unglycosylated wild-type HA in the absence of tunicamycin and was not altered in migration rate or steady-state amount by tunicamycin.

The effects of tunicamycin were previously mentioned. Tunicamycin acts

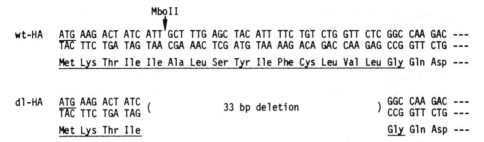

Figure 5 Nucleotide sequences of the wild-type and the derived d1-Ha deletion mutant. The nucleotide sequences and the encoded amino acids at the amino terminus are shown. The wild-type HA contains a signal peptide of 16 amino acids (underlined) which is cleaved by a protease. Also indicated is the MboII cleavage site where deletions are introduced to obtain this series of mutants. The mutant analyzed in this study lacks 11 of the 16 amino acids. The mutant DNA retains an MboII cleavage site because the GAAGA MboII recognition sequence is still present.

exclusively in the ER to inhibit glycosylation by suppressing the formation of the lipid-oligosaccharide donor (89). Kinetics of synthesis and degradation of viral proteins that are not translocated appear to be unaffected by the drug (19, 47, 58), but some viral glycoproteins, including the influenza A HA, have been shown to be unstable in the ER lumen in the unglycosylated state induced by tunicamycin (80). This has been shown in one instance to occur because the absence of mannose-rich carbohydrate renders the polypeptide susceptible to protease degradation in the ER lumen or in transit to the plasma membrane (58).

The results of the foregoing experiment suggested that the d1-HA polypeptide was unglycosylated, since it comigrated with unglycosylated wild-type HA made in the presence of tunicamycin (Fig. 5). Furthermore, failure to be glycosylated was the consequence of a translocation defect. If the d1-HA polypeptide had been translocated we would have expected to observe its degradation, as evidenced by reduced steady-state label in comparison to wild-type HA made in the absence of tunicamycin. The observed reduced steady-state amount of the wild-type HA polypeptide made in the presence of tunicamycin is a positive control for protease degradation occurring in the ER lumen.

The results of two additional experiments are consistent with the hypothesis that the d1-HA polypeptide is not translocated. Wild-type and d1-HA vector infected cells were labeled with [^{35}S] methionine and separated into "membrane" and "cytoplasmic" fractions by Dounce homogenization and differential centrifugation. Under these conditions, the d1-HA polypeptide was associated with the "cytoplasmic" (soluble) fraction, while wild-type HA was associated with the "membrane" (pelleted) fraction of infected cells (data not shown). In addition, indirect immunofluorescence of permeable

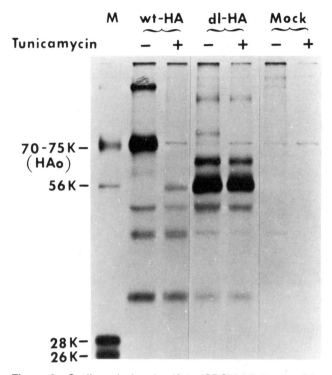

Figure 6 Sodium dodecyl sulfate (SDS)/polyacrylamide gel analysis of the wild-type and mutant HAs. Monolayers of AGMK cells were infected with the wild-type HA SV40 (wt-HA) and mutant HA SV40 (dl-HA) for 72 hours. Infected cells were pretreated with tunicamycin (1 μg/ml) for 60 minutes prior to labeling with [^{35}S]methionine (100 μCi/ml) in methionine-free medium in the presence of 1 μg of tunicamycin per milliliter medium. Cell lysates were immunoprecipitated with HA antiserum and analyzed on 15 percent SDS/polyacrylamide gels. M indicates the polypeptides markers (shown in 10^{-3} daltons) prepared from [^{35}S]-methionine-labeled influenza virions.

wild-type or dl-HA vector infected cells (Fig. 7) demonstrated that the wild-type HA was present in abundance within cells and at the plasma membrane, whereas dl-HA polypeptide was not associated with membranes.

Gething and Sambrook also demonstrated that the HA signal peptide is required for translocation (21). These investigators constructed an HA expression vector analogous to ours in that HA DNA sequences were cloned into the late region of SV40. Wild-type HA produced from their vector displayed properties completely analogous to those of HA produced during influenza virus infection. A deletion of DNA sequences encoding the signal peptide of HA was achieved by exonuclease digestion of the 5' end of HA DNA. Therefore, the wild-type HA translation initiation codon and downstream sequences encoding the signal peptide were deleted from their "signal" mu-

wt-HA

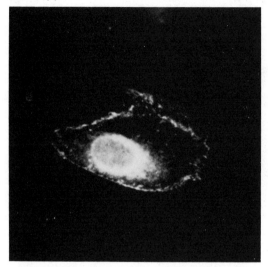

dl-HA

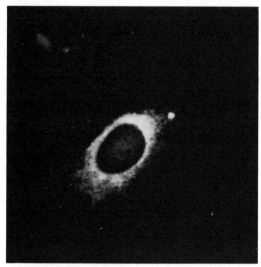

Figure 7 Indirect immunofluorescence assay of HA in infected cells. AGMK cells were infected with the SV40 ts helper and either the wild-type HA SV40 virus (top) or the mutant HA SV40 virus (bottom) for 72 hours. Cells were fixed and stained as described in the text. Accumulation of the wild-type HA on the outer membrane is discernible. The mutant HA stains as intensely as wild-type HA in the cytoplasm but does not accumulate on the membranes.

tant. This may account for the observed instability of "signal" HA polypeptide and/or its mRNA. However, the "signal" mutant DNA produced an HA with properties similar to those of d1-HA.

To investigate the specificity of the HA signal peptide amino acid sequence and the requirement for hydrophobicity in mediating translocation, we made point mutations in DNA encoding the wild-type signal peptide (unpublished data). Heteroduplex DNA was generated by denaturing and reannealing a mixture of pHA-SV40 and plasmid d1-HA DNA. Wild-type signal peptide DNA sequences that encode the 11 amino acids deleted from d1-HA were unpaired in this construction and were subjected to bisulfite mutagenesis, which causes base transitions predominantly in single-stranded DNA. We prepared a series of infectious SV40 vectors incorporating HA DNA bearing the desired point mutations in DNA sequences encoding the HA signal peptide. The procedure used to prepare the point mutants failed to generate HA DNA encoding a signal peptide in which a hydrophobic amino acid was replaced by a hydrophilic one. Eight mutants with one or more conservative amino acid changes in the signal peptide sequence not involving the -NH-Gly-Gln-COO- cleavage site were indistinguishable from wild type with respect to glycosylation, cell surface expression, and hemadsorption. One mutant polypeptide with an altered cleavage site (-NH-<u>Ser</u>-Gln-COO-) was translocated and glycosylated but failed to reach the cell surface. Immunofluorescence studies suggest that this mutant HA polypeptide was ultimately localized to the Golgi complex. Amino acid sequence analysis of its NH$_2$-terminus revealed that the signal peptide had not been removed during translocation.

The finding that conservative (hydrophobic) amino acid changes in the signal sequence of the influenza HA fail to perturb translocation is consistent with other available data that demonstrate that the only constant feature among the hydrophobic regions of signal peptides of different proteins is net hydrophobicity (29). In fact, there is diversity in sequence even among the signal peptides of influenza HAs of different antigenic subtypes and incomplete conservation of sequence among those of the same antigenic subtype (Fig. 2).

Mutant signal peptides were previously studied in prokaryotic cells that do not contain SRP or its analogue. Inouye (36, 56) has proposed a loop model for translocation in bacteria, wherein the short NH$_2$-terminal positively charged hydrophilic segment that precedes the hydrophobic region in prokaryotic and eukaryotic membrane and secretory proteins initiates an association with the (negatively charged) cytoplasmic side of the membrane. Once this association is established, translocation proceeds by looping the remainder of the protein through a pore in the membrane, hydrophobic signal peptide first. Cleavage of the signal peptide frees the NH$_2$-terminus of the body of the protein on the lumenal side of the membrane.

Escherichia coli mutants with altered signal peptide sequences have been described that fail to export maltose-binding protein (MBP) or λ receptor protein (5, 16). Point mutants with defects in the MBP signal peptide were selected for reduced utilization of maltose and the accumulation of MBP in cytoplasm. Sequence analysis revealed that four of five mutations introduced a single charged amino acid into the hydrophobic core of the MBP signal sequence. One mutation introduced a single conservative change from leucine to proline in the signal peptide and the resulting mutant was defective in MBP transport. In a study of λ receptor protein, DNA of 15 deletion or point mutants defective in export was sequenced. All deletion mutants lacked only the complete or nearly complete signal peptide DNA sequences. Of the four point mutants sequenced, all had mutations in the hydrophobic region of the signal peptide introducing aspartic acid, glutamic acid, or arginine into the sequence. These data suggest, in contrast to our observations, that nonconservative amino acid substitutions in the signal peptide prevent its function. The observation that a conservative change (leucine to proline) in the MBP signal peptide also interrupted translocation could be interpreted to indicate a constraint on the secondary structure of the signal peptide related to its interaction with the membrane. The fact that we did not make a similar observation for a eukaryotic membrane protein may relate to differences in the mechanism of translocation.

The signal peptide is necessary but not sufficient for translocation to occur. Phage λ receptor protein is coded for by the lam B gene. Moreno et al. (54) fused the 5′ terminus of the lam B gene, encoding NH_2-terminal λ protein sequences, to the upstream terminus of the gene encoding β-galactosidase, a cytoplasmic protein. Mutants synthesized proteins containing varying lengths of the NH_2-terminal end of the λ receptor protein fused to β-galactosidase. These fusion proteins were transported to the bacterial outer membrane, if most of the λ receptor sequences were encoded. Fusion proteins that included only the λ receptor signal peptide fused to full-length β-galactosidase remained in the cytoplasm. Since the signal peptide is cleaved before synthesis of C-terminal sequences is completed, it is unlikely that failure of translocation of the λ receptor signal peptide-β-galactosidase fusion protein(s) is due to direct interaction of β-galactosidase with the signal peptide of the lam B protein. It is more likely that sequences conferring unacceptable charge or secondary structure in β-galactosidase interfere with or interrupt translocation, despite the presence of a functional signal peptide.

The HA signal peptide cleavage mutant we studied had undergone a change from glycine (hydrophobic) to serine (neutral) at the C-terminus of the signal sequence. It is unclear why such a mutation results in failure of cleavage, but we speculate that the observation may indicate a conformational requirement for access of signal peptidase. Wild-type HAs of different antigenic subtypes contain either glycine or alanine at this position (Fig. 2), but serine occurs naturally at the C-termini of signal peptides in the secretory proteins pre-proalbumen and penicillinase, for example (29). As mentioned, the cleavage mutant glycoprotein did not reach the cell surface. Although it was glycosylated, it contained only endo H-sensitive carbohydrate, indicating that it was not subject to processing in the trans region of the Golgi. Aberrant or interrupted transport and glycosylation may be accounted for by the fact that the signal sequence was not cleaved. Therefore, transport of the mutant glycoprotein may have been altered, because it remained membrane-associated at both NH_2- and C-termini, or simply because of conformational differences from wild type. These data do suggest that removal of a signal peptide that is normally cleaved is not an obligatory step in translocation.

Mutations in the C-Terminal Membrane Insertion Site of the Influenza HA

That the C-terminal hydrophobic region of the mature influenza HA was membrane-associated was first shown by direct amino acid sequencing of a bromelain-cleaved fragment of HA2 (84) and confirmed by determination of the three-dimensional structure of HA (96). We studied the processing of mutant HA with deleted or altered C-terminal hydrophobic sequences (Table 1) (86, 87). The phenotype of such HA mutants was determined by SDS-PAGE of labeled immune-precipitated protein produced from mutant HA DNA in which deletions of sequences encoding the hydrophobic C-terminus had been made. Three mutants displayed an identical phenotype: they were glycosylated, and after 6 hours of labeling with no chase, discrete intracellular and extracellular (secreted) forms were detected. In each case, the secreted form of the mutant HA had a slower migration rate on SDS-PAGE than its corresponding intracellular species. In each case also, mutant HA was not detected at the plasma membrane of infected cells by IFA, and hemadsorption could not be demonstrated. When secreted and intracellular forms of these three mutant HAs were eluted from a preparative gel and digested with endo H, secreted species were entirely resistant and intracellular species were sensitive. In contrast, wild-type HA displayed partial endo H sensitivity, indicating it contained both mannose-rich and complex carbohydrate.

Tunicamycin treatment of mutant HA vector infected cells showed that migration rate differences on SDS-PAGE between intracellular and secreted forms of each mutant HA were accounted for by higher molecular weight complex carbohydrate present on secreted HAs. The tunicamycin experiment also suggested that secretion of mutant HA was achieved by transport through the same pathway utilized by wild-type HA, since tunicamycin inhibited not only glycosylation but also secretion (probably because of degradation in transit). Sequencing of mutant HA DNA in the region encoding the hydrophobic C-terminus of wild-type HA confirmed that two of three mutants had sustained large deletions in this region such that the predicted amino acid sequences of the mutant C-termini did not include an uninterrupted hydrophobic region of 20 amino acids or longer as required to span the membrane. The third mutant had sustained a random insertion of cellular DNA at this site with an analogous result.

In sum, mutant HA lacking a membrane anchorage domain was translocated, but the pathway of transport was modified. Secretion was thought to be a consequence of failure to be membrane-inserted. The finding of a discrete intracellular HA species bearing only mannose-rich carbohydrate after a long label with no chase may indicate that transport from ER to Golgi of these mutant polypeptides is slow. Alternately, these species may represent a proportion of the mutant HA that leaks out of transport vesicles because of failure to be anchored. Once transport to the Golgi is achieved, conversion of mannose-rich to complex carbohydrate and secretion would seem to occur rapidly. Additional complex carbohydrate added to secretory HA also suggested discrimination at the level of the Golgi complex in the transport process between glycoproteins destined for secretion and those destined for plasma membrane insertion.

A fourth mutant derived in the same procedure, dl-12, displayed a phenotype distinct from those just described (Table 1). dl-12 HA polypeptide was not secreted (87). A single immune-precipitated HA species was isolated from cells infected with the dl-12 HA vector that was not distinguishable from wild-type HA by SDS-PAGE before or after endo H treatment suggesting that dl-12 HA contained both mannose-rich and complex carbohydrate, like wild-type HA. However, in contrast to wild-type HA, this mutant was not detected at the cell surface by IFA. It was present in abundance in the cytoplasm in a distribution that suggested its association with the Golgi complex. Cells infected with the dl-12 vector did not display the property of hemadsorption, yet lysates of such cells were able to mediate hemagglutination. Sequencing of the DNA of dl-12 encoding the C-terminus of the mutant HA revealed that it had undergone a 5-bp deletion upstream from sequences encoding the wild-type C-terminus. This produced a frameshift mutation. The resulting downstream sequence fortuitously encoded an uninterrupted mutant hydrophobic or neutral sequence 20 amino acids in length, terminating in a mutant hydrophilic C-terminal tail. This construct was analogous in structure to the wild-type HA C-terminus but lacked any amino acid homology. A cell fractionation experiment indicated that the dl-12 polypeptide was membrane anchored.

These observations suggested that the hydrophobic membrane anchorage domain and/or the C-terminal hydrophilic (cytoplasmic) tail may interact with transporting membrane organelles to affect sorting. Corroborating evidence for this interpretation comes from the work of Rose and Bergmann (66, 67). These investigators introduced deletions into DNA clones encoding the C-terminus of the VSV G protein. G proteins that lacked the cytoplasmic domain and most of the trans-membrane domain were secreted slowly from cells, like the comparable HA mutants (86). Deletion mutants that retained the trans-membrane domain but had altered cytoplasmic (hydrophilic "tail") domains fell into two classes. One class exhibited arrested transport of the protein to the cell surface at a stage prior to the acquisition of complex oligosaccharides. The other class of mutants acquired complex oligosaccharides at a rate that was greatly reduced from wild type. Analysis by IFA suggested that both classes of "tail" or cytoplasmic domain mutants accumulated in the RER. The phenotype of these latter mutants compares to that of dl-12 HA, in that they exhibited modified sorting (failure to reach the plasma membrane). The latter observations indicate an effect on sorting mediated by the amino acid sequence of the hydrophilic C-terminus of the VSV G. Our observations suggest an effect on sorting of the HA may be mediated by mutation in the C-terminal hydrophobic region.

Other data suggest that sorting is determined not only by unique C- and/or NH_2-termini but also by internal sequences of the protein being sorted, perhaps mediated by charge properties or secondary structure. In prokaryotic cells, the failure of β-galactosidase bearing the signal peptide of the λ receptor protein (54) to be translocated is one example. In eukaryotic cells, the altered sorting of a secretory protein, rat growth hormone (rGH), bearing the wild-type hydrophobic C-terminus of the VSV G protein (28) is another. The mutant fusion rGH polypeptide is localized to the Golgi complex, rather than being inserted into the plasma membrane, as would be expected if the fused VSV G C-terminus were the sole determinant.

Yost et al. (99) demonstrated that NH_2-terminal signal peptides and C-terminal membrane anchorage domains have properties that distinguish them from one another independent of their topographical location in the linear protein sequence. In a cell-free transcription-translation system they showed that an (NH_2-terminal) β-lactamase-(C-terminal)-globin fusion protein was completely translocated across added microsomal membranes. If β-lactamase sequences, including those encoding its signal peptide, were deleted, translocation did not occur. When sequences encoding the immunoglobulin M transmembrane segment (M) were inserted in-frame between β-lactamase and globin sequences (NH_2-β-lactamase-M-globin-COOH), the M segment was fully functional. Only the normally exported β-lactamase segment of the fusion protein was translocated, and the M segment spanned the microsomal membrane with the globin moiety on the cytoplasmic side. However, the M segment failed to function as a signal peptide. The cell-free translation product of M segment sequences repositioned 5' to sequences encoding the translocation-defective β-lactamase-globin deletion mutant protein (NH_2-M-$\Delta\beta$-lactamase-globin-COOH) was not translocated.

Two Functions for the NH₂-Terminus of the Influenza NA

As mentioned previously, the influenza NA is membrane-inserted at its NH_2-terminus, and NH_2-terminal cleavage of the NA polypeptide does not occur during processing (10). To determine the requirement for the NH_2-terminal hydrophobic region of the NA for translocation, we first demonstrated the properties of NA produced from a full-length DNA copy of the NA virion RNA of influenza A/Udorn/72 (H3N2) cloned into the late region of the SV40 vector, pSV2330 (52). Wild-type NA-SV2330 vector produced a glycoprotein that was immune precipitable by heterospecific N2 NA antibody. The precipitated protein was glycosylated and comigrated on SDS-PAGE with NA produced from influenza virus infection. Vector-coded NA, like that produced during influenza virus infection, was shown to be present at the surface of infected cells. A very low level of neuraminidase enzymatic activity was demonstrated in lysates of such cells.

We then created deletions in the region of NA DNA encoding NH$_2$-terminal hydrophobic amino acid sequences. Deletion mutant NA DNA from three such clones was found to bear the desired deletions that were in-frame with the wild-type NA translation initiation codon (Fig. 8). We assessed the phenotype of polypeptides coded for by NA DNA that had sustained deletions of part (d1K) or all (d1I, d1Z) of the coding region for the hydrophobic NH$_2$-terminal.

d1K retained all but the seven carboxy-terminal amino acids of the

```
wt:   ATG AAT CCA AAT CAA AAG ATA ATA ACA ATT GGC TCT GTC TCT CTC ACC ATT GCA
      TAC TTA GGT TTA GTT TTC TAT TAT TGT TAA CCG AGA CAG AGA GAG TGG TAA CGT
      Met Asn Pro Asn Gln Lys Ile Ile Thr Ile Gly Ser Val Ser Leu Thr Ile Ala

      ACA ATA TGC TTC CTC ATG CAG ATT GCC ATC CAG GTA ACT ACT GTA ACA TTG CAT....
      TGT TAT ACG AAG GAG TAC GTC TAA CGG TAG GTC CAT TGA TGA CAT TGT AAC GTA
      Thr Ile Cys Phe Leu Met Gln Ile Ala Ile Gln Val Thr Thr Val Thr Leu His
                                                                          |
                                                                          36

d1 K: ATG AAT CCA AAT CAA AAG ATA ATA ACA ATT GGC TCT GTC TCT CTC ACC ATT GCA
      TAC TTA GGT TTA GTT TTC TAT TAT TGT TAA CCG AGA CAG AGA GAG TGG TAA CGT
      Met Asn Pro Asn Gln Lys Ile Ile Thr Ile Gly Ser Val Ser Leu Thr Ile Ala

                                    -174 bp
                                       ▽
      ACA ATA TCC TTC CTC ATG CAG ATT GCC ATC TGG TCA AAG CCG....
      TGT TAT AGG AAG GAG TAC GTC TAA CGG TAG ACC AGT TTC GGC
      Thr Ile Cys Phe Leu Met Gln Ile Ala Ile Trp Ser Lys Pro
                                           | |
                                           28 87

                                                       -192 bp
                                                          ▽
d1 I: ATG AAT CCA AAT CAA AAG ATA ATA ACA ATT GGC TCT GTC TCT CTC CCC AAA TTA GTG....
      TAC TTA GGT TTA GTT TTC TAT TAT TGT TAA CCT AGA CAG AGA GAG GGG TTT AAT CAC
      Met Asn Pro Asn Gln Lys Ile Ile Thr Ile Gly Ser Val Ser Leu Pro Lys Leu Val
                                                               | |
                                                               15 79

                                        -282 bp
                                           ▽
d1 Z: ATG AAT CCA AAT CAA AAG ATA ATA ACA ATT GGC TCA ATT CGG CTT TCT....
      TAC TTA GGT TTA GTT TTC TAT TAT TGT TAA CCG AGT TAA GCC GAA AGA
      Met Asn Pro Asn Gln Lys Ile Ile Thr Ile Gly Ser Ile Arg Leu Ser
                                                    |        |
                                                    11       106
```

Figure 8 5'-(+) Terminal nucleotide and predicted polypeptide sequences of WT NA DNA and of deletion mutant NA DNAs, d1K, d1I, and d1Z. Mutagenized NA DNA was cloned in pBR322, and DNA from 10 clones was sequenced from the 5' (+) end of the insert by the chemical method. d1K, d1I, and d1Z DNAs were shown to bear in-frame deletions of WT NA nucleotide sequences and were therefore selected for cloning in the expression vector, pSV2330. d1I and d1Z DNAs are lacking all or nearly all sequences coding for N-terminal hydrophobic region in NA. d1K is lacking sequences coding only for the seven carboxy-terminal amino acids of the hydrophobic region (residues 29 through 35). The downstream extent of the deletion in the predicted d1K polypeptide is eight amino acids greater than that in the d1I polypeptide.

hydrophobic region. In addition, the deletion in d1K extended 52 amino acids downstream from the hydrophobic region. However, the d1K polypeptide was glycosylated and was detected at the cell membrane by immunofluorescence assay of infected cells that had not been made permeable to antibody. Cell fractionation studies performed as previously described demonstrated that d1K accumulated in the "membrane" fraction (Fig. 9). In the presence of

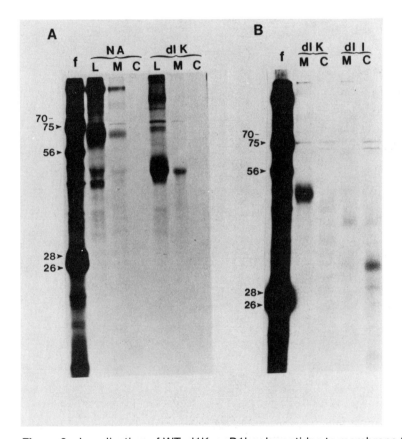

Figure 9 Localization of WT, d1K, or D1l polypeptides to membrane (M) or cytoplasmic (C) fractions of vector-infected cells. Cells were fractionated with a Dounce homogenizer in reticulocyte standard buffer with 1 mM phenylmethylsulfonyl fluoride. The postnuclear supernatant was made with 0.1 M NaCl and 1 mM EDTA and then centrifuged at 100,000 × g. The resultant supernatant (C) was made 1× with respect to RIPA buffer, and the pellet (M) was suspended in RIPA buffer for immunoprecipitation. (A) WT and d1K polypeptides localized to membrane fraction of infected, labeled cells. Migration of the glycosylated WT or d1K polypeptide is not altered by fractionation procedure from that of the respective polypeptides immunoprecipitated from whole cell lysate (L) in RIPA buffer. (B) Differential localization of the glycosylated D1K polypeptide to the M fraction and of the unglycosylated d1l polypeptide to the C fraction of vector-infected cells. Lanes "f" contained [^{35}S]-methionine-labeled influenza A/Udorn/72 viral proteins.

tunicamycin, d1K was not glycosylated, and its accumulation was greatly diminished (Fig. 10). By these criteria the d1K polypeptide was not distinguishable from wild-type NA. Both were translocated, glycosylated, and transported to the plasma membrane. In contrast, d1I and d1Z polypeptides were

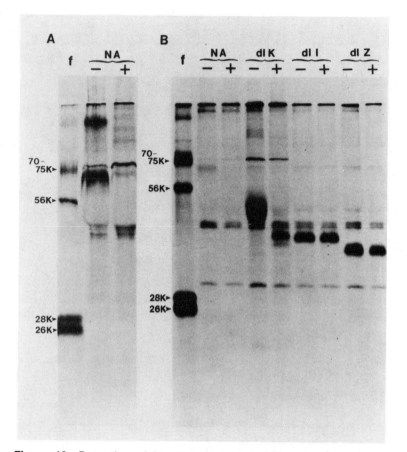

Figure 10 Detection of immunoprecipitable NA polypeptide in lysates of NA and △NA-pSV2330 vector-infected cells labeled with [^{35}S]methionine in the presence (+) or absence (−) of tunicamycin. Lysates were prepared in RIPA buffer [1 percent Triton X-100, 1 percent deoxycholate, 0.15 M NaCl, 0.1 percent SDS. 0.1 M Tris (pH 7.6)]6 hours after addition of label at 100 μCi/ml. NA protein was immunoprecipitated with sheep anti-Udorn/72 (N2) NA serum and staphylococcal protein A-Sepharose beads. Immunoprecipitates were run on a 15 percent SDS-polyacrylamide gel at 100 V for 16 hours. Lanes f contained total labeled cellular proteins from influenza A/Udorn/72 virus-infected cells labeled with [^{35}S]methionine: band × denotes a high molecular weight influenza virus-specific band that may represent a dimeric form of NA. (A) A 52-kd unglycosylated monomeric NA polypeptide in infected cells labeled in the presence of tunicamycin [lane wt (+)]. (B) Differential effect of tunicamycin on migration rate and level of glycosylated WT and polypeptides in contrast to unglycosylated D1I and d1Z polypeptides.

not detected at the cell surface, were not glycosylated, and their level was not detectably reduced by tunicamycin (Fig. 10). d1I polypeptide was detected predominantly in the "cytoplasmic" fraction of infected cells. These results indicated that d1I and d1Z polypeptides were not translocated. Although d1I lacked all but the initial 10 amino acids of the NH$_2$-terminal hydrophobic region in the wild-type NA protein, the C-terminal extent of the deletion in d1I was eight amino acids shorter than that in d1K (Fig. 9). Therefore, the observed differences in phenotype between d1K and d1I are not likely to be related to the deletion of sequences downstream from the NH$_2$-terminal hydrophobic region.

We concluded that the hydrophobic region at the NH$_2$-terminus of the NA protein includes a sequence that directs the translocation of the nascent NA polypeptide into the RER, which leads to glycosylation and cell surface expression. This is one functional definition of a signal sequence (9, 53, 75, 92). If the NH$_2$-terminus were only a transmembrane anchor, we would expect the phenotype of mutants d1I and d1Z to be similar to that of HA C-terminal deletion mutants, which are translocated, glycosylated, and secreted (86). We have not shown, however, that the NH$_2$-terminus of the NA fulfills a more rigorous definition of a signal sequence, i.e., that it mediates a translational arrest in the presence of a signal recognition particle, and deletion of hydrophobic sequences eliminates the translational arrest. Since the NA protein is not cleaved, a clear understanding of whether or how its NH$_2$-terminus interacts with the signal recognition particle during and after translocation would be of some interest. Also, our data do not rule out the possibility that the altered phenotype of mutant NA polypeptides d1I and d1Z is the result of a conformational change randomly induced by the sequence deletions. If so, our observations would demonstrate a conformational requirement for the deleted hydrophobic sequences.

If a signal sequence or an analog of one exists at the NH$_2$-terminus of the NA, it is likely to include the initial six hydrophobic residues that are conserved among NA genes (17, 31, 51). These were not deleted from mutant polypeptides (Fig. 8). Characteristic of a signal sequence, these residues are immediately downstream from the hydrophilic NH$_2$-terminal tail of the molecule. Sequence data suggests that a signal sequence needs to be at least 12 amino acids in length, whereas about 20 to 25 amino acids are necessary for transmembrane anchorage (68). If the signal sequence in the NA begins at lysine at position 6, it is interrupted after only 10 hydrophobic residues in d1I and after seven in d1Z. In contrast, the hydrophobic region in d1K polypeptide was shortened to 22 amino acids, and the resulting protein appears to be glycosylated and inserted in the outer cell membrane with an efficiency equivalent to that of wild type. It might be possible to distinguish the location and extent of the sequence crucial to translocation from that necessary for membrane insertion, for example, by extending the deletion of d1K DNA gradually upstream, progressively shortening the extent of the hydrophobic region but preserving a length sufficient to function in translocation. Ultimately, one might observe secretion of a glycosylated mutant NA protein analogous to HA mutant proteins that lack C-terminal hydrophobic sequences.

General References

DUNPHY, W. G. and ROTHMAN, J. E. 1985. Compartmental organization of the Golgi stack, *Cell* 42:13–21.

NUSSENZWEIG, V. and NUSSENZWEIG, R. S. 1985. Circumsporozoite proteins of malaria parasites, *Cell* 42:401–403.

SABATINI, D. D., KREIBICH, G., MORIMOTO, T., and ADESNIK, M. 1982. Mechanisms for the incorporation of proteins in membranes and organelles, *J. Cell. Biol.* 92:1–22.

WALTER, P., GILMORE, R., and BLOBEL, G. 1984. Protein translocation across the endoplasmic reticulum, *Cell* 38:5–8.

References

1. Air, G. M. 1979. Nucleotide sequence coding for the "signal peptide" and N terminus of the hemagglutinin from an Asian (H2N2) strain of influenza virus, *Virology* 97:468–472.

2. Alonzo, F. V., and Compans, R. W. 1981. Differential effect of monensin on enveloped viruses that form at distinct plasma membrane domains, *J. Cell. Biol.* 89:700–705.

3. Alonzo-Caplen, F. V., Matsuoka, Y., Wilcox, G. E., and Compans, R. W. 1984. Replication and morphogenesis of avian coronavirus in Vero cells and their inhibition of monensin, *Virus Res.* 1:153–167.

4. Bassford, P. J., and Beckwith, J. 1979. *Escherichia coli* mutants accumulating the precursor of a secreted protein in the cytoplasm, *Nature (London)* 277:538–541.

5. Bedouelle, H., Bassford, P. J., Fowler, A. V., Zabin, I., Beckwith, J., and Hofnung, M. 1980. Mutations which alter the function of the signal sequence of the maltose binding protein of *Escherichia coli, Nature (London)* 285:18–23.

6. Bergeron, J. J., Kotwal, G. J., Levine, G., Bilan, P., Rachubinski, R., Hamilton, M., Shore, G. C., and Ghosh, H. P. 1982. Intracellular transport of the transmembrane glycoprotein G of vesicular stomatitis virus through the Golgi apparatus as virualized by electron microscope radioautography, *J. Cell. Biol.* 94:36–41.

7. Bergman, J. E., Tokuyasu, K. T., and Singer, S. J. 1981. Passage of an integral membrane protein, the vesicular stomatitis virus G protein, through the Golgi apparatus en route to the plasma membrane, *Proc. Nat. Acad. Sci. U.S.A.* 78:1746–1750.

8. Blobel, G. 1980. Intracellular protein topogenesis, *Proc. Nat. Acad. Sci. U.S.A.* 77(3):1496–1500.

9. Blobel, G., and Dobberstein, B. 1975. Transfer of proteins across membranes. I. Presence of proteolytically processed and unprocessed nascent immunoglobulin light chains on membrane-bound ribosomes of murine myeloma cells, *J. Cell Biol.* 67:835–851.

10. Blok, J., Air, G. M., Laver, W. G., Ward, C. W., Lilley, G. G., Woods, E. F., Roxburgh, C. M., and Inglis, A. S. 1982. Studies on the size, chemical composition, and partial sequence of the neuraminidase (NA) from type A influenza viruses show that the N-terminal region of the NA is not processed and serves to anchor the NA in the viral membrane, *Virology* 119:109–121.

11. Bosch, F. X., Orlich, M., Klenk, H. -D., and Rott, R. 1979. The structure of the hemagglutinin, a determinant for the pathogenicity of influenza viruses, *Virology* 95:197–207.

12. Bretz, R., Bretz, H., and Palade, G. E. 1980. Distribution of terminal glycosyltransferases in hepatic Golgi functions, *J. Cell Biol.* 84:87–101.

13. Davis, A. R., Bos, T. J., and Nayak, D. P. 1983. Active influenza virus neuraminidase is expressed in monkey cells from cDNA cloned in simian Virus 40 vectors, *Proc. Nat. Acad. Sci. U.S.A.* 80:3976–3980.

14. Dobberstein, B., Garoff, H., and Warren, G. 1979. Cell-free synthesis and membrane insertion of mouse H-2D^d histocompatibility antigen and β_2-microglobulin, *Cell* 17:759–769.

15. Elder, K. T., Bye, J. M., Skehel, J. J., Waterfield, M. D., and Smith, A. E. 1979. *In vitro* synthesis, glycosylation, and membrane insertion of influenza virus haemagglutinin, *Virology* 95:343–350.

16. Emr, S. D., Hedgpeth, J., Clement, J. -M., Silhavy, T. J., Hofnung, M. 1980. Sequence analysis of mutations that prevent export of λ receptor, an *Escherichia coli* outer membrane protein, *Nature (London)* 285:82–85.

17. Fields, S., Winter, G., and Brownlee, G. G. 1981. Structure of the neuraminidase gene in human influenza virus A/PR/8/34, *Nature (London)* 290:213–217.

18. Gallione, C. J., and Rose, J. K. 1983. Nucleotide sequence of a cDNA clone encoding the entire glycoprotein from the New Jersey serotype of vesicular stomatitis virus, *J. Virol.* 46:162–169.

19. Garoff, H., and Schwarz, R. T. 1978. Glycosylation is not necessary for membrane insertion and cleavage of Semliki forest virus membrane proteins, *Nature (London)* 274:487–490.

20. Gething, M. -J., Bye, J., Skehel, J., and Waterfield, M. 1980. Cloning and DNA sequence of double-stranded copies of haemagglutinin genes from H2 and H3 strains elucidates antigenic shift and drift in human influenza virus, *Nature (London)* 287:301–307.

21. Gething, M.-J., and Sambrook, J. 1982. Construction of influenza hemagglutinin genes that code for intracellular and secreted forms of the protein, *Nature (London)* 300:598–603.

22. Gibson, R., Leavitt, R., Kornfeld, S., and Schlesinger, S. 1978. Synthesis and infectivity of vesicular stomatitis virus containing nonglycosylated G protein, *Cell* 13:671–679.

23. Green, J., Griffiths, G., Louvard, D., Quinn, P., and Warren, G. 1981. Passage of viral membrane proteins through the Golgi complex, *J. Mol. Biol.* 152:663–698.

24. Green, R. F., Meiss, H. K., and Rodriguez-Boulan, E. 1981. Glycosylation does not determine segregation of viral envelope proteins in the plasma membrane of epithelial cells, *J. Cell Biol.* 89:230–239.

25. Griffiths, G., Brands, R., Burlee, B., Louvard, D., and Warren, G. 1982. Viral membrane proteins acquire galactose in *trans* Golgi cisternae during intracellular transport, *J. Cell. Biol.* 95:781–792.

26. Griffiths, G., Quinn, P., and Warren, G. 1983. Dissection of the Golgi complex. I. Monensin inhibits the transport of viral membrane proteins from medial to *trans* Golgi cisternae in baby hamster kidney cells infected with Semliki forest virus, *J. Cell Biol.* 96:835–850.

27. Gruber, C., and Levine, S. 1985. Respiratory syncytial virus polypeptides. IV. The oligosaccharides of the glycoproteins, *J. Gen. Virol.* 66:417–432.

28. Guan, J.-L., and Rose, J. K. 1984. Conversion of a secretory protein into a transmembrane protein results in its transport to the Golgi complex but not to the cell surface, *Cell* 37:779–787.

29. Habener, J. F., Rosenblatt, M., Kemper, B., Kronenberg, H. M., Rich, A., and Potts, J. T. 1978. Pre-parathyroid hormone: amino acid sequence, chemical synthesis, and some biological studies of the precursor region, *Proc. Nat. Acad. Sci. U.S.A.* 75:2616–2620.

30. Harrison, S. C., and Kirchhausen, T. 1983. Clathrin, cages, and coated vesicles, *Cell* 33:650–652.

31. Hiti, A. L., and Nayak, D. P. 1982. Complete nucleotide sequence of the neuraminidase gene of human influenza virus A/WSN/33, *J. Virol.* 41:730–734.

32. Hiebert, S. W., Paterson, R. G., and Lamb, R. A. 1985. Hemagglutinin-neuraminidase protein of the paramyxovirus strain Simian virus 5: Nucleotide sequence of the mRNA predicts an N-terminal membrane anchor, *J. Virol.* 53:1–6.

33. Huang, R. T. C., Rott, R., Wahn, K., Klenk, H.-D., and Kohama, T. 1980. The function of the neuraminidase in membrane fusion induced by myxoviruses, *Virology* 107:313–319.

34. Huang, R. T. C., Wahn, K., Klenk, H.-D., and Rott, R. 1979. Association of the envelope glycoproteins of influenza virus with liposomes—A model study on viral envelope assembly, *Virology* 97:212–217.

35. Hubbard, S. C., and Ivatt, R. J. 1981. Synthesis and processing of asparagine-linked oligosaccharides, *Annu. Rev. Biochem.* 50:555–583.

36. Inouye, S., Soberon, X., Franceschini, T., Nakamura, K., Itakura, K., and Inouye, M. 1982. Role of positive charge on the amino-terminal region of the signal peptide in protein secretion across the membrane, *Proc. Nat. Acad. Sci. U.S.A.* 79:3438–3441.

37. Ishibashi, M., and Maizel, J. V. 1974. The polypeptides of adenovirus VI: Early and late glycopolypeptides, *Virology* 58:345–361.

38. Johnson, D. C., and Spear, P. G. 1983. O-Linked oligosaccharides are acquired by herpes simplex virus glycoproteins in the Golgi apparatus, *Cell* 32:987–997.

39. Jones, L. V., Compans, R. W., Davis, A. R., Bos, T. J., and Nayak, D. P. 1985. Surface expression of the influenza neuraminidase, and amino-terminally anchored viral membrane glycoproteins in polarized epithelial cells. (Submitted for publication)

40. Katz, F. N., Rothman, M. E., Knipe, D. M., Lodish, H. F. 1977. Membrane assembly: Synthesis and intracellular processing of vesicular stomatitis virus G, *J. Supramol. Structure* 7:353–370.

41. Katz, F. N., Rothman, M. E., Lingappa, V. R., Blobel, G., Lodish, H. F. 1977. Membrane assembly *in vitro:* Synthesis, glycosylation, and assymetric insertion of a transmembrane protein, *Proc. Nat. Acad. Sci. U.S.A.* 74:3278–3282.

42. Keil, W., Niemann, H., Schwarz, R. T., and Klenk, H. D. 1984. Carbohydrates of influenza virus. V. Oligosaccharides attached to individual glycosylation sites of the hemagglutinin of fowl plague virus, *Virology* 133:77–91.

43. Knipe, D. M., Lodish, H. F., and Baltimore, D. 1977. Localization of two cellular forms via the Golgi of the vesicular stomatitis virus glycoprotein, *J. Virol.* 21:1121–1127.

44. Kornfeld, R., and Kornfeld, S. 1980. Structure of glycoproteins and their

oligosaccharide units, in: *The Biochemistry of Glycoproteins and Proteoglycans* (W. J. Lennarz, ed.), pp. 1–34, Plenum Press, New York.

45. Lai, C.-J., Markoff, L. J., Zimmerman, S., Cohen, B., Berndt, J. A., and Chanock, R. M. 1980. Cloning DNA sequences from influenza viral RNA segments, *Proc. Nat. Acad. Sci. U.S.A.* 77(1):210–214.

46. Lazarowitz, S. G., Compans, R. W., and Choppin, P. W. 1971. Influenza virus structural and non-structural proteins in infected cells and their plasma membranes, *Virology* 46:830–843.

47. Leavitt, R., Schlesinger, S., and Kornfeld, S. 1977. Tunicamycin inhibits glycosylation and multiplication of Sindbis and vesicular stomatitis viruses, *J. Virol.* 21(1):375–385.

48. Li, E., Tabas, I., and Kornfeld, S. 1978. II. Characterization of the processing intermediates in the synthesis of the complex oligosaccharide units of the vesicular stomatitis virus G protein, *J. Biol. Chem.* 253:7762–7770.

49. Lingappa, V. R., Katz, F. N., Lodish, H. F., and Blobel, G. 1978. A signal sequence for the insertion of a transmembrane glycoprotein, *J. Biol. Chem.* 253:8667–8670.

50. Maeda, T., Kawasaki, K., and Ohnishi, S.-I. 1981. Interaction of influenza virus hemagglutinin with target membrane lipids is a key step in virus-induced hemolysis and fusion at pH 5.2, *Proc. Nat. Acad. Sci. U.S.A.* 78:4133–4137.

51. Markoff, L., and Lai, C.-J. 1982. Sequence of the influenza A/Udorn/72 (H3N2) virus neuraminidase gene as determined from cloned full-length DNA, *Virology* 119:288–297.

52. Markoff, L., Lin, B.-C., Sveda, M. M., and Lai, C.-J. 1984. Glycosylation and surface expression of the influenza virus neuraminidase requires the N-terminal hydrophobic region, *Mol. Cell Biol.* 4:8–16.

53. Milstein, C., Brownlee, G. G., Harrison, T. M., and Mathews, M. B. 1972. A possible precursor of immunoglobulin light chains, *Nature (London)* 239:117–120.

54. Moreno, F., Fowler, A., Hall, M., Silhavy, T. J., Zabin, I., and Schwartz, M. 1980. A signal sequence is not sufficient to lead β-galactosidase out of the cytoplasm, *Nature (London)* 286:356–359.

55. Nakamura, K., and Compans, R. W. 1977. The cellular site of sulfation of influenza viral glycoproteins, *Virology* 79:381–392.

56. Nakamura, K., Itakura, K., and Inouye, M. 1982. Role of positive charge on the amino-terminal region of the signal peptide in protein secretion across the membrane, *Proc. Nat. Acad. Sci. U. S. A.* 79:3438–3441.

57. Niemann, H., Boschek, B., Evans, D., Rosing, M., Tamura, T., and Klenk, H.-D. 1982. Post-translational glycosylation of coronavirus glycoprotein E1: Inhibition by monensin, *EMBO J.* 1:1499–1504.

58. Olden, K., Pratt, R. M., and Yamada, K. M. 1978. Role of carbohydrate in protein secretion and turnover: Effects of tunicamycin on the major cell surface glycoprotein of chick embryo fibroblasts, *Cell* 13:461–473.

59. Pollack, L., and Atkinson, P. H. 1983. Correlation of glycosylation forms with position in amino acid sequence, *J. Cell. Biol.* 97:293–300.

60. Porter, A. G., Barber, C., Carey, N. H., Hallewell, R. A., Threlfall, G., and Emtage, J. S. 1979. Complete nucleotide sequence of an influenza virus haemagglutinin gene from cloned DNA, *Nature (London)* 282:471–477.

61. Quinn, P. Griffiths, G., and Warren, G. 1983. Dissection of the Golgi complex II.

Density separation of specific Golgi functions in virally-infected cells treated with monensin, *J. Cell Biol.* 96:851–856.

62. Robbins, P. W., Hubbard, S. C., Turco, S. J., and Wirth, D. 1977. Proposal for a common oligosaccharide intermediate in the synthesis of membrane glycoproteins, *Cell* 12:893–900.

63. Rodriguez-Boulan, E., and Prendergast, M. 1980. Polarized distribution of viral envelope proteins in the plasma membrane of infected epithelial cells, *Cell* 20:45–54.

64. Rodriguez-Boulan, E., and Sabatini, D. D. 1978. Assymetric budding of viruses in epithelial monolayers, *Proc. Nat. Acad. Sci. U.S.A.* 75:5071–5075.

65. Rose, J. K., Adams, G. A., and Gallione, C. J. 1974. The presence of cysteine in the cytoplasmic domain of the vesicular stomatitis virus glycoprotein is required for palmitate addition, *Proc. Nat. Acad. Sci. U.S.A.* 81:2050–2054.

66. Rose, J. K., and Bergmann, J. E. 1982. Expression from cloned cDNA of cell-surface secreted forms of the glycoprotein of vesicular stomatitis virus in eucaryotic cells, *Cell* 30:753–762.

67. Rose, J. K., and Bergmann, J. E. 1983. Altered cytoplasmic domains affect intracellular transport of the vesicular stomatitis virus glycoprotein, *Cell* 34:513–524.

68. Rose, J. K., Welch, W. J., Sefton, B. M., Esch, F. S. and Ling, N. C. 1980. Vesicular stomatitis virus G is anchored in the viral membrane by a hydrophobic domain near the carboxy-terminus, *Proc. Nat. Acad. Sci. U.S.A.* 77:3884–3888.

69. Roth, M. G., Compans, R. W., Giusti, L., Davis, A. R., Nayak, D. P., Gething, M.-J., and Sambrook, J. 1983. Influenza virus hemagglutinin expression is polarized in cells infected with recombinant SV40 viruses carrying cloned hemagglutinin DNA, *Cell* 33:435–443.

70. Roth, M. G., Srinivas, R. V., and Compans, R. W. 1983. Basolateral maturation of retroviruses in polarized epithelial cells, *J. Virol.* 45:1065–1073.

71. Roth, M. G., Fitzpartrick, J. P., and Compans, R. W. 1979. Polarity of influenza and vesicular stomatitis virus maturation in MDCK cells: Lack of a requirement for glycosylation of viral glycoproteins, *Proc. Nat. Acad. Sci. U.S.A.* 76:6430–6434.

72. Rothman, J. E. 1981. The Golgi apparatus: Two organelles in tandem, *Science* 213:1212–1219.

73. Rothman, J. E., and Fine, R. E. 1980. Coated vesicles transport newly synthesized membrane glycoproteins from endoplasmic reticulum to plasma membrane in two successive stages, *Proc. Nat. Acad. Sci. U.S.A.* 77:780–784.

74. Rothman, J. E., Katz, F. N., and Lodish, H. F. 1978. Glycosylation of a membrane protein is restricted to the growing polypeptide chain but is not necessary for insertion as a transmembrane protein, *Cell* 15:1447–1454.

75. Sabatini, D. D., Kreibich, G., Morimoto, T., and Adesnik, M. 1982. Mechanisms for the incorporation of proteins in membranes and organelles, *J. Cell Biol.* 92:1–22.

76. Schachter, H., and Roseman, S. 1980. Mammalian glycosyltransferases, in: *The Biochemistry of Glycoproteins and Proteoglycans* (W. Lennarz, ed.), pp. 85–160, Plenum Press, New York.

77. Schlesinger, M. J., Magee, A. I., and Schmidt, M. F. G. 1980. Fatty acid acylation of proteins in cultured cells, *J. Biol. Chem.* 255:10021–10024.

78. Schmidt, M. F. G. 1982. Acylation of viral spike glycoproteins: A feature of enveloped RNA viruses, *Virology* 116:327–338.

79. Schmidt, M. F. G., and Schlesinger, M. J. 1980. Relation of fatty acid attachment to the translation and maturation of Vesicular Stomatitis and Sindbis virus membrane glycoproteins, *J. Biol. Chem.* 255:3334–3339.

80. Schwarz, R. T., Rohrschneider, J. M., and Schmidt, M. F. G. 1975. Suppression of glycoprotein formation of Semliki Forest influenza and avian sarcoma viruses by tunicamycin, *J. Virol.* 19:782–791.

81. Sefton, B. M. 1977. Immediate glycosylation of Sindbis virus membrane proteins, *Cell* 10:659–668.

82. Sekikawa, K., and Lai, C.-J. 1983. Defects in functional expression of an influenza virus hemagglutinin lacking the signal peptide sequences, *Proc. Nat. Acad. Sci. U.S.A.* 80:3563–3567.

83. Shida, H., and Dales, S. 1981. Biogenesis of vaccinia: carbohydrate of the hemagglutinin molecule, *Virology* 111:56–72.

84. Skehel, J. J., and Waterfield, M. D. 1975. Studies on the primary structure of the influenza virus hemagglutinin, *Proc. Nat. Acad. Sci. U.S.A.* 72:92–97.

85. Sveda, M. M., and Lai, C.-J. 1981. Functional expression in primate cells of cloned DNA coding for the hemagglutinin surface glycoprotein of influenza virus, *Proc. Nat. Acad. Sci. U.S.A.* 78:5488–5492.

86. Sveda, M. M., Markoff, L. J., and Lai, C.-J. 1982. Cell surface expression of the influenza virus hemagglutinin requires the hydrophobic carboxy-terminal sequences, *Cell* 30:649–656.

87. Sveda, M. M., Markoff, L. J., and Lai, C.-J. 1984. Influenza virus hemagglutinin containing an altered hydrophobic carboxy terminus accumulates intracellularly, *J. Virol.* 49:223–228.

88. Tabas, I., Schlesinger, S., and Kornfeld, S. 1978. Processing of high mannose oligosaccharides on the newly synthesized polypeptides of the vesicular stomatitis virus G protein and the IgG heavy chain, *J. Biol. Chem* 253:716–722.

89. Tkacz, J. J., and Lampen, J. O. 1975. Tunicamycin inhibition of polyisoprenyl N-acetylglucosaminyl pyrophosphate formation in calf liver microsomes, *Biochem. Biophys. Res. Commun.* 65:248–257.

90. Townsend, A. R. M., McMichael, A. J., Carter, N. P., Huddleston, J. A., and Brownlee, G. G. 1984. Cytotoxic T cell recognition of the influenza nucleoprotein and hemagglutinin expressed in transfected mouse L cells, *Cell* 39:13–25.

91. Varghese, J. N., Laver, W. G., and Colman, P. M. 1983. Structure of the influenza virus glycoprotein antigen neuraminidase at 2.9 Å resolution, *Nature (London)* 303:35–44.

92. Walter, P., Gilmore, R., and Blobel, G. 1984. Protein translocation across the endoplasmic reticulum, *Cell* 38:5–8.

93. Ward, C. W., Murray, J. M., Roxburgh, C. M., and Jackson, D. C. 1983. Chemical and antigenic characterization of the carbohydrate side chains of an Asian (N2) influenza virus neuraminidase, *Virology* 126:370–375.

94. Webster, R. G., Laver, W. G., Air, G. M., and Schild, G. C. 1982. Molecular mechanisms of variation in influenza viruses. *Nature (London)* 296:115–121.

95. White, J., Helenius, A., and Gething, M.-J. 1982. Haemagglutinin of influenza virus expressed from a cloned gene promotes membrane fusion, *Nature (London)* 300:658–659.

96. Wilson, I. A., Wiley, D. C., and Skehel, J. J. 1981. Structure of the hemagglutinin membrane glycoprotein of influenza virus at 3Å resolution, *Nature (London)* 289:366–373.

97. Winter, G., Fields, S., and Brownlee, G. G. 1981. Nucleotide sequence of the haemagglutinin gene of a human influenza virus H1 subtype, *Nature (London)* 292:72–75.

98. Wirth, D. F., Lodish, H. F., and Robbins, P. W. 1979. Requirements for the insertion of the Sinbis envelope glycoproteins into the endoplasmic reticulum membrane, *J. Cell Biol.* 81:154–162.

99. Yost, C., Hedgpeth, J., and Lingappa, V. R. 1983. A stop transfer sequence confers predictable transmembrane orientation to a previously secreted protein in cell-free systems, *Cell* 34:759–766.

100. Zilberstein, A., Snider, M. D., Porter, M., and Lodish, H. F. 1980. Mutants of vesicular stomatitis virus blocked at different stages in maturation of the viral glycoprotein, *Cell* 21:417–427.

Questions for Discussion with the Editors

1. *Viruses provide excellent model systems for analyzing intracellular glycoprotein transport, as you have detailed in your description of the strengths of this experimental system. What do you perceive, however, as the weaknesses of the "viral membrane" system?*

Our general approach to studying the processing and transport of viral glycoproteins has been discussed. We assume there is an analogy in vector-infected cells to events that occur during influenza viral infection. Vector-coded influenza hemagglutinin displays all the grossly detectable properties of influenza virus-coded hemagglutinin. However, one cannot be certain that vector-coded hemagglutinin achieves the trimeric three-dimensional form of the native molecule. It is conceivable that hemagglutination, hemadsorption, and fusion require only the accessibility of the requisite binding sites at the cell surface for these phenomena to be mediated by HA expressed from the SV40 vector. The finding that vector-coded hemagglutin is recognized by cytotoxic T cells is the best but not conclusive evidence that vector-coded hemagglutinin "looks" like the native molecule. If the three-dimensional structure of vector-coded hemagglutinin or neuraminidase at the cell surface is not precisely that of the native molecule, it may be due to a requirement for other influenza viral gene products in order for the surface antigens to attain their respective three-dimensional structures and spatial relationships to each other at the infected cell surface. If such a requirement exists, failure of vector-coded neuraminidase to attain its native three-dimensional structure may explain the markedly reduced enzymatic activity of vector-coded neuraminidase compared to that detectable at the surface of influenza infected cells. One of the products of the influenza matrix (M) gene, the M2 protein, has recently been shown to be plasma membrane associated and detectable at the cell surface. It contains an internal hydrophobic region sufficient in length to span the membrane. It would be of value to know more about the association, if any, of the M2 protein with the anchor termini of the hemagglutinin (and neuraminidase) that might affect the native configuration of these molecules on influenza virus infected cells.

In a much larger sense, a "weakness" of the viral membrane system has been i ts failure to shed more light on the processes of synthesis and maintenance of intracellular membraneous structures. This is because viruses that acquire glycoproteins at the plasma membrane, such as the alphaviruses, influenza virus, and vesicular stomatitis virus, have been principal subjects of study. The data do not tell us much about interactions between organelles responsible for glycoprotein formation (the endoplasmic reticulum and the Golgi) and other specialized organelles, such as the mitochondria and lysozomes.

2. *Please list some of the various cell types (e.g., pancreas acinar cells) in mammals that you would expect to display the highly exaggerated and specialized intracellular protein translocation processes similar to those you have described for virus-infected cells.*

It is probable that most functionally specialized glycoproteins produced by most specialized cell types are glycosylated and modified during passage through the endoplasmic reticulum and the Golgi apparatus in a manner analogous to the processing and transport of the viral glycoproteins we have discussed. As a specific example, the transport of amylase in and its secretion from pancreatic acinar cells has been followed by immune EM. Amylase traverses the ER and the Golgi before being packaged in Zymogen granules for transport to the cell surface and secretion. Therefore, pancreatic acinar cells do transport an endogenous gene product along the pathway taken by viral glycoproteins. Additional examples of cell-associated glycoproteins that probably are processed in a manner analogous to viral glycoproteins would include basement membrane protein, α_2-macroglobulin, collagen, histocompatibility antigens, membrane-associated enzymes (such as sucrase-isomaltase and leucine aminopeptidase localized to the small intestinal microvilli), and plasma membrane-anchored hormone receptors. Of current interest in the group of hormone receptors are those for broad-spectrum growth factors derived from tumors or normal cells. These include epidermal growth factor (EGF), platelet-derived growth factor (PDGF), and fibroblast growth factor (FGF).

The data suggest that cellular secretory glycoproteins are processed and secreted in a manner analogous to that of influenza hemagglutinin and vesicular stomatitis virus G protein mutants that lack the C-terminal membrane anchorage domain. In early studies that led to the exposition of the signal hypothesis, it was shown that the *in vitro* translocation of secretory immunoglobulin G expressed from mouse myeloma cell mRNA and that of the vesicular stomatitis virus G protein expressed from VSV mRNA were completely analogous. Recent data on the processing of important receptor and effector molecules of the immune system also support an analogy to that of viral glycoproteins and derived mutants lacking membrane domains, respectively. These include a number of glycoproteins that are detected in both membrane-bound and secreted forms, such as the class I and class II molecules of the major histocompatibility complex, immunoglobulin M, and the receptors for insulin, EGF, and interleukin 2 (IL-2). B and T lymphocytes seem to employ more than one strategy in order to generate alternate forms of a single functional protein that either retains or lacks a membrane anchorage domain. For example, different genes code for the alternate forms of MHC class I molecules, while a splicing event at the mRNA level results in alternate transcripts that retain or lack sequences that code for the membrane anchorage domain of immunoglobulin M.

Human Genetic Diseases

CHAPTER 14

Mammalian Expression of Genes Associated with Human Immunodeficiency Disease

R. Scott McIvor

RECENT TECHNICAL ADVANCES in the delivery of genetic material into mammalian cells, especially at the *in vitro* level, has contributed substantially to our understanding of gene expression (as seen in this volume). A variety of physical, chemical, and biological methods of gene transfer have been developed [reviewed (12)]. Information is also steadily accumulating on the role of specific gene sequences in the regulation and tissue specificity of gene expression. In many mammalian systems, the components of a sequence can now be optimized for expression in a particular cell type or for a particular function.

These developments have led to speculation that gene transfer technology might be applicable to the treatment of human genetic disease (2). Indeed, several reports have recently appeared concerning the introduction of genetic material into hematopoietic tissues of mice [(3, 5, 22, 37) and chap. 15], providing evidence for the competence of hematopoietic stem cells for gene transfer and a basis for animal studies that might ultimately lead to human gene therapy. Of the hundreds of identified genetic diseases (21), the immunodeficiencies associated with absence of purine nucleoside phosphorylase (PNP) or adenosine deaminase (ADA) activities (15, 17) are perhaps the most likely to be amenable to treatment by gene therapy (see below).

This chapter presents an approach to gene therapy, from the isolation of cDNA clones to the development of systems for gene transfer and expression, for treatment of PNP and ADA enzyme deficiencies. The enzymes and their genes are also described as well as the immunodeficiency diseases that result from the absence of enzyme activity.

Immunodeficiencies Associated with Absence of Purine Nucleoside Phosphorylase and Adenosine Deaminase

The absence of ADA activity in blood samples of certain children suffering from severe combined immunodeficiency (SCID) was a surprising discovery (9) which was followed shortly by a similar observation that PNP activity was lacking in samples obtained from individuals expressing a T-cell immunodeficiency (10). Both SCID and T-cell immunodeficiency are diseases inherited in an autosomal recessive manner and are extremely rare (15, 17), although more intense screening procedures have recently modified this view. Onset of the disease is within a year after birth (perhaps delayed for PNP deficiency,) and most patients succumb between the ages of 1 and 5 years.

The PNP enzyme is a trimer of molecular weight (MW) 31,000 as determined with the protein purified from human erythrocytes (25, 40) and as predicted from the cDNA coding sequence (38). PNP catalyzes the phosphorolysis of inosine, deoxyinosine, guanosine, and deoxyguanosine to the corresponding base and pentose-1-phosphate. Although the equilibrium constant for this reaction lies in the direction of nucleoside synthesis, phosphorolysis has been the primary physiological function associated with the enzyme, probably owing to high concentrations of intracellular inorganic phosphate and low levels of ribose- and deoxyribose-1-phosphate. All four substrates accumulate in the plasma and are excreted in the urine of individuals lacking PNP activity. However, the one substrate that appears to cause immunotoxicity is deoxyguanosine (Fig. 1). In T-cells, the combination of nucleoside kinases and nucleotidases are such that deoxyguanosine (but not the other PNP substrates) can be phosphorylated to deoxyGMP, due largely to the substrate specificity of deoxycytidine kinase. Further phosphorylation of deoxyGMP yields deoxyGTP, which at high levels inhibits ribonucleotide reductase (by a feedback mechanism) in a manner that depletes affected cells of deoxyCTP. The resultant deoxynucleotide imbalance impedes DNA synthesis and, ultimately, the immunoproliferative response (17).

This model for the metabolic basis of immunotoxicity in PNP deficiency was formulated largely from somatic cell genetic studies using cultured mouse S49 T-lymphoma cells (30). A PNP-deficient cell line was isolated that was sensitive to deoxyguanosine. Secondary mutants resistant to deoxyguanosine retained the PNP deficiency but were unable either to transport or phosphorylate deoxyguanosine (with mutations in the nucleoside carrier or deoxycytidine kinase, respectively) or contained an altered ribonucleotide reductase that was resistant to feedback inhibition by deoxyGTP.

ADA is a 42,000-MW protein that catalyzes the apparently irreversible deamination of adenosine to inosine and of deoxyadenosine to deoxyinosine. Both products accumulate in the blood of deficient patients (17). A hypothesis similar to that described above for PNP has been presented to explain why the absence of ADA activity so drastically affects both arms of the immune system, based on the lymphocyte-specific phosphorylation of deoxyadenosine ultimately leading to feedback inhibition of ribonucleotide reductase by deoxyATP

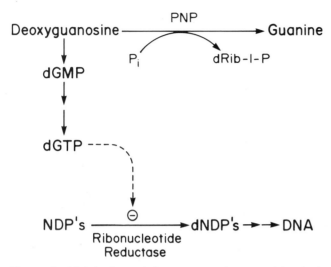

Figure 1 Metabolism of deoxyguanosine, resulting in immunotoxicity. PNP, purine nucleoside phosphorylase; dRib-1-P, deoxyribose-1-phosphate; P$_i$, inorganic phosphate; dGMP, deoxyguanosine-5'-monophosphate; dGTP, deoxyguanosine-5'-triphosphate; NDP's, ribonucleoside-5'-diphosphates; dNDP's, deoxyribonucleoside-5'-diphosphates.

(17). However, adenosine is also generally toxic for many different cell types. Inhibition of methylation reactions associated with S-adenosylmethionine (15), inhibition of pyrimidine biosynthesis (15), as well as several other mechanisms (3, 28), have also been proposed as explanations for immunotoxicity in ADA deficiency.

A number of approaches have been taken in the treatment of these diseases, including enzyme therapy (infusion of irradiated, normal erythrocytes), deoxycytidine or uridine administration for PNP deficiency, and bone marrow transplantation from matched donors (17). The latter has proved to be successful in many but not all cases where it has been attempted.

The possibility that ADA- and PNP-associated immunodeficiencies might be amenable to treatment by gene therapy is apparent for a number of reasons: (a) They are diseases of the hematopoietic system, and hematopoietic tissue is readily available in the form of bone marrow, including pluripotent hematopoietic stem cells with a demonstrable capacity for self-renewal (35). Bone marrow transplantation procedures have been well-studied in both animals and humans, and the transfer of genetic material into bone marrow stem cells of mice has been demonstrated (3, 5, 22, 37). Treatment of genetic diseases affecting other tissues (e.g., Lesch-Nyhan syndrome) will probably require gene transfer systems that are more elaborate than what is presently available. (b) Displacement of the endogenous, inactive gene product will not be necessary, defining the therapy is gene supplementation. (c) The specific cell lineage into which the functional ADA or PNP gene is transferred and expressed would be of no consequence, just as long as the overall amount of enzyme activity in hematopoietic tissues is sufficient to lower the concentration of deoxynucleoside in the circulation to nontoxic levels. Nucleosides and nucleic acid bases are generally transported across the plasma membrane of mammalian cells by rapid facilitated diffusion processes (27), so that substrate entry into the cell and product efflux from the cell would not be rate-limiting for PNP and ADA reactions. (d) A relatively small amount of enzyme activity should suffice in relieving immunotoxicity, while a large excess of activity would probably not be harmful. Individuals who retain only a small percentage of the normal level of ADA activity in their blood remain immunocompetent (17), and infusion of irradiated normal erythrocytes into enzyme-deficient patients has been clinically efficacious in some cases (17). On the other hand, cultured mammalian cells have been isolated that contain amplified ADA genes and produce ADA at a level of up to 75 percent of cytosol protein [(13, 39) and R. Kellems, personal communication]. In summary, the biological requirements are not stringent for a putative therapeutic procedure in terms of target tissue, cell type, or regulation of gene expression. Finally, these diseases are lethal if untreated.

Sequence and Organization of the Human PNP and ADA Genes

The isolation of cDNA clones containing sequences encoding PNP and ADA has recently been reported by a number of laboratories (7, 11, 13, 32, 36, 39). Availability of PNP- and ADA-specific sequences permits analysis of the structure and expression of the ADA and PNP genes (35, 38) and the molecular genetic anomalies associated with immune dysfunction (1, 31). These cDNA clones have also provided starting material for the construction of systems for gene transfer and expression of PNP and ADA (see below).

Goddard et al. isolated cDNA clones containing the entire coding sequence for human PNP by complementation in PNP-deficient *Escherichia coli* (11). The use of such a selective procedure in isolating molecular clones is advantageous since it ensures the presence of a complete, functional coding sequence (i.e., guards against truncated sequences and cloning artifacts) and obviates the need for extensive screening by hybridization. The PNP cDNA sequence was subsequently used to screen a lambda phage library and isolate the PNP gene (38). Southern analysis of genomic clones using various portions of the cDNA sequence as probe indicated the presence of coding sequence on five exons within a gene spanning approximately 12 kb of DNA. The ability of sequences upstream of the first exon to initiate transcription of the PNP cDNA has been demonstrated in mouse cells, and this promoter is presently being further characterized (S. R. Williams, R. S. McIvor, and D. W. Martin, Jr., unpublished results).

cDNA clones encoding ADA have been isolated from human (7, 32, 36), mouse (39), and rat (13) sources. The complete human ADA coding sequence has been corroborated by three separate groups (1, 7, 32) and one of these clones was recombined for expression in mouse cells, verifying the presence of a functional sequence (32). Wiginton et al. (36) also reported an interesting cDNA clone, evidently derived from an incompletely processed messenger since it contained a 76-bp intron sequence complete with splice signals. A complete, functional mouse ADA coding sequence has been cloned by complementation in *E. coli* (R. Kellems, personal communication). The human ADA gene has been isolated from a human cosmid library and consists of 12 exons dispersed across 32 kb of DNA (34). An upstream sequence has also been identified that functions as a promoter (i.e., results in the production of human ADA when fused to the human ADA cDNA sequence and transfected into mouse L cells) and contains a number of interesting features in common with promoters of other eukaryotic "housekeeping" genes (34).

Using ADA-specific sequences as probe, defects in the production of active enzyme have been traced in several cases to apparent errors in RNA splicing in lymphoblast cell lines derived from ADA-deficient patients (1). Splicing errors detected by S1 nuclease mapping were subsequently shown to be consistent with the intron-exon structure of the ADA gene (34). In other cases, amino acid substitutions most likely render the ADA protein either catalytically inactive or unstable (31). Specific mutation sequences have been identified in clones isolated from a cosmid library prepared from one of these cell lines and are presently being functionally characterized (D. Valerio, personal communication).

Gene Transfer and Expression of Human PNP and ADA

The isolation of cDNA clones containing the entire coding sequences for human ADA and PNP (see previous section) makes possible the construction of plasmids designed for expression of these enzymes in mammalian cells after gene transfer. Several different promoter/host cell combinations (detailed below) have been used to test for PNP expression. These results directed studies on ADA expression to a specific promoter/host cell combination. After the initial demonstration that the PNP cDNA sequence could be successfully expressed in mammalian cells, subsequent experiments were designed to address two major questions through gene transfer studies in cell culture:

1. What sequence combinations result in successful expression of PNP and ADA in mouse cells? This is important because of ongoing attempts to transfer and express human ADA and PNP in hematopoietic tissues of mice as an animal model.
2. What sequence combinations result in successful expression in human cells? This addresses the ultimate goal of transfer and expression in human hematopoietic cells.

PNP and ADA Expression Plasmids

Before the sequence of the PNP gene was determined, it had been observed that the coding sequence was most likely contained in its entirety in a BamHI-AvaI fragment (11). These restriction sites were present in pPNP1 (Fig. 2), but not in pPNP2, while *both* clones were able to complement the PNP⁻ phenotype of bacterial host cells. The location of the BamHI site has since been confirmed in a genomic clone (38) approximately 50 bp downstream from the transcriptional initiation site (S. R. Williams and D. W. Martin, Jr., unpublished results).

The 1242-bp BamHI-AvaI fragment (38) was used to construct pEPD (20), a plasmid that contains PNP and methotrexate (MTX)-resistant DHFR (29) transcriptional units, each flanked on the 5′ side by an SV40 early promoter

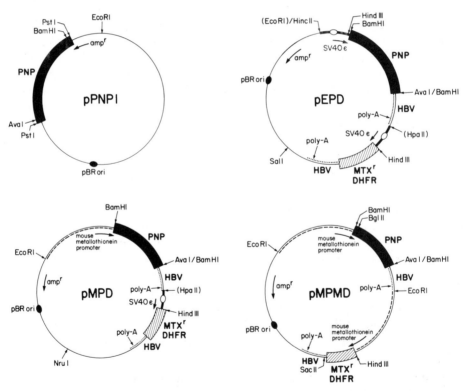

Figure 2 Plasmids designed for expression of human PNP in mammalian cells. PNP and DHFR coding regions are designated by solid and striped boxes, respectively. A thin line represents bacterial vector sequence. Parallel dashed and dotted lines indicate mouse metallothionein promoter sequence and hepatitis B virus surface antigen sequence (including the polyadenylation signal), respectively. A thick line denotes SV40 sequence including the early promoter region.

and on the 3′ side by a sequence containing the polyadenylation signal of the hepatitis B virus surface antigen gene (Fig. 2). In subsequent plasmid constructions the SV40 early promoter was replaced by the mouse metallothionein promoter in the 5′ flanking sequences of PNP (pMPD) and then DHFR (pMPMD). The methotrexate-resistant DHFR gene functions as a dominant-selectable marker in gene transfer experiments and can be amplified by culturing transfectant cells at elevated concentrations of MTX (see Chap. 5).

The plasmid pLL was isolated from a cDNA library prepared from human Molt-4 cell mRNA by screening with a fragment previously shown to contain ADA-specific sequences (32). Sequence data indicated the presence of what was most likely the translational start site within a unique NcoI restriction site (CCATGG). The presence of a unique NcoI restriction site at the translational start site of the PNP coding sequence facilitated its replacement in pMPMD by the ADA coding sequence to construct pMAMD. The sequences flanking the ADA coding region in pMAMD on the 5′ side are thus identical to the sequences flanking the PNP coding sequence on the 5′ side in all of the PNP expression plasmids.

PNP Expression in Rodent Cells

PNP expression plasmids introduced into Chinese hamster ovary (CHO) cells or mouse LTK⁻ APRT⁻ (LTA) cells by using the DNA-calcium phosphate coprecipitation technique (20). After an expression period of 1 to 3 days, host cells were harvested and either plated onto selective medium (containing 0.25 µM MTX) or extracted and assayed for transient expression of human PNP. Isolated clones were expanded, harvested, and extracts analyzed for the presence of human PNP by isoelectric focusing (pH 4 to 6.5) and staining for PNP activity (20).

A typical result from this kind of experiment is shown in Fig. 3. PNP activity from human (HeLa) cells is clearly distinguishable (pI = 6.1) from mouse or hamster PNP (pI = 5.1 to 5.3). Rodent cells stably transfected with the plasmids indicated (CHO-EPD2, LTA-MPMD5) contained a substantial amount of PNP activity focusing at the same pI as human PNP. The presence of two intermediate bands of PNP activity in these gels is explained by the trimeric structure of PNP, giving rise to two hybrid molecules containing different combinations of human and rodent PNP subunits. Transient expression of human PNP was also detectable in L cells 2 to 3 days after transfection with pMPD and in both in L cells and CHO cells 1 to 2 days after transfection with pEPD, facilitating a rapid evaluation of these plasmids for PNP expression (20). However, hybrid forms of PNP were not detected in transient analyses. One explanation for this is that transiently expressed human PNP was perhaps synthesized in a cellular compartment separate from the site of endogenous rodent PNP synthesis. Alternatively, human PNP synthesis might have occurred in a very small number of cells,

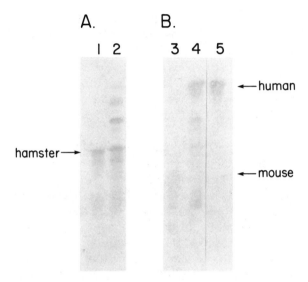

Figure 3 PNP isozyme analysis of extracts from mammalian cells. Samples were electrofocused between pH 4 and 6.5, then the gel was stained *in situ* for PNP activity. (A) Lane 1, untransfected CHO cells; lane 2, CHO-EPD 2 (an isolate of CHO transfected with pEPD; see Fig. 2). (B) Lane 3, untransfected mouse LTA (TK⁻, APRT⁻) cells; lane 4, LTA-MPMD5 (an isolate of LTA transfected with pMPMD; see Fig. 2); lane 5, HeLa cells.

allowing formation mainly of the human homotrimer and masking the presence of mouse enzyme subunits.

The ability of a plasmid construction to provide human PNP expression in a given cell type appeared to be dependent on the promoter which was used to initiate transcription of the PNP gene (Table 1). While approximately half of the pEPD transferrents of CHO cells contained detectable human PNP levels, there was none detected in any of four pEPD transferrents of mouse LTA cells on initial isolation. Human PNP expression was ultimately detected in one of the latter transferrents but only after adaption to growth in 500 μM MTX (i.e., amplification of PNP sequence along with DHFR). The mouse

TABLE 1 PNP expression in rodent cells

Plasmids[a]	CHO cells		LTA cells	
	Transient	Stable	Transient	Stable
pEPD	+	+	±	±
pMPD	−	N.D.[b]	+	+
pMPMD	−	N.D.	+	+

[a]See Fig. 2 for plasmid constructions.
[b]N.D., not determined.

metallothionein promoter effected substantial levels of PNP expression in mouse L cells and was inducible with zinc, even though human PNP was undetectable in transient analyses of CHO cells transfected with pMPD. L-cell lines stably transfected with pMPD were difficult to isolate, perhaps because the SV40 early promoter initiates transcription of DHFR in this plasmid. This promoter appeared to be of limited value in effecting transcription of the PNP gene in this L-cell line (see above). A plasmid containing two mouse metallothionein promoter sequences to initiate transcription of both PNP and DHFR genes (pMPMD, Fig. 2) provided transfection frequencies of approximately 10^{-4} in L cells, and, in contrast to transferrents of pEPD or pMPD, isolated pMPMD transferrent clones grew out well and gave consistently positive results in assaying for human PNP activity by IEF-isozyme analysis (20).

PNP Expression in Human Cells

The mouse metallothionein promoter thus seemed to be a most valuable sequence for effecting PNP expression in mouse cells. In order to determine the ability of this transcriptional unit to function in a human system, pMPD was transfected into HeLa cells and MTX-resistant colonies were isolated (20).

The mouse metallothionein promoter has been shown to function in human cells (18), so one would expect pMPD to provide PNP expression in MTX-resistant HeLa cells transferrents. Southern analyses indicated that HeLa cells possessed a peculiar ability to retain the DHFR function necessary for survival but to exclude the superfluous PNP sequences in about half of the transferrents analyzed. An example is shown in Fig. 4A. Genomic DNA from HeLa-MPD2 digested with BglII contained a fragment not found in the parental HeLa cell line hybridizing to both PNP and HBV probes. HeLa-MPD1 contained no such PNP-hybridizing BglII fragment, but was positive when reprobed for HBV sequences. The IEF-isozyme analysis used to detect human PNP expression in rodent systems was not applicable to HeLa cell transferrents, due to the identity of endogenous human PNP. However, Northern analysis of HeLa-MPD-2 indicated the presence of several transcripts which hybridized to both PNP and HBV probes (Fig. 4B). One of these (indicated by arrow) was of an appropriate size to extend from the transcriptional initiation site of the mouse metallothionein promoter through the PNP gene to the polyadenylation signal of the HBV sequence. These results were interpreted as evidence for the ability of this sequence combination to function at least through the level of transcription in effecting PNP expression in a human system (20).

Expression of ADA in Rodent Cells

Experience in obtaining PNP expression in rodent cells led to the supposition that human ADA might also be expressed in mouse cells if the mouse me-

A.

B.

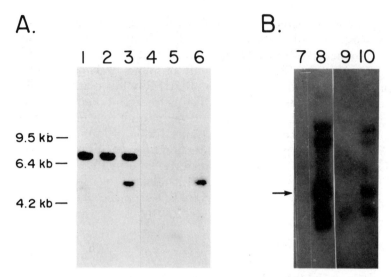

Figure 4 Blot hybridization analysis of HeLa cell isolates transfected with pMPD (see Fig. 2). (A) Southern blot of Bg1II-digested genomic DNA, probed first for PNP sequences (lanes 1 to 3) and then for HBV sequences (lanes 4 to 6). Lanes 1 and 4, untransfected HeLa cells; lanes 2 and 5, HeLa MPD1; lanes 3 and 6, Hela-MPD2. (B) Northern blot of oligo-dT-selected RNA from untransfected HeLa cells (lanes 7 and 9) and from HeLa-MPD2 (lanes 8 and 10) probed first for HBV sequences (lanes 7 and 8) and then for PNP sequences (lanes 9 and 10).

tallothionein promoter was used to initiate transcription of the ADA coding sequence. Mouse L cells were exposed to a calcium phosphate coprecipitate of pMAMD (32). Two days later, the cells were harvested and extracts were prepared and subjected to isoelectric focusing between pH 4 and 5. The gel was stained for ADA activity (Fig. 5). The transfected cells contained human ADA activity that was electrophoretically distinct from mouse ADA (32). Human ADA activity has also been demonstrated in clones of L cells stably transfected with pMAMD (19).

Expression of human ADA in rodent cells has recently been used as a means of evaluating activity of a human ADA promoter sequence (34). An ADA "minigene" (promoter fragment fused to cDNA) transfected in mouse L cells resulted in transient expression of human ADA. Expression of mouse (R. Kellems, personal communication) and human (24) ADA sequences has also been demonstrated in COS cells using SV40-based vector systems. The ability of a rat ADA sequence to function as a selectable marker in medium containing deoxycoformycin has also been assessed using a plasmid which contains the long terminal repeat of Friend spleen focus forming virus to initiate transcription of the ADA gene (26).

Figure 5 ADA isozyme analysis to detect expression of human ADA in mouse cells. Recipient cells were exposed to a calcium phosphate coprecipitate of pMAMD (34), and then harvested 2 days later. Cell extracts were electrofused between pH 4 and 5 and then the gel was stained *in situ* for ADA activity. Lane 1, HeLa cells; lane 2, L cells; lane 3, L cells transfected with pMAMD and incubated with 50 μM ZnCl$_2$; lane 4, L cells transfected with pMAMD and incubated without Zn.

Conclusion: Possibilities for Expression in Animals

The evidence presented above shows that cDNA sequences encoding human ADA and PNP are expressed in cultured mammalian cells when introduced in the form of appropriately constructed plasmid vectors. However, the use of plasmid vector systems for introducing genetic material into hematopoietic tissues of animals remains technically problematic in a general sense and has been reported by only one laboratory [Chap. 15 and (5)]. The crucial problem in such experiments is how to genetically transform a small [0.1 percent (35)] proportion of cells (i.e., pluripotent hematopoietic stem cells) in a large, heterogeneous population (whole bone marrow) using methods that provide relatively low transfer efficiencies. This problem is further compounded by the unknown competence of stem cells for the uptake, integration, and expression of exogenously provided DNA. Attempts to introduce pMPMD and pMAMD into mouse bone marrow using the DNA-calcium phosphate coprecipitation technique in this laboratory have so far been unsuccessful. Perhaps the ability to use plasmid vectors for gene transfer into bone marrow cells in a reproducible manner will require investigating other means of delivery, such as electric-field-induced gene transfer or protoplast fusion (12).

A more reliable method of gene transfer into mouse bone marrow cells has been provided by the development of retroviral delivery systems (14, 22, 37). Williams et al. recently reported the successful and reproducible transfer of a defective Moloney murine leukemia virus genome into hematopoietic tissue using a helper-free system (37). The presence of a newly introduced genetic material was demonstrated using a clonal analysis system (Southern analysis of spleen colonies derived from transplanted stem cells that had been infected with virus prior to transplantation) in which 15 percent of the spleen colonies analyzed contained transferred sequences. Transplantation of cells obtained from a single infected spleen colony into secondary recipients and regeneration of the entire hematopoietic system of those mice proved that a truly pluripotent stem cell population had been infected with the virus. Transfer into other hematopoietic progenitor cells has been demonstrated *in vitro* by Joyner et al. where colonies of a granulocyte-macrophage type, resistant to the antibiotic G418, were obtained after infection with a retrovirus encoding the bacterial neomycin resistance gene (14).

Retrovirus expression has also been demonstrated after transfer into mouse bone marrow cells. Miller et al. (22) transplanted bone marrow cells that had been infected with a defective Moloney leukemia virus encoding human hypoxanthine phosphoribosyltransferase (HPRT) into lethally inrradiated recipients along with replication competent helper virus. The defective HPRT virus was subsequently rescued from spleen and bone marrow cells of these mice, showing the presence of the virus genome in hematopoietic tissues. The spleen tissue also contained human HPRT which was electrophoretically distinguishable from endogenous mouse HPRT.

In a number of laboratories, retroviruses have been constructed that are capable of transducing sequences encoding human PNP and ADA [(8, 33); S. Orkin, personal communication; W. F. Anderson and E. Gilboa, personal communication; R. S. McIvor, J. Johnson, A. D. Miller, D. Valerio, S. R. Williams, I. M. Verma, and D. W. Martin, Jr., unpublished results]. The presence of enzymatically active gene product in virus-infected rodent cells is demonstrable using isozyme analyses similar to those described above. Experiments are now underway to infect hematopoietic stem cells with these viruses and to determine whether or not they express the human enzyme activities *in vivo*.

Development of mouse model systems for gene transfer into hematopoietic tissues will certainly facilitate human therapy. Recent work on amphotropic packaging systems (6, 23), generating helper-free defective viruses capable of infecting a wide variety of host cell types including human cells, has been an important advance.

Before human treatment might be attempted there will undoubtedly be many questions to address concerning the stability and safety of sequences introduced by using a retrovirus or by any other means. Retroviruses are notorious for their ability to recombine (16), and the specific defective viruses used to transduce the desired expression sequences must be evaluated in terms of the homogeneity of virions in producer cultures (e.g., freedom from replication-competent virus) and maintenance of the sequence structure transduced into target cells and tissues. An obvious concern is that newly integrated sequences might initiate or activate transcription of an endogenous oncogene or have some other detrimental effect on gene function. These possibilities can be addressed through experiments in cell culture and in animals and vectors modified to deal with specific problems that arise. Once treatment becomes feasible, however, the risk inherent in any such procedure should be weighed carefully against the patient's chances for survival by receiving treatment otherwise available at that time.

Acknowledgments

Patricia Hoffee, Rod Kellems, and Stuart Orkin are gratefully acknowledged for providing preliminary results and preprints of their work, as are David Martin, Jr., and Dinko Valerio for critical reading of the manuscript. This work was performed in the laboratory of David Martin, Jr., and was supported in part by USPHS grants AI17435 and AM20428 from the National Institutes of Health.

General References

BERNSTEIN, A., BERGER, S., HUSZAR, D., AND DICK, J. 1985. Gene transfer with retrovirus vectors, in: *Genetic Engineering: Principles and Methods*, vol. 7 (J. K. Setlow and A. Hollaender, eds.), pp. 235–261, Plenum Press, New York.

MARTIN, D. W., Jr., and GELFAND, E. W. 1981. Biochemistry of diseases of immunodevelopment, *Annu. Rev. Biochem.* 50:845–877.

References

1. Adrian, G. S., Wiginton, D. A., and Hutton, J. J. 1984. Structure of adenosine deaminase mRNAs from normal and adenosine deaminase-deficient human cell lines, *Mol. Cell. Biol.* 4:1712–1717.
2. Anderson, W. F. 1984. Prospects for human gene therapy, *Science* 226:401–409.
3. Carr, F., Medina, W. D., Duke, S., and Bertino, J. R. 1983. Genetic transformation of murine bone marrow cells to methotrexate resistance, *Blood* 62:180–185.
4. Carson, D. A., Wasson, D. B., Lakow, E., and Kamatani, N. 1982. Possible metabolic basis for the different immunodeficiency states associated with genetic deficiencies of adenosine deaminase and purine nucleoside phosphorylase, *Proc. Nat. Acad. Sci. U.S.A.* 79:3848–3852.
5. Cline, M. J., Stang, H., Mercola, K., Morse, L., Ruprecht, R., Browne, J., and Salser, W. 1980. Gene transfer in intact animals, *Nature (London)* 284:422–425.
6. Cone, R. D., and Mulligan, R. C. 1984. High-efficiency gene transfer into mammalian cells: Generation of helper-free recombinant retrovirus with broad mammalian host range, *Proc. Nat. Acad. Sci. U.S.A.* 81:6349–6353.
7. Daddona, P. E., Shewach, D. S., Kelley, W. N., Argos, P., Markham, A. F., and Orkin, S. H. 1984. Human adenosine deaminase: cDNA and complete primary amino acid sequence, *J. Biol. Chem.* 259:12101–12106.
8. Friedman, R. L. 1985. Expression of human adenosine deaminase using a transmissable murine retrovirus vector system, *Proc. Nat. Acad. Sci. U.S.A.* 82:703–707.
9. Giblett, E. R., Anderson, J. E., Cohen, F., Pollara, B., and Meuwissen, H. J. 1972.

Adenosine deaminase deficiency in two patients with severe impaired cellular immunity, *Lancet* 2:1067–1069.

10. Giblett, E. R., Ammann, A. J., Wara, D. W., Sandman, R., and Diamond, L. K. 1975. Nucleoside phosphorylase deficiency in a child with severely defective T-cell immunity and normal B-cell immunity, *Lancet* 1:1010–1013.

11. Goddard, J. M., Caput, D., Williams, S. R., and Martin, D. W., Jr. 1983. Cloning of human purine-nucleoside phosphorylase cDNA sequences by complementation in *Escherichia coli*, *Proc. Nat. Acad. Sci. U.S.A.* 80:4281–4285.

12. Gorman, C., 1985. High efficiency gene transfer, in: *DNA Cloning* (D. Glover, ed.), pp. 143–168, IRL Press, Oxford, Washington.

13. Hunt III, S. W., and Hoffee, P. A. 1983. Amplification of adenosine deaminase gene sequences in deoxycoformycin-resistant rat hepatoma cells, *J. Biol. Chem.* 258:13185–13192.

14. Joyner, A., Keller, G., Phillips, R. A., and Bernstein, A. 1983. Retrovirus transfer of a bacterial gene into mouse haematopoietic progenitor cells, *Nature (London)* 305:556–558.

15. Kredich, N. M., and Hershfield, M. S. 1983. Immunodeficiency diseases caused by adenosine deaminase deficiency and purine nucleoside phosphorylase deficiency, in: *The Metabolic Basis of Inherited Disease* (J. B. Stanbury, J. B. Wyngaarden, D. S. Frederickson, J. L. Goldstein, and M. S. Brown, eds.), pp. 1157–1183, McGraw-Hill, New York.

16. Linial, M., and Blair, D. 1984. Genetics of retroviruses, in: *RNA Tumor Viruses* (R. Weiss, N. Teich. H, Varmus, and J. Coffin, eds.) pp. 648–783, Cold Spring Harbor Laboratory, Cold Spring Harbor, N.Y.

17. Martin, D. W., Jr., and Gelfand, E. W. 1981. Biochemistry of diseases of immunodevelopment, *Annu. Rev. Biochem.* 50:845–877.

18. Mayo, K. E., Warren, R., and Palmiter, R. D. 1982. The mouse metallothionein-I gene is transcriptionally regulated by cadmium following transfection into human or mouse cells, *Cell* 29:99–108.

19. McIvor, R. S., Valerio, D., Williams, S. R., Goddard, J. M., Simonsen, C. C., Duyvesteyn, M. G. C., Van Ormondt, H., Van der Eb, A. J., and Martin, D. W., Jr. 1985. Mammalian expression of cloned cDNA sequences encoding human purine nucleoside phosphorylase and adenosine deaminase, *Ann. N.Y. Acad. Sci.* 451:245–249.

20. McIvor, R. S., Goddard, J. M., Simonsen, C. C., and Martin, D. W., Jr. 1985. Expression of a cDNA sequence encoding human purine nucleoside phosphorylase in rodent and human cells, *Mol. Cell. Biol.* in press.

21. McKusick, V. A. 1978. *Mendelian Inheritance in Man*, 5th ed., Johns Hopkins Univ. Press, Baltimore.

22. Miller, A. D., Eckner, R. J., Jolly, D. J., Friedman, T., and Verma, I. M. 1984. Expression of a retrovirus encoding human HPRT in mice, *Science* 225:630–632.

23. Miller, A. D., Lau, M.-F., and Verma, I. M. 1985. Generation of helper-free amphotropic retroviruses that transduce a dominant-acting, methotrexate-resistant dihydrofolate reductase gene, *Mol. Cell. Biol.* 5:431–437.

24. Orkin, S. H., Goff, S. C., Kelley, W. N., and Daddona, P. E. 1985. Transient expression of human adenosine deaminase cDNAs: Identification of a nonfunctional clone resulting from a simple amino acid substitution, *Mol. Cell. Biol.* 5:762–767.

25. Osborne, W. R. A. 1980. Human red cell purine nucleoside phosphorylase: Purifi-

cation by biospecific affinity chromatography and physical properties, *J. Biol. Chem.* 225:7089–7092.

26. Pfeilsticker, J., and Hoffee, P. A. 1985. Construction of a mammalian expression vector containing a cDNA for rat adenosine deaminase, *Fed. Proc.* 44:667.

27. Plagemann, P. G. W., and Wohlhueter, R. M. 1980. Permeation of nucleosides, nucleic acid bases, and nucleotides in animal cells, *Curr. Top. Membrane Transp.* 14:225–330.

28. Seto, S., Carrera, C. J., Kubota, M., Wasson, D. B., and Carson, D. A. 1985. Mechanism of deoxyadenosine and 2-chlorodeoxyadenosine toxicity to nondividing human lymphocytes, *J. Clin. Invest.* 75:377–383.

29. Simonsen, C. C., and Levinson, A. D. 1983. Isolation and expression of an altered mouse dihydrofolate reductase cDNA, *Proc. Nat. Acad. Sci. U.S.A.* 80:2495–2499.

30. Ullman, B., Gudas, L. J., Clift, S. M., and Martin, D. W., Jr. 1979. Isolation and characterization of purine-nucleoside phosphorylase-deficient T-lymphoma cells and secondary mutants with altered ribonucleotide reductase: Genetic model for immunodeficiency disease, *Proc. Nat. Acad. Sci. U.S.A.* 76:1074–1078.

31. Valerio, D., Duyvesteyn, M. G. C., Van Ormondt, H., Meera Khan, P., and Van der Eb, A. J. 1984. Adenosine deaminase (ADA) deficiency in cells derived from humans with severe combined immunodeficiency is due to an aberration of the ADA protein, *Nucleic Acids Res.* 12:1015–1024.

32. Valerio, D., McIvor, R. S., Williams, S. R., Duyvesteyn, M. G. C., Van Ormondt, H., Van der Eb, A. J., and Martin, D. W., Jr. 1984. Cloning of human adenosine deaminase cDNA and expression in mouse cells, *Gene* 31:137–143.

33. Valerio, D., Duyvesteyn, M. G. C., and Van der Eb, A. J. 1985. Introduction of sequences encoding functional human adenosine deaminase into mouse cells using a retroviral shuttle system, *Gene* 34:163–168.

34. Valerio, D., Duyvesteyn, M. G. C., Dekker, B. M. M., Weeda, G., Berkvens, T. M., Van der Voorn, L., Van Ormondt, H., and Van der Eb, A. J. 1985. Adenosine deaminase: Characterization and expression of a gene with a remarkable promoter, *EMBO J.* 4:437–443.

35. Wagemaker, G. 1985. Hemopoietic cell differentiation, in: *Bone Marrow Transplantation: Biological Mechanisms and Clinical Practice*, vol. 3 (D. W. Van Bekkum and B. Lowenberg, eds.), Marcel Dekker, New York.

36. Wiginton, D. A., Adrian, G. S., and Hutton, J. J. 1984. Sequence of human adenosine deaminase cDNA including the coding region and a small intron, *Nucleic Acids Res.* 12:1015–1024.

37. Williams, D. A., Lemischka, I. R., Nathan, D. G., and Mulligan, R. C. 1984. Introduction of new genetic material into hematopoietic stem cells of the mouse, *Nature (London)* 310:476–480.

38. Williams, S. R., Goddard, J. M., and Martin, D. W., Jr. 1984. Human purine nucleoside phosphorylase cDNA sequence and genomic clone characterization, *Nucleic Acids Res.* 12:5779–5787.

39. Yeung, C. Y., Frayne, E. G., Al-Ubaidi, M. R., Hook, A. G., Ingolia, D. E., Wright, D. A., and Kellems, R. E. 1983. Amplification and molecular cloning of murine adenosine deaminase gene sequences, *J. Biol. Chem.* 258:15179–15185.

40. Zannis, V., Doyle, D., and Martin, D. W., Jr. 1978. Purification and characterization of human erythrocyte purine nucleoside phosphorylase and its subunits, *J. Biol. Chem.* 253:504–510.

Questions for Discussion with the Editors

1. *You emphasize that a key step in constructing a plasmid that will be successfully expressed is the choice of a suitable promoter. Do you feel that the metallothionein promoter is optimal? Or do you predict that by testing other promoters even more effective ones can be identified?*

This chapter described experiments that addressed the ability of promoters to provide human PNP expression in a species-specific manner. Further imposition of tissue-specific constraints on promoter/enhancer activity will also dictate which elements should be included to provide expression in hematopoietic cells. In this regard, the 5' regulatory sequences derived from the mouse metallothionein I gene are probably not the best choice. In initial experiments, it may be necessary to use viral regulatory sequences, which in many cases have not appeared to be subject to the same expressional constraints as cellular genes. Ultimately one might predict that natural regulatory elements from the human ADA and PNP genes should provide appropriate expression levels. However, the effects of vector (e.g., retroviral) sequences and the overall efficiency of gene transfer in the total cell population must also be considered.

2. *To improve the effectiveness of transforming hematopoietic cells it appears that the small subpopulation of pluripotent stem cells will have to be isolated from whole bone marrow. Are there any differentiated characteristics (e.g., size or surface properties) of those stem cells that can be exploited for their isolation?*

There are two approaches that can be taken to increase the efficiency of gene transfer into hematopoietic stem cells, and these are not mutually exclusive. Pluripotent stem cells of the mouse can be enriched from total marrow populations by using antibody staining to a surface marker (Qa-2m)[1] in conjunction with cell sorting techniques. However, similar purification of human hematopoietic stem cells is not currently feasible, and in general it will probably be more straightforward to use a gene transfer vector system that has the capability of transforming every cell in a given population. Recombinant retroviruses provide such a system, and in conjunction with partial purification of stem cells and advanced stem cell culture techniques may result in the ability to achieve close to 100% gene transfer efficiency.

[1]Harris, R. A., Hogarth, P. M., Wadeson, L. J., Collins, P., McKenzie, I. F. C., and Penington, D. G. 1984. An antigenic difference between cells forming early and late haematopoietic spleen colonies (CFU-S). *Nature (London)* 307:638–640.

CHAPTER 15

Gene Transfer into Hematopoietic Cells

Martin J. Cline

BONE MARROW is a natural target tissue for gene transfer in living mammals. It contains multipotent and committed stem cells that may continue to proliferate throughout the lifespan of the animal. Bone marrow cells are readily obtained; they can be extensively manipulated *in vitro* and can still reconstitute hematopoiesis when injected into marrow-depleted animals. In addition, there is already extensive experience with bone marrow transplantation in humans. For these reasons, bone marrow was the first target tissue selected for genetic manipulation in intact animals (6, 10).

Hematopoiesis

In most mammals the bone marrow and, to a lesser extent, the spleen and liver are the principal sites of blood cell formation (5, 9). Multiple cell lineages are produced in the marrow including erythrocytes, granulocytes, mononuclear phagocytes, lymphocytes, eosinophils, and megakaryocytes. There is good evidence that all these cell lines arise from common ancestral multipotent stem cells. These stem cells are capable of proliferative self-renewal as well as differentiation to committed stem cells which are lineage-specific. Thus the same multipotent stem cell can differentiate to an erythroid stem cell, to a granulocytic stem cell, to a megakaryocytic stem cell, and so forth. These committed precursor cells are unipotent and are capable of further differentiation only within a single cell lineage. In postnatal animals the pluripotent stem cells are relatively rare (about 1 in 10^4 nucleated cells in the marrow) and are normally in a nondividing, quiescent state. They enter the proliferative cycle when there are potent demands for increased hematopoiesis, as for example after marrow depletion by irradiation or chemotherapeutic drugs such as methotrexate (MTX) or vinblastine. The committed stem cells of granulo-poiesis and erythropoiesis are more common (one stem cell in 10^2 to 10^3 nucleated marrow cells) and, under ordinary conditions, are continually dividing. Adjustments in the numbers of circulating blood cells are usually made from within the compartments of committed stem cells.

Pluripotent stem cells have a high but not unlimited self-renewal capacity; committed progenitor cells have a limited self-renewal capacity. When pluripotent stem cells are injected intravenously into an irradiated, marrow-depleted, genetically-compatible recipient they proliferate to reconstitute the pluripotent compartment but also differentiate to committed stem cells.

Pluripotent stem cells (CFU-S) can be assayed in mice by their ability to form hematopoietic colonies in the spleen of irradiated genetically compatible recipients (20). Committed progenitors of erythrocytes (BFU-E, CFU-E), granulocytes and macrophages (CFU-GM), eosinophils (CFU-eos), T lymphocytes (CFU-TL), and megakaryocytes (CFU-M) are assayed by their ability to form cell-specific colonies *in vitro* in a semisolid supporting matrix such as agar under appropriate hormonal stimulus (5, 9).

From the scheme of hematopoiesis illustrated in Fig. 1, one can postulate the following: (a) genes transferred into pluripotent stem cells should be detectable among several cell lineages; (b) genes introduced into primitive pluripotent stem cells or the early committed progenitors of specific lineages should persist for long periods of time if they are stable in the genome of the transformed cells; and (c) genes introduced into committed stem cells with a limited self-renewal capacity should have a limited survival in the animal irrespective of whether or not they are stable in the genome of the host cells. The scheme in Fig. 1 further suggests that drug-resistance genes might be useful dominant selectable markers if the selective drugs influenced *either* cell survival or the rate of cell proliferation. A transformed hematopoietic stem cell containing a drug-resistance gene should continue to proliferate in an uninhibited manner despite drug administration to the animal, whereas the much larger of untransformed cells should be relatively inhibited. Ultimately the progeny of the transformed stem cells should dominate the proliferating hematopoietic tissues of the animal.

Inherited Diseases of the Bone Marrow as Targets

A number of monogenic diseases affecting the bone marrow have been identified as potential targets for gene transfer. These include sickle cell disease, the severe thalassemias, Gaucher's disease (glucocerebrosidase deficiency), certain immunodeficiency states (e.g., adenosine deaminase deficiency) and red cell enzymopathies resulting in severe hemolytic anemias (e.g., pyruvate kinase deficiency). Current therapy for these diseases consists mainly of supportive treatments, such as transfusions and folic acid supplementation for the hereditary anemias, antibiotics and γ-globulin for immunodeficiency diseases, and splenectomy and orthopedic procedures for Gaucher's disease.

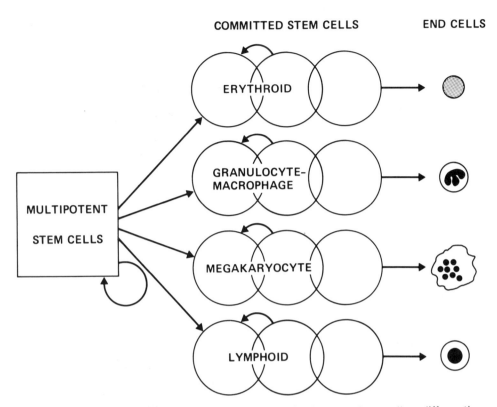

Figure 1 Multipotent stem cells are capabable of self-renewal as well as differentiation to lineage-specific stem cells. These committed stem cells have a limited self-renewal capacity. They divide to generate functional, fully differentiated end cells.

These treatments are palliative rather than curative and the possibility of using gene transfer techniques to attack the basic genetic defect has broad, if not universal, appeal. Alternative approaches to severe genetic diseases of the hematopoietic system are prenatal detection and abortion, and bone marrow ablation and transplantation from a genetically matched donor. The potential benefits and limitations of these procedures have been extensively evaluated (16, 19). Suffice it to say that both alternatives have important limitations including a high morbidity and mortality among bone marrow transplant recipients. These limitations make the possibility of gene transfer an attractive alternative. It is likely, therefore, that the serious genetic disorders affecting bone marrow cells will be the first disease targets of gene transfer, both because of the accessibility and proliferative kinetics of the bone marrow tissue and the serious and widespread nature of the genetic diseases that affect the hematopoietic system.

Gene Transfer Techniques Used with Hematopoietic Cells

A large number of gene transfer techniques of variable efficiency and practicality have been applied to mammalian cells in culture (4) (see Chap. 9). These include:

1. Chromosome transfer
2. Microcell-mediated gene transfer
3. Cell fusion
4. Spheroplast fusion
5. Liposome fusion
6 Direct injection
7. Transfection (DNA-mediated gene transfer)
8. DNA viral vectors
9. RNA viral vectors
10. "Electroporation"

Of these, only the last four techniques would at present appear to have potential applicability to gene transfer into marrow cells of intact animals or ultimately of humans.

Transfection has been the technique most widely applied to a variety of mammalian cell systems *in vitro*. With surface-adherent mammalian cells and a variety of selectable vectors, efficiency of genetic transformation by transfection has varied between 1 in 10^3 to 1 in 10^7 cells. Efficiency has varied with the cell type, the selective gene, and with the conditions of the transfection procedure. Incubation time, DNA concentration, and the use of glycerol shock are all potentially important variables. Transfection of nonadherent hematopoietic cells has not been systematically studied; however, anecdotal evidence suggests that efficiency of stable gene transfer is low relative to adherent cells. Another disadvantage of transfection in hematopoietic cells is the unknown stability of cotransferred, nonselectable genes. The apparent advantages of transfection are its simplicity and its utility with characterizable noninfectious reagents, which can include enhancing elements and modulatable promoters limited only by the ingenuity of the constructor. Furthermore, transfected genes, even when introduced into cells inefficiently and at low copy number, can often be amplified under appropriate selective conditions. In practice, at the present time, some genes can be introduced into the bone marrow cells of animals by transfection but the relatively rough estimates that are possible in this complex system suggest that the kinetics are suboptimal and that only a small fraction of stem cells (perhaps 1 to 10 percent) are genetically altered after an intense selective procedure. If efficiency is too low (less than approximately 10^{-6}) then even stringent selective conditions may be inadequate to get clinically effective expression of new genes introduced into a population of marrow cells.

Retroviruses

Because of the limitations of transfection techniques, considerable interest has been generated in retroviral vectors (13). Potential advantages of these vectors for gene transfer in hematopoietic tissues are: (a) efficient internalization, integration and expression of genes of interest. Up to 100 percent of cells are transformed in some tissue culture systems. (b) Large numbers of cells can conveniently be "infected." (c) Integration of genetic information into the host's genome is generally stable. (d) The host range and target cell populations are often manipulable. The single most obvious disadvantage of these vectors (in addition to a fortuitous, disastrous integration resulting in activation of an undesirable host gene) is the possibility of recombinational events that could convert a replicating defective viral vector into an infectious agent.

Recently, retroviral vectors have begun to be applied to bone marrow systems *in vitro* and *in vivo*. The first attempt along these lines was one in which we utilized a modified spleen necrosis virus containing a Herpes virus thymidine kinase (TK) insert (17). This modified avian virus could infect but not replicate in rat cells. It efficiently transformed TK$^-$ rat fibroblasts to TK$^+$ status *in vitro*. Unfortunately it was inefficient as a vector for rat bone marrow cells *in vitro*. As an alternative approach (which was based on host range of an engineered virus), has been to examine packaging-defective viral vectors. The ψ-2 cell line has been used to introduce a neomycin-resistance gene into hematopoietic cells of mice via a retroviral vector (8). This line of 3T3 cells contains a packaging-defective proviral DNA permanently integrated. ψ-Helper viral RNA is constitutively produced, yielding the proteins required to make a viral particle, but it does not package its own RNA. When the cell line is transfected with DNA of an appropriate retroviral vector only (or probably only) the vector particles are produced. These packaged vectors can be used to infect bone marrow cells. In this case, the retroviral vectors contain dihydrofolate reductase (DHFR) and a neomycin-resistance gene. Using this approach, Williams et al. (21) have recently introduced new genetic material into multipotent hematopoietic stem cells of mice (see below).

Another example of a retroviral vector is that constructed by Verma and colleagues based on Moloney murine leukemia and sarcoma viruses and containing human hypoxanthine phosphoribosyltransferase (12). In this system a packageable helper virus was not excluded and infectious particles could be expected to persist *in vivo*. Other groups of investigators have also developed retroviral vector systems that can introduce new genetic information into murine hematopoietic stem cells, but in which infectious viruses are not eliminated.

It presently appears that if the problem of infectious particles generated either in vector production or by recombinational events could be eliminated, then retroviral vectors would be most promising agents for gene transfer into hematopoietic and other somatic tissues. However, there is always a possibility for recombinational events that will generate infectious virus. Temin and colleagues have observed homologous and illegitimate recombinations involving transfected viral DNA (1). If this proves to be a general phenomenon, then retroviral vectors may always have intrinsic limitations.

DNA Viral Vectors

Many investigators have utilized SV40 vectors to introduce and express various genes in mammalian cells in tissue culture. By its nature this system involving viral replication with ultimate injury of the host cell has no ready applicability to the hematopoietic system of intact animals.

Isolated studies have considered the use of modified vaccinia and polyoma viruses as potential gene vectors. On the other hand many investigators have suggested that some of the papilloma viruses may be useful for gene transfer. In cells of some mammalian species including humans, papilloma viral DNA replicates as a stable episome with approximately 50 to 200 copies per cell. Nonviral genes introduced by this means have been expressed in tissue culture cell lines, but there is as yet no extensive experience with blood-forming cells. Obvious problems in considering papilloma viruses as vectors for use in intact animals are host range of virus, adaptation of the vectors to the hematopoietic system, and elimination of transforming potential.

Electroporation

An intriguing technique for gene transfer into mammalian cells involves the imposition of a very brief, high-voltage electric field on a suspension of cells. Because transient pores are thought to form in the cell membrane as a consequence of depolarization, the technique has been called "electroporation" by some investigators or "electric shock" by the more literal minded. The technique was described more than a decade ago but has only recently had extensive application to problems of cell fusion (as in hybridoma formation) and to the introduction of macromolecules into cells (22). Although little of this later information has been published, anecdotal reports from several investigators suggest that it is appreciably more efficient than transfection in transferring genes into nonadherent cells, including cells of hematopoietic lineage (15). The general range of voltage and pulse duration specified to yield a high efficiency of gene transfer and acceptable levels of cell death is 2 to 4 kV and 0.1 to 1.0 μsec.

Dominant Selectable Markers

For gene transfer systems with less than 1 to 10 percent efficiency, one must consider selectable markers and the imposition of selective pressure to spread the inserted gene throughout the target cell population. Drug resistance genes as selectable markers have generally been used with "wild-type" cells that are not deficient in an exploitable enzyme system. A wide variety of selectable systems have been used *in vitro,* including genes for resistance to aminoglycoside antibodies, DHFR inhibitors, aspartate transcarbamylase inhibitors, chloramphenicol, mycophenolic acid, and membrane-transport inhibitors. Most of these agents have little potential for application *in vivo* because of their toxicity to intact animals. In my opinion only one of the currently available selective systems—the DHFR system—has potential for clinical application to hematopoietic cells. Of the other drug-selection systems that might be useful *in vivo,* the chloramphenicol acetyltransferase (CAT) system deserves consideration since chloramphenicol can produce marrow suppression. However, this effect is infrequent, unpredictable, and not clearly related to drug dosage. Chloramphenicol at high doses can be cardiotoxic.

Methotrexate (MTX), the most widely used inhibitor of DHFR, has been employed in clinical medicine for nearly 30 years as an antiproliferative agent for treatment of leukemia, certain solid tumors, and certain skin diseases. In animals and man the antiproliferative effects of MTX have been well characterized and are manifested primarily in the bone marrow and in the gastrointestinal tract. The degree of suppression of marrow can be titrated with an acceptable level of precision by adjustment of the dosage of drug. MTX is therefore, in principle, a useful drug for gene transfer *in vivo.* Mechanisms of resistance to MTX include amplification of the DHFR gene, transport system mutations, and mutations that alter MTX binding by the DHFR gene product. Both eukaryotic (18) and prokaryotic (14) genes or cDNAs have been cloned and placed in mammalian expression vectors that can transform both DHFR and wild-type mammalian cells *in vitro.* A mutant mammalian DHFR gene is of particular interest since it codes for a gene product that is reasonably active, yet resistant to 10^{-5} M MTX (2, 18). The mutant DHFR protein results from a single amino acid change, which alters its binding of both natural substrate and MTX.

Animal Model Systems

DHFR

We previously described attempts at introducing MTX-resistance genes into intact mice utilizing hematopoietic progenitor cells and their ability to repopulate the bone marrow of lethally irradiated recipients (2, 6, 10). Initially we used transfection of high molecular weight DNA from an MTX-resistant cell line containing a mutant mouse DHFR gene and utilized indirect assays of transformation, including karyotype analysis and analysis of the hematological status of MTX-treated recipient mice (6). The strategy of the experimental design is illustrated in Fig. 2. Subsequently we utilized assays that would detect the mutant gene product by its MTX-resistance and by its abnormal electrophoretic mobility (2, 7). To briefly summarize our observations:

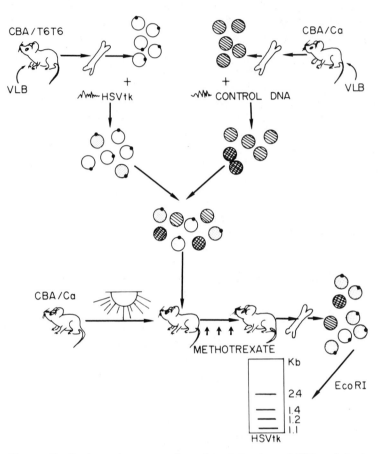

INSERTION OF HSVtk INTO MAMMALIAN CELLS

Figure 2 Design of an experiment to introduce MTX-resistant genes into mouse bone marrow cells. T6T6 cells contain a distinctive chromosomal marker. In these studies Herpes virus thymidine kinase genes were introduced *in vitro* and the cells transferred to syngeneic irradiated recipient mice. These were treated with MTX. After a period of time the presence of the HSV TK gene was detected by predominance of the chromosomally marked cell population and Southern blot analysis.

1. Evidence of transformation was obtained in more than 50 percent of animals utilizing indirect and direct assays of transformation.
2. MTX administration was required for expression of the transferred mutant DHFR genes.
3. Expression of the transferred gene took at least 4 weeks from the administration of marrow to irradiated recipients.
4. Once established, the karyotype of the transformed cells persisted even when MTX was discontinued.

5. Mutant DHFR protein was expressed in hematopoietic tissues (Fig. 3) but not in nonhematopoietic tissues of experimental animals.
6. Bone marrow cells of the genetically altered animals were relatively resistant to MTX.
7. The recipient mice remained clinically well and without evidence of toxicity for a large portion of their life spans after gene transfer (Fig. 4).

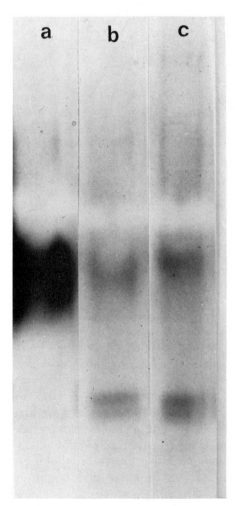

Figure 3 Mutant mouse DHFR was detected by electrophoretic mobility in nondenaturing gels. (a) Enzyme from 3T6R1 cells containing mutant enzyme; (b and c) partially purified enzyme from spleens of mice transfected with mutant genes.

Figure 4 CBA mouse 15 months after irradiation injection with transfected bone marrow.

After running a successful series of experiments with DNA from a cell line containing a mutant DHFR gene, we consistently failed to observe gene transfer when we used DNA from cell lines containing amplified wild-type DHFR genes.

We concluded that the transferred gene had to convey a high level of MTX resistance in order to effective. Observations similar to those outlined above were subsequently made by Carr et al. (3). Some negative observations from these experiments are also pertinent. We attempted cotransfection into mouse hematopoietic cells of genomic mutant DHFR genes and cloned genomic human β-globin genes. We then looked for human β-globin expression in mouse erythrocytes utilizing specific antibody and a radioimmunoassay. Test animals receiving bone marrow transfected with both DHFR and human β-globin genes were treated with phenylhydrazine in order to induce a hemolytic anemia and expand the size of the spleen and to stimulate erythropoiesis. Once we had sorted out an artifact in the radioimmunoassay induced by oxidative denaturation of mouse globin, we concluded that there was no evidence of expression of the cotransfected human β-globin gene in mouse hematopoietic tissues.

In a recent series of studies we have utilized a vector containing an SV40 promoter and cloned cDNA from the mutant mouse DHFR constructed by

Simonsen et al. (18). We have preliminary evidence that gene sequences are detectable in a high percentage of recipient animals between 4 and 12 weeks after transfection, that these genes have a mammalian methylation pattern consistent with gene replication in host cells, that the introduced genes are expressed, and that between 1 and 10 percent of the target progenitor cells demonstrate resistance to MTX at concentrations between 10^{-7} and 10^{-5}. We regard these results as encouraging and have defined our next set of objectives as follows:

1. Examine other promoters that may be more appropriate to hematopoietic cells or that may be modulated by drug administrations
2. To use more efficient gene insertion techniques for use with bone marrow cells, and
3. To insert vectors that contain both DHFR and a nonselectable utilitarian gene.

Our perspective at present is that this is a useful system for intact animals and possibly for clinical application, but that its efficiency will have to be improved.

Herpes Virus Thymidine Kinase in Animals

Simultaneously with our initial attempts to insert mutant DHFR genes into hematopoietic cells of mice, we undertook studies of herpes virus thymidine kinase (HSV TK) as a potential selectable marker (10, 11). Several features distinguished the transfecting system using HSV TK from the DHFR animal model. Insertion and expression of the HSV TK gene was less regularly observed than in the DHFR systems; with HSV TK in plasmid and with ligated HSV TK genes, only 2 of 6 and 8 of 14 animals, respectively, demonstrated the phenotype of the transformed cell population. With Herpes TK vectors, transformation was generally unstable, and less than 15 percent of animals receiving transfected bone marrow demonstrated retention of the viral genes for several months. Although the phenotype of cells transfected with HSV TK was more frequently observed in the presence of MTX selection, expression of this phenotype was also observed in the absence of MTX administration to experimental animals. This observation was regarded as important for subsequent clinical studies. We observed over 100 mice for periods of 3 to 18 months for evidence of toxicity of the HSV TK vectors. No clinical toxicity was observed.

Retroviral Vectors

The most extensively documented *in vivo* retroviral vector system is that recently described by Williams and his colleagues (21). They used the ψ2 MSV-DHFR-NEO producer cell line described above which generates over

10^7 CFU/ml free of wild-type virus. These investigators cocultivated mouse marrow cells with this producer cell line for either 48 hours or 6 days before returning the hematopoietic cells to irradiated syngeneic mice. These cells formed spleen colonies (CFU-S) which were used as a source of DNA for Southern analysis. The analysis revealed that about 10 to 25 percent of hematopoietic cells had a detectable NEO-pBR hybridizing fragment. Generally, multiple sites of integration were observed in a colony derived from a single cell. Such spleen colony cells carrying new genes were used to repopulate a secondary irradiated marrow recipient. No evidence of transferred gene expression was reported in these studies and the introduction of ψ2 MSV-DHFR-NEO cells was not vigorously excluded. Nevertheless, this approach appears to be promising for efficient gene transfer into hematopoietic cells. Long-term observation of animals for possible recombination of the introduced vectors with endogenous retroviruses is obviously necessary.

Applications to Human Diseases

Thalassemia and Sickle Cell Disease

Many investigators active in the field of human genetics believe that it is only a matter of time before systematic clinical trials of gene therapy are undertaken in humans. Assuming the technical problems of gene transfer and efficient gene expression can be surmounted, then the first logical application of gene insertion techniques in humans will be in genetic diseases of the hematopoietic system including thalassemia and sickle cell disease. These are logical targets because they arise as a consequence of isolated defects of single genes or small gene clusters whose expression is limited to blood forming cells. These diseases are therefore theoretically amenable to correction by insertion of functional single genes into hematopoietic cells. Furthermore, they are important diseases with a high associated morbidity and mortality that affect millions of people throughout the world. Every investigator in this field has his or her own concept of the important obstacles that must be overcome before gene therapy of marrow diseases can become a reality. My list of these obstacles includes the following:

1. Development of an efficient system for inserting genes into marrow cells *in vitro*
2. Development of methods for introducing genetically altered cells back into the blood forming system of the recipient patient
3. Construction of gene vectors that can be confidently expected not to generate infectious agents or to insert at multiple sites in a single target cell
4. Development of selective techniques (for low-efficiency gene-transfer methods) that have acceptable levels of clinical toxicity
5. Enhancement of levels of expression of introduced nonselectable, utilitarian genes such as β-globin, so that clinically useful levels of

gene product can be generated in developing blood cells

6. Assurance of stability of the transferred genes so that the process need not be repeated more often than every few months
7. Development of a method of destroying cells containing the transferred genes so that such cells can be eliminated should any undesirable side effects of transferred genes become apparent.

In planning the strategy for accomplishing these goals and in planning for future clinical investigation, it is worth considering what has been learned thus far in regard to introducing genes into the hematopoietic system of humans.

In 1980, my laboratory undertook studies aimed at gene insertion into marrow cells of two patients with severe β-thalassemia. Both patients had evidence of heart involvement by iron deposition as part of the disease process. HSV TK was selected as the dominant selectable marker because insertion of this gene in aminal model systems had been observed in the absence of MTX administration, and use of this cytotoxic drug was deemed an undesirable feature of initial clinical studies. The objective of the study was to define conditions necessary for transfection of human blood-forming cells and return these cells to active hematopoietic sites. The experimental design was essentially that used in the previous animal model (2, 3, 6, 10) except that irradiation of 300 rads was limited to a short length of the femur of each patient. Betwen 9×10^8 and 11×10^8 nucleated cells were transfected *in vitro* with the vectors shown in Fig. 5 and then returned to the patients intravenously. Blood and marrow samples were obtained over a period of more than 1 year. Our observations may be briefly summarized as follows: Analysis of bone marrow and blood DNA for the presence of HSV TK and pBR sequences revealed similar patterns in the two patients. Neither of these foreign genes was detected in blood samples obtained before the procedure. Between 1 and 2 weeks after transfection, the internal 2.4 kb EcoRI digested fragment of the HSV TK vectors were found in blood cells of both patients at low copy number (less than one copy per haploid genome). Between 3 and 10 weeks after the procedure, multiple TK bands were observed in Southern blots of DNA from hematopoietic cells. The predominant band was still the internal 2.4 kb EcoRI fragment, but smaller (0.9 to 1 kb) and larger (3.8 to 11 kb) fragments were found in blood cells of both patients (Fig. 6). After 10 weeks no additional high-molecular-weight TK sequences were observed, and these sequences gradually disappeared from DNA of blood cells and were undetectable after 3 months in one patient and after 9 months in the other. Plasmid sequences were transiently detected in blood cell DNA between the first and tenth week.

Although the high-molecular-weight bands could have resulted from partial digestion with EcoRI, this is unlikely to be the case since the same pattern was observed in several independent digestions, and parallel reactions with marker DNA were digested to completion. The relative intensity of the bands suggested integration at different sites into a major and several minor hematopoietic cell clones. The intensity of the major high-molecular-weight bands

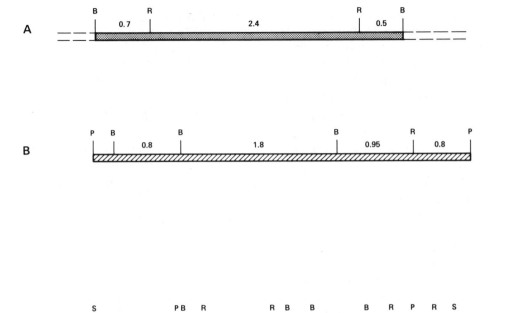

Figure 5 Restriction enzyme maps of vectors used to transfect bone marrow cells (units are in kilobases and are approximations). (A) Herpes virus gene sequences containing TK gene within the EcoRI fragment; BAM HI fragments were ligated prior to transfection. (B) Human gene sequences containing β-globin gene (Hβ1). (C) Recombinant vector containing both HSV TK (crosshatched area) and Hβ1 (hatched area) sequences in pBR 322. The plasmid was cleaved with Sal-I and ligated prior to transfection. S, Sal-I; P, Pst-I; R, EcoRI; B, BamHI.

(equivalent to about one gene copy per haploid genome) suggests either that many blood cells carried the HSV TK gene at low copy number or that a minor population carried many copies of the foreign gene. The presence of predominant and minor TK gene inserts in genomic DNA at this time-point was supported by Sal-I digestion of DNA and analysis with a radiolabeled TK probe. A predominant band of 4 kb was observed with fainter bands of 3.0 and 6.2 kb.

These observations suggest that Herpes TK as an identifiable marker was introduced into blood cells of two thalassemic patients. The time of appearance of the TK gene in human blood cells at about 5 weeks after transfection of human hematopoietic precursor cells is similar to that observed in the mouse model. In the mouse, viral gene product was detected in the blood; however, limited materials prevented a similar analysis in the human studies. The transitory nature of the inserted TK gene also resembles the pattern seen in mice, where instability of this foreign gene in hematopoietic cells over a period of 3 to 4 months is the rule. Instability might reflect (a) carriage of the

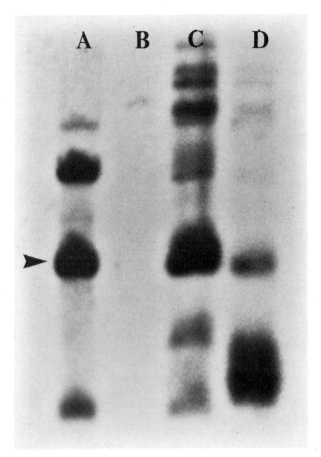

Figure 6 Samples of 10 μg of DNA were digested with EcoRI (lanes A and C) or EcoRI plus MboI (lane D), hybridized with ³⁵p-HSVtk, and filmed for 24 hours. (A) Patient's sample at 10 weeks; (B) blank; (C and D) mouse LTK⁻ cells transfected with HSV TK and selected in HAT medium. A single TK⁺ colony was grown up but not recloned before DNA extraction. The TK bands in A are (from the top) 7.3, 4.8, 3.7 (faint), 2.4 (arrow) and 1.0 kb.

gene as as unstable episome, (b) unstable integration of the gene, or (c) its insertion into committed stem cells that divide but do not renew the proliferative cell compartment.

No toxicity was observed from the procedure. We regard these observations as encouraging that genes can indeed be introduced into human hematopoietic precursor cells and that these cells can, under the appropriate circumstances, return to patients' bone marrows to resume proliferation and generation of mature blood cells.

Conclusions

Several different techniques appear to have applicability to gene transfer into hematopoietic cells of intact animals. Transfection of precipitated DNA and retroviral vectors has already achieved some measure of success in animal model systems. Each of these methods have some inherent limitations. Other potentially useful techniques are still untried in intact animals. Early studies in animals have also identified potentially useful selectable gene sequences that could be used in a clinical setting. The first studies in humans are encouraging with regard to the possibility of inserting useful gene sequences into hematopoietic cells and reintroduction of the modified cells into the hematopoietic system. The obvious technological hurdles before clinical application are the efficiency of the transfer technique, proving freedom from infectious agents or the induction of neoplasia, and modulating the level of expression of useful genes. None of these obstacles seems insurmountable at present.

General References

CLINE, M. J. 1984. Gene therapy, *J. Mol. Cell. Biochem.* 59:3–10.

METCALF, D., and MOORE, M. A. S. 1971. *Haematopoietic Cells,* North-Holland, Amsterdam.

References

1. Bandyopadhyay, P. K., Watanabe, S., and Temin, H. M. 1984. Recombination of transfected DNA in vertebrate cells in culture, *Proc. Nat. Acad. Sci. U.S.A.* 81:3476–3480.

2. Bar-Eli, M., Stang, H. D., Mercola, K. E., and Cline, M. J. 1983. Expression of a methotrexate-resistant dihydrofolate reductase gene in transformed hematopoietic cells of mice, *Somatic Cell Genet.* 9:55–67.

3. Carr, F., Medina, W. D., Dube, S., and Bertino, J. R. 1983. Genetic transformation of murine bone marrow cells to methotrexate resistance, *Blood* 62:180–185.

4. Cline, M. J. 1984. Gene therapy, *J. Mol. Cell. Biochem.* 59:3–10.

5. Cline, M. J., and Golde, D. W. 1979. Cellular interactions in haematopoiesis, *Nature (London)* 277:177–181.

6. Cline, M. J., Stang H., Mercola K., Morse, L., Ruprecht, R., Browne, J., and Salser, W. 1980. Gene transfer in intact animals, *Nature (London)* 284:422–425.

7. Haber, D. A., Beverly, S. M., Kiely, M. L., and Schimke, R. T. 1981. Properties of an altered dihydrofolate reductase encoded by amplified genes in cultured mouse cells, *J. Biol. Chem.* 256:9501–9506.

8. Mann R., Mulligan, R. C., and Baltimore, D. 1983. Construction of a retrovirus packaging mutant and its use to produce helper-free defective retrovirus, *Cell* 33:153–159.

9. Metcalf, D., and Moore, M. A. S. 1971. *Haematopoietic Cells*, North-Holland, Amsterdam.

10. Mercola, K. E., Stang, H. D. Browne, J., Salser, W., and Cline, M. J. 1980. Insertion of a new gene of viral origin into bone marrow cells of mice, *Science* 208:1033–1035.

11. Mercola, K. E., Bar-Eli, M., Stang, H. D., Slamon, D. J., and Iine, M. J. 1982. Insertion of new genetic information into bone marrow cells of mice: Comparison of two selectable genes, *Ann. N. Y. Acad. Sci.* 397:272–280.

12. Miller, A. D., Jolly, D. J., Friedmann T., and Verma, I. M. 1983. A transmissible retrovirus expressing human hypoxanthine phosphoribosyltransferase (HPRT): Gene transfer into cells obtained from humans deficient in HPRT, *Proc. Nat. Acad. Sci. U.S.A.* 80:4709–4713.

13. Miller, A. D., Eckner, R. J., Jolly, D. J., Friedmann, T., and Verma, I. M. 1984. Expression of a retrovirus encoding human HPRT in mice, *Science* 225:630–632.

14. O'Hare, K., Benoist, C., and Breathnach, R. 1981. Transformation of mouse fibroblasts to methotrexate resistance by a recombinant plasmid expressing a prokaryotic dihydrofolate reductase *Proc. Nat. Acad. Sci. U.S.A.* 78:1527–1531.

15. Potter, H., Weir, L., and Leder, P. 1984. Enhancer-dependent expression of human kappa immunoglobulin genes introduced into mouse pre-B lymphocytes by electroporation, *Proc. Nat. Acad. Sci. U.S.A.* 81:7161–7165.

16. Rowley, P. T. 1984. Genetic screening: Marvel or menace? *Science* 225:138-144.

17. Shimotohno, K., and Temin, H. M. 1981. Formation of infectious progeny virus after insertion of Herpes simplex thymidine kinase gene into DNA of an avian retrovirus, *Cell* 26:67-77.

18. Simonsen, C. C., and Levinson, A. D. 1983. Isolation and expression of an altered mouse dihydrofolate reductase cDNA, *Proc. Nat. Acad. Sci. U.S.A.* 80:2495–2499.

19. Storb, R., Prentice, R. L., Buckner, C. D., Clift, R. A., Applebaum, F., Deeg, J., Doney, K., Hansen, J., Mason, M., Sanders, J. E., Singer, J., Sullivan, K. M., Witherspoon, R. P., and Thomas, E. D. 1983. Graft-versus-host disease and survival in patients with aplastic anemia treated by marrow grafts from HLA-identical siblings, *New Engl. J. Med.* 308:302–307.

20. Till, J. E., and Mc Culloch, E. A. 1961. A direct measurement of Ru radiation sensitivity of normal mouse bone marrow cells, *Radiat. Res.* 14:213–222.

21. Williams, D. A., Lemishka, I. R., Nathan, D. G., and Mulligan, R. C. 1984. Introduction of new genetic material into pluripotent hematopoietic stem cells of the mouse, *Nature (London)* 310:476–480.

22. Zimmermann, U. 1982. Electric field-mediated fusion and related electrical phenomenon, *Biochem Biogphys. Acta.* 694:227–277.

Questions for Discussion with the Editors

1. *Is it conceivable that a bifunctional gene vector system could be developed that includes, in addition to structural genes for hematopoiesis, genes that induce rapid cell proliferation in successfully transformed cells?*

Yes, it is conceivable that bifunctional gene vectors may be developed. However, it is unclear which genes are involved in the control of cellular proliferation.

Clearly some of the oncogenes are candidates for this role; however, definitive identification of "proliferation" genes has not been achieved.

2. *What are the prospects for extending the techniques you described for gene transfer into hematopoietic cells to other cells, for example, insulin deficient pancreatic cells?*

The prospects of extending the techniques of gene transfer to hematopoietic cells are very good for the future. It is likely that viral vectors with specific tissue tropism will be developed. These vectors may be used in combination with tissue-specific enhancer elements and may well allow expression of transferred genes in appropriate tissues.

Index